"Conner has constructed an elegant correlation between the natural sciences and other dimensions of human culture—philosophical, ethical, aesthetic, and religious—through a holistic naturalism that recognizes the influence of both material and non-material energies. Drawing on physics, special relativity, quantum mechanics, information theory, process philosophy, and religious naturalism, he shows how values and meaning emerge from the physical world itself and exposes the delusion of any simplistic divide between 'objective' and 'subjective' dimensions of reality."

—Jennifer G. Jesse, Professor Emerita of Philosophy and Religion, Truman State University

"In *Physics, Information, and Meaning*, David Conner successfully defends several audacious claims. Drawing on quantum physics, he demonstrates that all of reality consists of a non-material wholistic dynamic process. He then uses this metaphysics along with feminist and liberation theologies and process philosophy to support his definition of God as the 'total causal matrix' in which all of nature, including human beings, exists. I am convinced by his arguments and moved by his religious interpretations."

—Constance Wise, author of *Hidden Circles in the Web: Feminist Wicca, Process Thought and Occult Knowledge*

"This is the best book I know that shows the richness and vitality of relativity theory and quantum mechanics. David Conner presents us with a new philosophy of nature. Influenced by A. N. Whitehead, Conner goes far beyond Whitehead in depicting the contemporary, naturalistic roots of 'meaning.'"

—Nancy K. Frankenberry, author of *The Faith of Scientists*

"Physics and physical science are in a great transition, as the widely accepted paradigm of gravitational cosmology has receded and is being replaced by new models and scientific thinking about the universe. David Conner's book will be perfect for those who want an accurate, detailed, sophisticated, but not overly technical, picture of how process thinkers and other progressive researchers see the recent changes in contemporary physics and cosmology. For those who see the physical universe as a clue to greater questions of meaning, and deeper concerns of value, this book is a sober and reliable guide to nature, writ large. It is especially relevant to those who want an informed account of how our studies of the natural world should form our thinking about the social, cultural, and philosophical order of our 'world,' and not just the universe itself."

—Randall Auxier, Professor of Philosophy and Communication Studies, Southern Illinois University

"In *Physics, Information, and Meaning,* David Conner provides a compelling alternative to the mechanistic materialism that pervades modern scientific theories and erects artificial divisions between matter and mind, particles and purposes, empirical facts and personal meanings, natural history and human history. Drawing on experiments in special relativity and quantum mechanics as well as the process philosophies of Heraclitus, Bergson, James, and Whitehead, Conner argues that the physical world is a complex, holistic, and organismic network that traffics in information and is productive of feelings, decisions, goals, and values. Faith, theology, and experiences of the divine also arise naturally in the world. The final two chapters make a vigorous and impassioned case for a religious naturalism, a wholly naturalistic but spiritually evocative vision of reality that hallows the sacred depths of the universe and imagines God not as an anthropomorphic being beyond nature but as the 'total causal matrix' of nature itself. This book is a tour de force—and a must-read for anyone interested in nonsupernaturalistic interpretations of religion and nonreductionistic accounts of science."

—Demian Wheeler, Sophia Associate Professor of Religious and Theological Studies, United Theological Seminary of the Twin Cities

Physics, Information, and Meaning

Physics, Information, and Meaning

How Nature Creates Complexity, Purpose, and Values

DAVID E. CONNER

WIPF & STOCK • Eugene, Oregon

PHYSICS, INFORMATION, AND MEANING
How Nature Creates Complexity, Purpose, and Values

Wipf & Stock
An Imprint of Wipf and Stock Publishers
199 W. 8th Ave., Suite 3
Eugene, OR 97401

www.wipfandstock.com

PAPERBACK ISBN: 979-8-3852-5948-9
HARDCOVER ISBN: 979-8-3852-5949-6
EBOOK ISBN: 979-8-3852-5950-2

VERSION NUMBER 03/24/26

Contents

Preface

THE AIM OF THIS book is to describe the physical world in such a way as to show that physical nature is the source of goals, purposes, valuation, and meaning. In what follows, nature in its entirety is understood as holistic and organismic. The source of nature's complexity and holism is depicted not as some supernatural Being or spiritual principle. Rather, nature is said to be all there is. There is no supernatural source behind or beyond the universe. On the other hand, nature does turn out to be a great deal more creative and complex than the reductionistic, mechanistic conception of particles and forces to which modern science has often limited its attention. Instead, the premise of this book is that science itself—particularly special relativity and quantum mechanics—points to the principles upon which holism and meaning are based.

Relativity and quantum mechanics are not simple topics. Nevertheless, though some of the ideas presented here are complicated, this book is intended for a general audience. The discussion is not technical in the sense that it is meant only for scientists or philosophers.

In this book, the process philosophy of Alfred North Whitehead is a frequent resource. Whitehead (1861–1947) was a British mathematician who later in life produced several books about what he called "the philosophy of organism." Though Whitehead is sometimes portrayed as a panpsychist or even as a theologian, Whitehead himself stated directly that the source of his philosophy was mathematical physics,[1] and he was skeptical about the value of academic theology. But in any case, Whitehead's philosophy is not the primary subject in what follows. His philosophy is simply one resource among others.

The final two chapters consider the subject of religion. Inasmuch as the book's central argument is about the origins of meaning, value, and

1. Whitehead, *Science and the Modern World*, ch. 9.

life itself, religion seems appropriate as a concluding topic. However, the type of religion described here is not the meticulously defined system of mandatory beliefs and behaviors that has so often been associated with Western monotheism. I think the type of religion presented here should seem plausible to most readers, and it may even be engaging. But it is not viewed as compulsory.

The topics are presented in what I believe is a logical order. However, it is conceivable that, depending on their existing fields of study, some readers might want to skip certain chapters. For example, scientists might choose to skip chapters 4 and 6. Philosophers might similarly scan through sections dealing with traditional epistemology and metaphysics.

Acknowledgments

I OWE AN INCALCULABLE debt to my teachers—especially Charles Milligan, Harvey Potthoff, John Cobb, Alton Templin, John Spangler, and David Griffin. Scholarly friends and conversation partners such as William Dean, Donald Crosby, Nancy Frankenberry, Demian Wheeler, Jennifer Jesse, Constance Wise, Tom Howe, and Randy Auxier have been a blessing. My wife, Claudia Schmitt, has listened patiently to countless impromptu speeches about fascinating subjects such as the Pauli exclusion principle and hybrid physical prehensions. I must express heartfelt thanks to Jim and Kathy Bartsch for their consistent interest and generous support as this book was being written. Finally, the members of the churches who have allowed me to serve as their pastor have kept me mindful of the fact that life is essentially an excursion in pursuit of meaning, value, self-transcendence, and faith. Intellectual achievement—whether scientific, philosophical, theological, or otherwise—must finally be held accountable to these larger ends.

1

Introduction

In my library there are several books about quantum mechanics. A few of those books are actual textbooks filled with equations and other technical information. In college I majored in chemistry, and you can't understand chemistry—chemical bonding, for example—unless you understand something about quantum mechanics. But after college I did not become a professional scientist, and therefore over the years most of my books about quantum mechanics have tended to be nontechnical descriptions written for laypersons.

Among these nontechnical volumes, one of my favorites is *Quantum Strangeness: Wrestling with Bell's Theorem and the Ultimate Nature of Reality*, by astronomer-physicist George Greenstein. Greenstein's book is enjoyable partly because it is concise, understandable, and engaging, but, most of all, it is enjoyable for its honesty. Greenstein openly admits that, despite all of his years as a physicist, there are still aspects—major aspects—of quantum mechanics that, to him and to most physicists, just don't make sense. What is even more praiseworthy is that instead of taking the approach of most science writers—which is to assert that quantum mechanics is inherently paradoxical and cryptic, and we are simply stuck with that fact—Greenstein says that what quantum science really calls for is an entirely new outlook on nature, what Greenstein calls "*an experimental metaphysics*."[1] In other words, it's not that quantum physics is inherently weird. The problem is that our old, familiar beliefs about nature are, well, *wrong*. What are these old, familiar beliefs? Of course

1. Greenstein, *Quantum Strangeness*, 5–6, 102.

the details vary depending on the explainer, but the worldview familiar to most scientists, and to most laypersons also, can be described as mechanistic materialism. This is the idea that nature is composed of tiny material particles that interact with each other according to predictable, mechanistic, mathematical laws.

The trouble is that this view doesn't work with quantum mechanics. In quantum mechanics, reality sometimes seems to be composed of particles, but at other times it behaves like waves. Matter is sometimes located in a specific place, but at other times matter seems to be "smeared out" in a way that prevents its precise location from being stated. Furthermore, even empty space isn't really empty. It is filled with "virtual particles" that seem to pop in and out of existence. Because of these and other quantum riddles, Greenstein suggests that we need a radically new depiction of nature that will treat all of these quantum phenomena as normal, as routine, while still also allowing for the ordinary modes of physical interaction that we observe in everyday life.

The book that you are now reading is a response to this need for a new philosophy of nature, an "experimental metaphysics."[2] The chapters that follow describe both the special theory of relativity and quantum mechanics. After each descriptive chapter, I offer interpretive chapters in which I spell out basic principles (four drawn from relativity and five drawn from quantum mechanics) that summarize a new philosophy of nature—something, I hope, like what Greenstein has in mind. No doubt this sounds audacious, especially since, as I have admitted, I am not a scientist. Therefore I can only ask for indulgence from you, the reader. Even if I have not done a perfect job of explaining them, the ideas that are examined here are fascinating.

They are also important. The ideas are not only scientific; they are philosophical. They depict a world that cannot be reduced to physical forces and mechanistic principles, a world that is alive and productive of value—a world like the one that we actually experience personally and socially, a world in which outcomes actually *matter*. It turns out that the philosophy that, I am proposing, emerges from relativity and quantum mechanics abolishes the absolute dichotomy between fact and value. It

2. By *metaphysics*, Greenstein and I both mean the branch of academic philosophy that examines the nature of reality, including the relation between mind and matter, and between fact and value. *Metaphysics* in this book does not signify pseudo-intellectual claims about mystical knowledge or spiritual forces that are said to inhabit physical objects.

offers a solution to the mind–body problem, or to the "hard problem" of consciousness.[3] The worldview that is developed in this book facilitates a greater sense of continuity between the natural sciences, the social sciences, and the humanities. In other words, it proposes a path to conciliation to end the perennial culture war between reductionist-materialist and holistic-humanist outlooks about the functioning of nature and the meaning of life. It reveals the natural place of human nature within nature as a whole.

So, this book has been written partly for scientists who are interested in a less weird, more sensible interpretation of quantum mechanics. It is also written for nonscientists who are interested in a worldview that reconciles contemporary science with ordinary experience and human values. It is written for people who are concerned about ethics (human and environmental), about values, and even about spirituality or religion, but who have grown dissatisfied with both contemporary ethical relativism and traditional religious supernaturalism. And finally, it is written for anyone who is tired of or discouraged by the "scientific" assumption that reality can be boiled down to an enormous, purposeless collection of particles and forces.

One afternoon a few years ago I was watching a documentary about the birth of the universe, which we commonly call the Big Bang. Before the Big Bang—if the word *before* can even be applied—there were no preexisting parts of the universe sitting around in space, waiting to be utilized. Rather, the Big Bang created everything, all at once.[4] Therefore, there is a sense in which everything that now exists is a result of the Big Bang. Apparently thinking of this, the narrator of the documentary, a well-known astronomer, commented with a certain air of amazement that everything that now exists was, in some sense, already contained within that primordial burst of energy. Then the narrator went on to list, as outcomes of the Big Bang, not only the appearance of atoms, stars,

3. The "hard problem of consciousness" asks how physical activities in the brain can be the underlying source of human awareness and create our sense of personal identity. The three answers usually cited—physicalist reductionism, mind–matter dualism, and panpsychism—are inadequate and do not include the approach advocated in this book.

4. Especially since the launch of the James Webb Space Telescope in December 2021, the idea of the Big Bang is being reevaluated. Some suggest that the Big Bang may not provide the best explanation for present conditions in the universe. In any case, the argument that I am now developing does not depend on the Big Bang as a single, unique event but only on the idea that the universe as it now exists has somehow evolved from a primordial state consisting of energy and particles.

galaxies, and planets, but examples from human history as well, such as the rise and fall of various civilizations, the development of governments and their legal systems, the industrial revolution, and cultural achievements such as the Enlightenment, the symphonies of Beethoven, and the plays of Shakespeare. The Big Bang somehow contained the material ingredients and the "laws of nature" that were necessary for all of these things to come into being—not only physical things but cultural artifacts. I remember that in that moment it seemed remarkable to me to hear a scientist making this connection between natural history and human culture.[5]

The narrator continued, but I was distracted by what he had already said. I am an ordained minister, and the thought came to me that the products of the Big Bang would therefore also include *religion*. Religious experience and religious history certainly are aspects of human culture, so, *a fortiori*, religion, too, must be traceable all the way back to the formation of the cosmos. For a moment I felt surprised by the thought that the Big Bang had, in a sense, produced religion—and then I was surprised about the fact that I felt surprised. I am accustomed to thinking of religion as a natural, ordinary activity because religion is an integral aspect of human behavior, and human behavior is an integral aspect of the natural world. (The claim that religion has a supernatural basis is, in my view, untenable.) Nevertheless, the coincidental connection between a documentary on astronomy and the existence of religion seemed surprising.

In a way, this element of surprise is what this book seeks to examine and render unnecessary. Why—this book asks—should we feel surprised when we recognize a continuity between scientific cosmology and human culture? Why do we typically think of nature as a subject to be studied with systematic objectivity and of human experiences such as religion, literature, history, and the arts as the outcome of mere personal preferences and social trends? We live in an era when the supposed impartiality of the scientific method is lionized as the paradigmatic path to truth—but this method, it is assumed, is irrelevant to personal feelings and social mores. Is this categorical contrast between science and

5. A discussion of the fact the natural universe is suited for the origination of human culture may seem to be related to what is known as "the anthropic principle," explicated by theoretical physicist Brandon Carter (b. 1942) and numerous others. However I do not rely on the anthropic principle in any of the arguments advanced in this book. See Wikipedia, "Anthropic Principle."

culture really appropriate? Or necessary? This book argues that the answer to those questions is *no* and that the fixed partition between science and the humanities is not only unnecessary but inappropriate. In fact, as I have said, this book consults science itself, in the form of relativity physics and quantum mechanics, as a way of discovering the origins of life, purpose, culture, and human feeling, within the sheer, primordial processes of physical creation. My final goal is to provide a worldview in which the relation between the physical causes described by science and the phenomena of human experience—culture, consciousness, and even faith—are integrally related, cut from the same cloth. I argue that scientific ideas—such as frame of reference in relativity, and superposition and entanglement in quantum mechanics—can be interpreted in ways that support the existence of values, meaning, and purposiveness, though I think I do this in ways that avoid being simplistic, naïve, unscholarly, or anti-intellectual.

I.

Before we go any further, I would reassure the reader that this book is not an attempt to show that science and the humanities are, when analyzed, merely different versions of the same thing—some monolithic reality called "human knowledge." I do not argue that human culture can be explained away as a result of purely physical principles or, more specifically, that human consciousness is really just a manifestation of biology. Nor, on the other hand, do I argue that the sciences and the humanities can peacefully coexist because they are about radically different subjects and rely on radically different methods. I do not claim, for example, that science and religion are not in conflict because science is about facts and religion is about values. Indeed, one of the goals of this book is to disestablish the facts–values dichotomy that is so widely taken for granted.

Other books have sought to reconcile science and culture or to overcome the separation between facts and values. Usually, however, these books take what is essentially a philosophical approach, relying on what are held to be new insights and clearer reasoning. This book takes what I believe is a more unusual tactic. Essentially, I rely on an analysis of the discoveries of twentieth-century physics in order to arrive at certain philosophical conclusions—probably *new* conclusions, for most readers—about the nature of reality. Instead of assuming that science has already arrived at a relatively settled worldview, I attempt to start fresh, describing certain experiments, especially related to relativity and quantum mechanics, then offering interpretations of the results. To some readers the attention devoted to science may seem excessive or prolonged—but there is a reason for this. My argument is that religion and science are not intrinsically in opposition. It is *supernaturalistic* religion and *mechanistic-reductionistic* science that are in opposition. For years the notion has been widely accepted that philosophies of determinism are the necessary consequence of an honest interpretation of science. This is not so. Determinism is, rather, the consequence of an inadequate interpretation of science.

The central argument of this book is that nature itself can provide a basis for understanding the emergence of goal-oriented behavior in animals and even plants, and that scientific, natural processes can also account for human experiences of self-awareness and culture. To be sure, this conclusion is already entailed by our initial statement that everything that now exists was made possible by the "Big Bang." Consciousness and culture really *are* parts of nature, not mere illusions. But as we have noted, this simple comment about the Big Bang is difficult to defend when examined more closely. The problem, as I have said already, is that the intellectual milieu of our own time has fostered a pernicious division between natural history and human history, between mechanism and purposiveness, between facts and values, between our physical bodies and our inner thoughts and feelings, between outer occurrences and inner meanings, between scientific truth and personal conviction. This polarized way of viewing the world and interpreting our lives has tended to alienate us from our very selves, it has served as a specious justification for our exploitive attitudes about nature, and it is has made religion and social traditions seem to be nothing more than a matter of arbitrary, unfounded personal opinions.

This book counteracts our habitually divided way of looking at things with an argument that is not often attempted—namely, that twentieth-century physics itself can plausibly be interpreted in ways that lead to a reconciliation between all of the polarities just listed.[6] The challenge

6. I acknowledge the similarities that exist between my argument in this book and the overall thesis of Paul Davies in *The Mind of God* and of Davies and John Gribbin in *The Matter Myth*. For example, Davies credits quantum mechanics as a source of clarity in his analysis of various philosophical options (Davies, *Mind of God*, 183). However, Davies's work generally tends to be more descriptive and less constructive than mine, and our proposals differ substantially at critical points.

we face is that an argument based on science sounds naïve and untenable if stated casually or without convincing examples. There are many books that interpret twentieth-century physics philosophically, but in most of them the general principles derived from the physics are simply stated as accepted conclusions. The seminal problems and experiments of the pioneering scientists are minimized or left out entirely.

In this book, the twentieth-century experiments that are most significant are examined in some detail. The results of the experiments are explicated in a way intended to substantiate a new understanding of nature that can serve as a corrective to the reductionism and materialism that still lurk in most scientific accounts of natural phenomena. However, I certainly do not rely on science alone. In fact, philosophy, more than science, has provided the general basis for this book's central thesis, for the interpretations of relativity and quantum mechanics that I propose could not have been initially envisioned without the benefit of prior acquaintance with certain philosophical perspectives. Nevertheless, I do argue that the science itself stands on its own as a source of a new outlook on nature and on life.

This use of science makes it necessary to include an explicit repudiation of what has been called *scientism*, the attitude that science is our most objective and reliable path to truth and that the scientific method should serve as a model for all other modes of inquiry. As Randy Auxier and Gary Herstein point out, an overreliance on science, or an uncritical trust in it, ironically poses a grave threat to any attempt (such as the present one) to derive a new philosophy of nature from scientific discoveries.[7] Science must be balanced with philosophical insight. On the other hand, as I reject scientism, it is also necessary to repudiate traditional metaphysics, if metaphysics is understood as an attempt to formulate a grand worldview that is either all-inclusive, logically necessary, or incontestable. I view metaphysics as a pragmatic activity rather than as a perfectible one. The postmodernism that is pervasive in today's intellectual climate reminds us, quite rightly, (1) that when we attempt to describe reality, the words we use are unavoidably ambiguous and, ultimately, self-referential; (2) that our perceptions of the world exist as mental images that do not necessarily depict the physical world accurately (because, for one reason, the world is actually composed of atomic

7. Auxier and Herstein, *Quantum of Explanation*, 233.

particles about which we have only indirect knowledge);[8] and (3) that there is in any case no neutral, objective foundation for statements that are allegedly about ultimate reality since all worldviews are propounded in some particular historical context and in terms of some personal or social perspective.[9] In every age, intellectuals have displayed a predilection for bestowing the status of orthodoxy upon certain worldviews so that those worldviews then restrict the ways in which nature itself may be understood. In other words, worldviews accrue various methods of self-authentication. Though the particular worldview of any era may cite numerous observations and experiments, the attitude entailed by that worldview eventually surrenders its original empiricism and becomes rationalistic, and then doctrinaire. This observation pertains not only to philosophers and theologians but to most scientists. In our own time quantum mechanics has repetitiously been called "weird" or "inexplicable" largely because scientists themselves cannot bring themselves to relinquish the venerable conviction that the world *should* behave according to the principles of Newtonian mechanics.

In this book, my emphasis is on the worldview-changing differences between classical physics and the discoveries of twentieth-century physics. I believe that we can learn a great deal from these differences, not only scientifically but philosophically, without relapsing into modernism[10] or succumbing to scientism. Despite the admonitions of postmodernists, there are a good many statements that may simply be regarded as true. It is still true that combustion is caused not by phlogiston but by rapid oxidation; that diseases are caused not by unbalanced "humors" but by germs;[11] and that atoms are not tiny, solid, indivisible corpuscles but a patterned arrangement of electrons moving about a relatively massive nucleus. It should be recognized that the insights of postmodernism can be applied so ardently that postmodernism itself becomes as grandiose as the modernism that it abjures.

8. The philosophical theory of perception is analyzed more thoroughly in chapter 13.

9. See *postmodernism* in the online Duignan et al., "Postmodernism and Relativism."

10. *Modernism* is generally associated with the Enlightenment belief that truth is universal and can be discovered by factual evidence and impartial reason (logic). To some extent, no doubt, these criteria really are objective, but on the other hand, context, perspective, and personal and societal motives always play some role in the selection and interpretation of evidence, and even logic is not always employed neutrally.

11. From antiquity through the Middle Ages, it was believed that illnesses were caused by unbalanced "humors" in the body. See Wikipedia, "Humorism."

The final chapters of this book move from science to a discussion of human experiences and values. I reject the traditional philosophy that human experience consists of subjective "representations" that must be tested by their "correspondence" to the "reality" of the "objective world." Based on my explication of relativity and quantum theory, I argue that experience itself is a form of reality—though, to be sure, we must still pay attention to the difference between experiences that originate in physical circumstances and experiences that originate in acts of imagination. Very often the two are blended. An essential aspect of human experience occurs in the context of *social conventions*—patterns of perception and behavior that are adopted by an entire group. These conventions include activities as diverse as the use of language, traffic signals and road signs, political and legal standards, and religious practices and beliefs.

The final two chapters deal with ideas about God and religion that, I think, are compatible with science. But I avoid treating religion as it is sometimes presented, as a system of beliefs and doctrines accompanied by certain ritual practices and moral precepts. I hold that the doctrines and the liturgies of religions are, mainly, the *results* of *religious feelings and experiences* rather than being the other way around. Of course, once religion has been embraced, the liturgy and even the doctrines can also become sources of inspiration. But my point is that religion is essentially a response to various inner needs and feelings, experienced both individually and socially, and not a collection of ideas and doctrines. In life we encounter a perennial, though often repressed, need to be grasped by a source of inspiration that is felt to be utterly transcendent, ultimate, and eternal, and yet experienceable in the context of ordinary living. It is the search for, and the experience of, this source that motivates religious devotion.

The conclusions in this book are comprised of certain major assertions. The terms and concepts mentioned here are discussed more fully in the chapters that follow. For now, stated as briefly as possible, the major points are these: (1) The discoveries of twentieth-century physics—notably the theory of special relativity and quantum mechanics—have implications for a radically new outlook on nature. (2) This new understanding includes the recognition that events are caused not only by material collisions and physical forces but also by the influence of transmitted information, received *as* information without being reduced to purely material forms. I argue that we observe this process in quantum superposition and quantum entanglement, for example. (3) The interaction of physical and informational factors allows us to develop a theory of reality in which subjectivity and goal-oriented functioning can be integrated with traditional mechanistic explanations of causation. (4) This reconciliation between physical causes and as-yet-unrealized goals is analyzed and supported by referring to the ideas of several philosophers, including Heraclitus of Ephesus (535–475 BCE), Henri Bergson (1859–1941), William James (1842–1910), and Alfred North Whitehead (1861–1947).

The philosophy of Whitehead is particularly valuable, and most of the key ideas in this book come from him. In Whitehead's philosophy, nature is held to be composed neither of particles nor waves but of *events*—what Whitehead called "actual entities" or "actual occasions." Whitehead's actual occasions, considered singly, are very much like the virtual particles that are described by quantum mechanics as occurring in empty space; indeed, Whitehead speaks of them this way.[12] Whitehead held that events are constantly coming into being by means of a process that he called *concrescence*. Concrescence involves not only mechanistic causes (collisions and physical forces) but abstract or conceptual information, such as mathematical data. In concrescence the constant blending of physical and conceptual elements means that nature is not purely mechanistic or deterministic. This nondeterminism allows for the functioning of purposes, aims, and goals throughout nature. However, the claim that quantum indeterminacy somehow "explains" consciousness or mentality lacks nuance and is misleading. Whitehead's philosophy has widely been viewed as an existing system of thought that can be applied to the interpretation of relativity and quantum mechanics, but I believe it is more nearly the case that his philosophy was *inspired by* relativity and quantum theory. If the conclusions I reach resemble Whitehead's ideas, it is because Whitehead's philosophy was prompted by the same scientific discoveries that I discuss in chapters 4 through 8. In any case, this book is not merely an exposition of Whitehead's philosophy, and I do not accept the doctrines of any existing philosophical outlook uncritically. I must also say emphatically that I do not interpret either Whitehead's philosophy or the principles of quantum mechanics as a basis for the claim, which has been made surprisingly often, that the universe is conscious,

12. Whitehead, *Process and Reality*, 56, 99, 177, 199. It is remarkable that Whitehead seems to have been aware of something like "virtual particles" in the mid-1920s when his Gifford Lectures (which became *Process and Reality*) were being written. This was a period when the basic principles of quantum mechanics were still being formulated by Bohr, Schrödinger, Sommerfeld, Heisenberg, Pauli, and others.

that the universe has something like a "soul," or that the universe is evolving towards some specific, ultimate goal. To the extent that the doctrine of *panpsychism* endorses any of these ideas, I think panpsychism is mistaken. These issues are discussed in greater detail in chapters 9 and 13.

(5) The inclusion of purposive elements in the various processes of nature helps us to understand biological evolution. In the evolution from primitive organisms to more complex life-forms, the presence of purposive functioning encourages the formation of social groups and the origination of established patterns of behavior, including instincts and the habits of herds, packs, hives, and so on. (6) Among human beings, these conventional patterns include systems of values and the sacred conventions of religion. Here I rely on the idea of conventions developed by theologian William Dean. The usefulness of sacred conventions is explained and undergirded by embracing the pragmatic theory of truth as opposed to the correspondence theory. (These theories are presented in chapter 10.) (7) This understanding of social conventions leads to *historicism*, the recognition that all personal values and religious beliefs (if any) occur in the context of some specific personal and social history. All statements made about actual facts are context-related. Therefore no creed or religious tradition can simply be true objectively for all persons, places, and times. On the other hand, there *are* objective criteria for assessing moral, political, and religious beliefs and behaviors.

I have sought to keep this book from being too technical or pedantic—though, considering the subject matter, this avoidance has often been difficult to achieve. Abstract argumentation is frequently relieved—and, I hope, clarified—by the regular use of illustrations and thought experiments. In the scientific sections I have provided information about the historical contexts in which the scientists were working, and I have described their experiments in some detail, not only because the experiments are interesting in themselves but because *an understanding of experimental results imparts clarity and credibility to the attendant new theories*—theories which by now have become recognized as scientific principles. Most books about relativity or quantum mechanics that are intended for a general audience minimize any explanation of the experiments themselves and jump to a presentation of general principles. This creates an impression that special relativity and quantum theory were developed mainly by theoreticians engaged in calculations and thought experiments. It is important to avoid that impression. Both special relativity and quantum mechanics originated in laboratory experiments that then demanded interpretation. Many people—perhaps the majority—find the principles of relativity and quantum mechanics enigmatic and elusive. The best way to counteract these difficulties is to understand the actual experiments that led unavoidably to the principles that were eventually adopted. Thus chapters 4 and 6 describe, respectively, the foundational experiments related to special relativity and quantum mechanics.

It is not necessary to read this book from cover to cover in order to assess its basic claims. Depending on the background and the interests of the reader, certain sections might be avoided entirely, or at least skipped and then returned to after further reading has made the general ideas more familiar. There is an unavoidable hazard inherent to this type of writing. It is that scientists may find the explanations of relativity and quantum mechanics overly simple, and philosophers and theologians may find the discussions in their areas similarly rudimentary. One supposes that this hazard is unavoidable in multidisciplinary writing. The advantage gained is that important ideas are all contained in one volume so that a valuable level of integration is achieved.

II.

I began by discussing the idea that the humanities are somehow a part of nature. Obviously human history is an outgrowth of natural history—as is everything in the world—but for centuries the prevailing attitude in Western culture has been that *human* history is somehow distinct or separated from *natural* history. There are several reasons for this disconnection.

The main reason is derived from something simple and obvious: our own intuitive differentiation between our body and our mind. As children we intuitively learn to make a distinction between bodily existence and mental awareness. We are able to think *about* our bodies, but we think *with* our minds. If a medical condition requires that your foot be amputated, you are able to think of your foot as an object that will be detached from *you*, yourself. This is not to say that this fact can be considered dispassionately but rather that you do not equate your foot with your personal identity. Your foot is separable. It is inherent to the process of thinking that we learn to distinguish objects themselves from our thoughts about those objects. Sigmund Freud referred to this act of distinguishing thoughts from objects as a contrast between "the primary process" (imagining something) and the "secondary process" (efforts to produce an actual result involving what was imagined).[13]

This differentiation between mind and body rose to the status of a philosophical doctrine in the dualism of French philosopher René Descartes (1596–1650), who posited two fundamental and totally dissimilar types of reality: "thinking substance" and "material substance." Today most intellectuals disavow Cartesian dualism, but the gap between body and mind still lives on in the so-called *hard problem* of consciousness. This problem arises because we have an immediate awareness of our own sensations, thoughts, and emotions, and yet we are told that this sense of personal awareness is actually caused by physical, neurological signals in the brain. The attempt to reduce our experience of personal consciousness to neurobiological processes is, for most people, not entirely convincing. This is the core of the hard problem.

The hard problem is often exacerbated by the scientism that is endorsed by many intellectuals—sometimes more frequently by nonscientists than by scientists. Scientism amounts to an unscientific endorsement—tantamount to a religious conviction—that scientific methods, centering upon the mathematical analysis of physical conditions and measurements, produce an accurate depiction of reality. Scientism is often accompanied by the parallel but not identical belief that our inner feelings and preferences are merely fleeting, privately held experiences that are somehow not fully real and therefore not actually able to cause physical events. Thus we separate human history from natural history. I discuss this problem more fully in chapters 2, 3, and 12, proposing a solution to the hard problem of consciousness in chapter 12.

Religion is not discussed in this book until the final two chapters. However, readers may find it interesting now, at the beginning, to understand this book's general outlook on religion. In the past there have been several reasons why the customary split between natural history and human history has been applied especially to religion. One reason is the parochial nature of many conventional religious teachings themselves. The monotheistic religions have historically taught that reality is divided into two realms—the natural and the supernatural. It is said that God, a supernatural, eternal being, made the world for the specific purpose of redeeming the human race. Nature provides a physical stage upon which the action in this story takes place, but nonhuman life-forms (plants and

13. Hall, *Primer*, 24–29 (ch. 2, secs. 1 and 2).

animals) are little more than furnishings—stage props, as it were—in the drama of salvation. Human beings alone are said to have souls. Natural history is thus relegated to a supporting role in a created realm in which the human pilgrimage is believed to be paramount. Monotheistic religions have opposed the idea that religion is simply a natural phenomenon, favoring instead the assertion that religion—or, rather, faith—is based on supernatural revelation, regarded as a gift from God.

More recently Christian theologians—especially Protestants—have avoided the natural-supernatural terminology in favor of a postmodern, "post-liberal" confessionalism in which religious beliefs are said to hinge ultimately not on empirical evidence but on a sheer commitment to "confess" (profess or endorse) the gospel. This outlook may be traced to Danish theologian Søren Kierkegaard (1813–1855), Swiss theologian Karl Barth (1886–1968), American theologian George Lindbeck (1923–2018), and many others.[14] But whether phrased in terms of supernaturalism or confessionalism, in many religious perspectives there is a clear rejection of efforts to support religious beliefs by appealing to evidence from nature. This attitude of detachment from the physical world is found also in religions other than Christianity. To some extent it can be said to pertain to Hindu and Buddhist conceptions of *maya*, the doctrine that the physical world is ceaselessly changing and is therefore unreal or that the human mind is chronically ensnared in thoughts about bodily feelings which conceal the true nature of the self. In these ways religions have deliberately distanced themselves from nature.

The other side of this coin is that a good many people have wanted to distance themselves from religion because of the perceived implausibility, the moral rigidity, or the manifest irrelevance of many religious

14. Lindbeck, for example, wrote that "though there can be no single best theology . . . yet 'dogmatic' methods are preferable to 'apologetic' ones in constructive or systematic work" (Lindbeck, *Nature of Doctrine*, 12). *Apologetic* in this context refers to the effort to utilize empirical and logical methods of argumentation in order to demonstrate that Christian beliefs are reasonable and credible. Apologetics are found even in early Christian history in theologians such as Justin Martyr (ca. 100—ca. 165). Liberal Christian theology has typically embraced apologetics, whereas orthodox and conservative theologians have minimized or avoided it. In theology, apologetics may be contrasted to *confessionalism*, the view that faith statements are appropriately established not by rational argumentation or experiential evidence but by the explicit endorsement by some denomination or other community of believers—that is, on the "confession" of some carefully formulated creedal statement. In theology, the word *dogmatic* refers not to an attitude of doctrinaire inflexibility but to teachings that have officially been approved by a historical church council or some other ecclesiastical body.

precepts and practices. Persons who accept scientific explanations of natural events are understandably reluctant to believe that the first two human beings were named Adam and Eve and that they lived in a paradisiacal garden called Eden (Gen 2:5—3:24), or that Elisha caused an iron axe head to float in the Jordan River (2 Kgs 6:1–7), or that Christ rose from the dead and is seated in heaven at the right hand of God (Heb 1:3, 1 Pet 3:22, Acts 7:55–56). Furthermore, tradition-bound religions have frequently clung obstinately to draconian stances such as the refusal to ordain women[15] and the condemnation of homosexuality. Such moral blunders in the name of religion may not at first seem to be at odds with natural history or natural science, but a de facto disengagement from natural history is present in the fact that this rigid approach to morals is based implicitly on an authoritarian understanding of truth that is at odds with the methodological reliance upon reason and experience that the scientific method exemplifies.

As indicated already, this book embraces the perspective that religion is ultimately a product of nature. This idea is couched in a philosophical outlook known as *naturalism*—the belief that nature is all there is and that there is no going behind or beyond the natural universe in order to explain how events take place or how things got to be the way they are. This said, however, this book also embraces the conviction that nature cannot adequately be explained—as many have attempted to explain it—by restricting explanatory principles solely to material bodies, the transfer of energy, and physical mechanisms. Nature is composed not only of matter and physical forces. It also includes information that is partly nonphysical (for example, mathematical) in nature. In this book the world's constant traffic of information is viewed as an essential precursor to very rudimentary forms of goal-seeking or teleology. This book holds that, because of the primeval flow of information, it is natural—in fact, virtually inevitable—that social, organismic structures would eventually emerge among plants, animals, and human beings. It is also natural that human beings, as purposive, self-aware creatures, would spontaneously search for sources of goodness, meaning, and hope in life's overall context. This is how religions originate.

I have said that religion is not primarily a matter of cognitive assent to specified truth-claims; religion is not merely the idea that God exists or that people's souls continue after death in some sort of afterlife.

15. This pertains not only to Roman Catholicism but to a good many right-wing Protestant churches.

Religion is often caricatured this way: first, by those who oppose religion, because popular, oversimplified depictions of religious beliefs create an easy target for ridicule; second, by fundamentalist Christians, because short lists of simplistic dogmas are appealing to defensive-minded zealots; and third, by most media outlets, because uncomplicated stereotypes are easier to portray and sensationalize in news reports. But after forty years as a pastor, I remain convinced that most persons who are sincerely religious—who participate in a religious community for years rather than engaging in what turns out to be a form of short-term dabbling—remain committed to religion because they experience an innate personal need for spiritual wholeness, a need that is addressed by what in the monotheistic religions is symbolized by the metaphor of communion with God. I do not hold that religion somehow ought to be mandatory, but I do argue that religion can be approached sympathetically and embraced rationally not only by the naïve but by the intellectual and the scholarly. The fact that you think critically does not mean that you have to be a skeptic.

III.

I have stated that this book begins with an analysis of special relativity and quantum mechanics as the basis for the development of a new view of nature. In fact, though chapters 2 and 3 provide an initial discussion about philosophical materialism and an examination of the nature of information, most of the first half of this book deals with twentieth-century physics. Philosophy and the humanities are mentioned only in passing. Readers who are not very interested in science are asked to be patient about this seeming preoccupation with physics. An understanding of this science is necessary in order to formulate an idea of nature in which feelings, values, and aspirations are real and functional as the core of life. The purpose here is to overcome the perennial compartmentalization that separates natural history and human history into two cultures.

As relativity and quantum theory are being introduced, I have avoided any extensive focus on mathematical techniques and technical details. Where mathematical formulas are included in the text, they are presented for the benefit of readers for whom the math may be interesting. Readers who find equations unhelpful may skip over the math without feeling that they are missing an essential step in the discussion. The focus is rather on the unanswered questions and conceptual difficulties that gave rise to twentieth-century discoveries in physics, and on the actual experiments that eventually led to answers. This approach, I hope, will be not only more interesting and easier to understand but more relevant to our overall aims. Chapters on relativity and quantum mechanics are followed by a discussion of their philosophical implications. Then I outline an overall interpretation of nature entailed by these philosophical implications.

I move from science to a consideration of the meaning of truth by contrasting the *correspondence theory of truth* and the *pragmatic theory of truth*. The correspondence theory is the intuitive belief that our mental ideas and convictions are true when they correspond to the material facts of the external world. The pragmatic theory rejects this implied division between internal mind and external world, taking instead the position that experience itself is an empirical fact and that all understandings of truth must incorporate a subjective aspect. Truths are regarded more nearly as a tool for guiding decisions and actions than as objective descriptions of reality, for, in fact, statements about reality are always to some extent perspectival, and moreover, the "objective" world itself contains realities that are in fact not objective but subjective.

The chapters in this book focus on certain topics. The present chapter is, obviously, an introduction. Chapter 2 considers the historic notion of matter (material existents) in some detail, with special attention to the philosophical and scientific problems that have arisen because of this notion. Chapter 3 deals with the nature of information—what it is and how it functions not only in human thinking but in natural processes. I affirm repeatedly that *nature traffics in information*. To some extent, chapter 3 makes clarifying observations about the nature of information but without proffering a full theory of information, which is provided in later chapters. Chapter 4 presents special relativity, focusing on the Michelson–Morley experiment and on Einstein's analysis of its results. Chapter 5 presents four general principles inferred from the axioms of special relativity, and it outlines the meaning of these principles for our overall view of nature.

Chapter 6 presents the roots of quantum mechanics, attending especially to black-body radiation, Rutherford's discoveries about atomic structure, and the Bohr model of the atom. Chapter 7 describes the conclusions drawn by Schrödinger, Heisenberg, and others. Although nonlocality and entanglement are conceivably the most conspicuous examples of the transmission of pure information, instead of appealing to nonlocality I discuss atomic structure, including electron orbitals, superposition, the Pauli exclusion principle, excited and stable states, and the basis of the periodic table. All of these phenomena provide compelling evidence for the paramount role of mathematical information throughout nature. Chapter 8 formulates five fundamental philosophical principles based on the basic ideas of quantum mechanics.

Chapter 9 shifts from physics to philosophy, returning to the idea that nature traffics in information. This assertion incorporates the notion of information-sharing into an overall perspective on nature in which Aristotelian formal and final causes are restored to their rightful place in nature. References to Aristotle are explained as we go along. I emphasize the processive-relational atomism of events that has already been mentioned. Process, atomism, relationality, and information are integrated into a worldview that affirms the reality of a primordial form of experience and purposiveness. This notion may be referred to as "radical empiricism" or "panexperientialism." This position entails a rather emphatic refutation of reductionistic materialism; in my view, holism has been on the defensive for far too long. It is time to go on the attack.

In chapter 10 I summarize the ideas of philosophers Heraclitus, Henri Bergson, and William James, whose concepts contribute to my interpretation of relativity and quantum theory. Chapter 11 considers the philosophy of Alfred North Whitehead. Chapter 12 moves from the atomic realm to the role of creative holism in the formation of macroscopic objects and social groups, shifting then to human patterns of relation. Chapter 13 analyzes human perception and the role of symbols in all forms of mentality. I argue that the separation of facts from values is a false dichotomy and that ethical norms related to environmentalism and human rights are not mere, subjective preferences. In chapter 14 I discuss the unavoidable, though often downplayed, role of values in science and the role of facts in the arts and humanities. In chapter 15 I analyze various difficulties pertaining to traditional ideas of God. In chapter 16 I describe four different intellectual resources or methods that are helpful in developing new ideas about God.

It may be asked why God is even mentioned in this book. My answer is that we must examine the claim that science itself suggests some notion of God. Consider, briefly, the following. We have the strong force, effective only at a very short range, to hold the nucleus together. We have the weak force, which enables various important nuclear transformations. We have the electrostatic force to hold the electrons in a cloud around the nucleus. We have gravity, relatively the weakest of the forces, to hold matter in clumps, thus causing the formation of stars and planets. And we have the Pauli exclusion principle, which mysteriously organizes the electrons into orbitals, making chemical structure and bonding possible. The outcome of all this is stable matter that can exist in three states: solid, liquid, and gas. All three states are necessary for the emergence of life. Even a devout skeptic might well wonder about the seemingly remote probability of these factors working together as they do. The various forces and outcomes occur as if a purposive being had deliberately planned all of this at the dawn of creation. On the other hand, there are also considerations that make this idea of a Creator tenuous or simply implausible. Pertaining to all of this, several traditional theistic proofs are analyzed in chapter 15. But in the end, this book is an attempt to describe the source of the universe's tendency towards complexity and meaning in terms that are religiously suggestive but without resorting to religious anthropomorphism or theological personalism.

I have appealed to nonscientists to be patient with the chapters on science. It is only fair also to appeal to scientists to be patient with the chapters dealing with philosophy. I believe that, to some degree, this book can be read by scientists solely for its ideas about the interpretation of relativity and, perhaps especially, of quantum mechanics, which even today is still regarded by many scientists as a murky subject. But in any case, the chapters on philosophy and religion are not intended only for philosophers and religionists. As I have said already, I hope that these chapters, and indeed the whole book, may be helpful to any person who has grown weary of Western culture's intransigent rending of human experience into two antagonistic poles—one populated with facts, figures, and technology, and the other with feelings, values, aspirations, and, sometimes, faith. In 1932 Alfred North Whitehead wrote,

> It is the business of philosophical theology to provide a rational understanding of the rise of civilization, and of the tendernesses of mere life itself, in a world which superficially is founded upon the clashings of senseless compulsion. I am not disguising my belief that in this task, theology has largely failed. . . . The defect of the liberal theology of the last two hundred years is that it has confined itself to the suggestion of minor, vapid reasons why people should continue to go to church in the traditional fashion.[16]

16. Whitehead, *Adventures of Ideas*, 218.

Since the time when Whitehead wrote those words, countless books—including more than a few that were inspired by his own ideas—have attempted to remedy the defect that he cited: the failure to reconcile science with the existence of purposiveness and values. Many of those books have been highly creative, deeply meaningful, and widely influential. However, in today's world, no one can claim that Western culture's view of nature has, on the whole, gotten any less exploitive, or that the breach between the natural sciences and the humanities is diminishing, or that a vital, wholesome spiritual vision is at the forefront of our social awareness. And yet, in spite of all of our blind spots, whether cultural or personal, *the natural world is in fact not founded upon mere, mechanical necessity*, upon "the clashings of senseless compulsion." In Whitehead's words, the world is not lifeless. It is alive. This book attempts to contribute to a new vision of the world and of our life's place in that world—a vision of the universe in which the pursuit of justice, compassion, hope, and faith can naturally be at home.

2

What's the Matter with Matter?

In this chapter and in chapter 3 I develop several ideas pertaining to two topics: the nature of matter and the nature of information. These two chapters serve as a prelude to our consideration of two subjects: the theory of relativity and quantum mechanics. The claim that relativity and quantum theory can and should inform our thinking about the humanities and the social sciences may seem odd, but it is based on three underlying postulates. The first is that, from a philosophical standpoint, relativity and quantum mechanics are not merely an update or an addendum to prior science, as they have sometimes been understood. Rather, *relativity and quantum mechanics call upon us to adopt an entirely new understanding of reality*—a profoundly innovative worldview or metaphysics.

Parenthetically it should be noted that in this book *metaphysics* does not refer to the counter-rational ideas of the New Age movement but to the area of philosophy dealing with the analysis of reality and the basic principles of existence. Metaphysics is regarded as one of the four main branches of philosophy, the other three being epistemology, ethics, and logic. Since this book embraces philosophical naturalism—the belief that nature is all there is and that there is no supernatural realm or supernatural type of influence—when I use the word *metaphysics*, the word *cosmology* can usually be substituted, if cosmology is understood in the philosophical sense and not as a subject of specialized study in astronomy.

The second postulate is that *relativity and quantum mechanics require us to discard, once and for all, the age-old notion of matter*. Of course the traditional concept of matter continues to have practical value in many settings, but as a key to understanding the nature of reality, matter must be abandoned.

The third postulate is that *nature traffics in information*. It is impossible to understand the natural processes that constitute the world without acknowledging that information is constantly being transmitted, received, and utilized. In what follows I explore the question of what information is and how it is put to use—not only by human beings but throughout nature.

Though metaphysics is a major focus, in this book I do not offer a fully developed metaphysical system. Rather, I begin by describing and critiquing the scientific-physicalist worldview that is dominant not only in the West but in technology-using societies all over the world. As we proceed I provide comments about the ideas and perceptions that a more updated, integrative worldview might entail. Then we will consider the discoveries and conclusions of the theory of relativity and quantum mechanics, with the goal of being guided by those discoveries and conclusions as we imagine a new outlook on nature and our place within it.

I.

The first postulate affirms the need for a new metaphysics or cosmology. The tenets of the fresh worldview that is needed have been discussed thoroughly by certain scholars, but, unfortunately, the most important ideas have not gained general recognition. Though there are significant exceptions, the needed consultation between science and philosophy has generally been lacking.

One reason for this lack is that even when attending only to their chosen field, both scientists and philosophers encounter more than enough material to occupy their time. The sciences produce a constant stream of new information. Faced with the demands of their own discipline, scientists can hardly be blamed for failing to study philosophy. Unfortunately, an additional obstacle is the result of intellectual prejudice. More than a few scientists dismiss philosophy summarily, regarding it as little more than an interminable exercise in inconclusive verbal legerdemain. During my senior year in college I decided to take an elective course in the history of philosophy. My class schedule had to be approved by my faculty advisor, who taught physical chemistry and inorganic theory. (I majored in chemistry.) When my advisor saw my choice for an elective he shook his head in mock sadness and said, "Philosophy? Ah. My condolences." This comment was lighthearted, but many scientists' dismissals of philosophy are not.[1] Even when scientists do not reject philosophy overtly, their lack of familiarity with philosophy is problematical, for it keeps them from recognizing that in their own work they cannot help but make philosophical assumptions—assumptions about the nature of reality, about the nature of knowledge and its sources, and about the ways in which various methods of investigation tend to bias results. When scientists are unaware of philosophical issues, it truncates their recognition of viable alternatives to their own working presuppositions. Philosophers, also, have often allowed their attention to be restricted by the traditional limits of their own field. This has led to a neglect of the deeper implications of twentieth-century science. It is reckless to generalize, but academic philosophy for several decades has often focused primarily on either an ongoing expansion of linguistic analysis or on ethics and various forms of liberation thought. When philosophers do turn their attention to metaphysics, it is often in the form of attempts, either implicit or explicit, to reformulate nineteenth-century Hegelian idealism in contemporary terms. But metaphysics is usually not at the forefront of contemporary philosophical discussion. Beginning with Kant's *Critique of Pure Reason* (1781/1787), many philosophers have viewed attempts to describe the nature of reality as inherently dubious, if not simply unfeasible.[2] During the twentieth century, disdain for metaphysics expanded under the influence of logical positivism and neopragmatism.[3] Even

1. For example, the well-known physicist Stephen Hawking (1942–2018) asserted that "philosophy is dead" because "it has not kept up with modern developments in science," adding that "scientists have become the bearers of the torch of discovery in our quest for knowledge" (Hawking, *Grand Design*, 5). For a critical response to Hawking and other science-based dismissals of philosophy, see Auxier and Herstein, *Quantum of Explanation*, 197–208.

2. Immanuel Kant (1724–1804) systematically described the intrinsic limits of human thought, arguing rigorously that it is impossible for rational analysis to ascertain the nature of ultimate reality.

3. *Logical positivism* maintained that, in order to be meaningful, statements must be either logically necessary or empirically verifiable. Though not always explicitly anti-metaphysical, positivism tends to undermine traditional metaphysical systems because metaphysical systems entail all-encompassing claims about being itself that are difficult to verify. *Neopragmatism* holds that words and language are meaningful because of the ways they are used in specific contexts; they can never fully succeed as purely objective descriptions of the "real" objects and situations that they are intended to represent. That is because words themselves are arbitrary human inventions, and the experiences

philosophies such as historicism and empirical naturalism, with their implicitly metaphysical doctrine that there is no type of reality beyond history or beyond nature, are frequently unwilling to engage in the construction of any detailed metaphysical system. Such systems, it is felt, are too speculative and too unempirical to be convincing.

It is easy to sympathize with those who would like to evade the complexities and puzzles of philosophical metaphysics—but, nevertheless, I argue that it is a mistake to try to neglect this task. After all, as complicated and abstruse as metaphysical philosophy may be, metaphysics arises out of a universal human need—the need to distinguish between what is real and what is imaginary. As I've just noted, some modern philosophers have argued that the word *real* itself is misleading or unproductive in philosophy and that discussions about reality are pointless. I disagree. There is a sense in which we begin talking about reality with small children when we need to convince them that there really is no boogeyman under the bed or that the movie they have just seen is "only pretend." A moment's thought suggests that this distinction between the real and the imaginary cannot help but remain important, if in more nuanced terms, in the lives of adults.

In this book I argue that a consideration of questions about metaphysics is actually unavoidable when we try to understand relativity theory and quantum mechanics. I also argue that the new metaphysics that is suggested by these scientific discoveries can lead to a very helpful and integrative way of understanding nature as a whole. Today there can be no question that old-fashioned claims to metaphysical certitude and apodictic truth are simply unsustainable. In this book, what is being claimed for metaphysics is not philosophical finality of statement but simply that relativity and quantum mechanics, when properly understood and interpreted, offer a path to a new picture of reality that is plausible enough and useful enough to serve as a new point of departure for the central argument of this book.

---

to which they refer are always interpreted in terms of our own mental categories, which have no direct relationship with the physical world. For example, we perceive a granite rock as a solid object, but in fact it is composed of molecules that are impossible to see individually.

II.

The second postulate is that *relativity and quantum mechanics require us to discard, once and for all, the ancient notion of matter.* This assertion entails the conviction that matter, as it has traditionally been understood and as it is still understood by most people, can no longer be regarded as having any type of metaphysical primacy. As Davies and Gribbin say, "The old Newtonian mechanistic paradigm has been superseded" and is being replaced by a new paradigm in which "mind, intelligence, and information are more important."[4] Ordinary sense perception may make it appear that the objects around us are solid and materially static, but relativity and, especially, quantum mechanics depict the physical world as something radically more dynamic, interactive, and unpredictable than matter as it has historically been conceptualized. In other words, the traditional view of matter turns out to be mistaken, even from the point of view of physics.

> Descartes founded the image of the human mind as a sort of nebulous substance that exists independently of the body. Much later, in the 1930s, Gilbert Ryle derided this dualism in a pithy reference to the mind part as "the ghost in the machine." Ryle articulated his criticism during the triumphal phase of materialism and mechanism. The "machine" he referred to was the human body and the human brain, themselves just parts of the larger cosmic machine. But already, when he coined that pithy expression, the new physics was at work, undermining the world view on which Ryle's philosophy was based. Today, on the brink of the twenty-first century, we can see that Ryle was right to dismiss the notion of the ghost in the machine—not because there is no ghost, but because there is no machine.[5]

In 1938 Alfred North Whitehead wrote,

> The notion of empty space, the mere vehicle of spatial interconnections, has been eliminated from recent science. The whole spatial universe is a field of force, or in other words, a field of incessant activity. The mathematical formulae of physics express the mathematical relations realized in this activity.[6]

4. Davies and Gribbin, *Matter Myth*, 283. It is noteworthy that Davies and Gribbin are not professional philosophers, but physicists.

5. Davies and Gribbin, *Matter Myth*, 309.

6. Whitehead, *Modes of Thought*, 136.

> [The moral of this story] is the extreme superficiality of the broad generalizations which mankind acquires on the basis of sense perception. The continuous effort to understand the world has carried us far away from all those obvious ideas. Matter has [now] been identified with energy, and energy is sheer activity: the passive substratum composed of self-identical enduring bits of matter has been abandoned, so far as concerns any fundamental description. Obviously this notion [of matter] expresses an important derivative fact. But it has ceased to be the presupposed basis of theory. The modern point of view is expressed in terms of energy, activity, and the vibratory differentiations of space-time. Any local agitation shakes the whole universe. The distant effects are minute, but they are there. . . . [In] the modern concept the group of agitations which we term matter is fused into its environment. There is no possibility of a detached, self-contained local existence.[7]

The relinquishment of matter as a foundational concept cannot help but feel counterintuitive. Our notion of matter is not merely an intellectual dogma that we learn in school. As Whitehead noted, matter is at the forefront of our experience from the moment of our birth. Infants learn to follow objects with their eyes. A toy can be seen, grasped, shaken, thrown, or dropped, but in most instances the toy itself is unaffected. From moment to moment and day to day the objects around us look the same and feel the same. Their apparent stability is also observable in their *inertia*—that is, unless they are acted upon by some external force, they sit still without spontaneous motion. Material surfaces such as walls, tables, cups, and hardwood floors look and feel solid and smooth, so we naturally believe that they *really are* solid and smooth. From the time of our earliest experiences, the evident stability of solid objects becomes the basis for understanding the world.

Our personal impressions and beliefs about the qualities of matter have been reinforced by 2,500 years of Western intellectual history. Aristotle (384–322 BCE) listed four explanatory factors or causes that are operative in the world. Readers who are not already acquainted with these four factors are asked to pay close attention, for Aristotle's terminology will remain useful in the chapters that lie ahead. The four factors or types of causation are: (1) the "material constituent," (2) the "form or pattern," (3) the "agent whereby change or state of rest is first produced," and (4) the "end or wherefore"—that is, the purpose

7. Whitehead, *Modes of Thought*, 137–38.

or outcome of the event or action.[8] In modern language we use the word *cause* mainly to refer to Aristotle's third explanatory factor—the acting agent or physical influence, which is also known as the *efficient* cause or the *mechanistic* cause.[9] Aristotle wanted to explain causation comprehensively, and so he listed all of the prerequisites that contribute to the occurrence of an event.

Thinking specifically about the material cause, Aristotle reasoned that whenever any change occurs, there must first be something already present that "persists throughout the change." "This thing," Aristotle said, "is the material."[10] As examples Aristotle cited "the bronze of a statue, the silver of a cup."[11] Aristotle's examples and most other examples of the material cause emphasize solidity, weight, durability, and physical volume.

René Descartes (1596–1650) famously, or notoriously, analyzed reality in terms of two fundamental "substances": thinking substance (more or less what we would call *mind*) and physical substance (what we call *matter*).[12] Descartes held that the defining property of material substance is *extension*—by which he meant that matter takes up space whereas thinking does not.[13] It is not always recognized that Descartes's concept of matter represents a novel departure from the ideas of his predecessors. Where Aristotle's analysis affirms the integration of the physical (in the efficient cause) and the purposive (in the final cause), Descartes posited a categorical, ontological breach between mind and matter.

As an aside, for the benefit of readers who are not familiar with philosophy, I will mention that *ontology* and *ontological* refer to inquiry into *the nature of being*. A simple ontology might distinguish between things

8. Aristotle, *Metaphysics* 5.2, 1013a (88). Citations for Aristotle's *Metaphysics* here and throughout are referenced by book, chapter, then Bekker number, with the page reference to the Hope translation in parentheses.

9. Regarding the efficient cause, it is worthy of note that Aristotle's comment "whereby change or state of rest is first produced" anticipates Newton's first law of motion, which states that if an object is at rest or moving at a constant speed in a straight line, it will remain at rest or keep moving in the same direction at the same speed unless it is acted on by an external force (Aristotle, *Metaphysics* 5.2, 1013a–1013b [88]).

10. Aristotle, *Metaphysics* 12.2, 1069b (249).

11. Aristotle, *Metaphysics* 5.2, 1013a (88).

12. The technical term *substance* was inherited by Descartes from the medieval philosophers and ultimately from Aristotle, who used *substance* (*ousia* in Greek), with varying nuances, to refer to a theorized enduring essence underlying all forms of existence (see Peters, *Greek Philosophical Terms*, s.v. "ousia" [150]). For Descartes, *substance* can be understood to mean "fundamental type of being."

13. See Copleston, *Descartes to Leibniz*, 119.

that are real—the Grand Canyon—and things that are imaginary—the tooth fairy. A slightly more elaborate ontology could include things that are not real but are destined to become real, such as the state of affairs in the city of Chicago one month from today. A fourth category could include things that can never become real—a photograph of you that was taken in 1844. There could also be a fifth category: things that are not real but that may possibly become real. In chapter 3, which deals with the ontological status of information, I will devote special attention to this fifth category, observing that one of the most important functions of information is to guide events in such a way that certain outcomes that do not yet exist are brought into existence.

Returning to Descartes, the ontological gap between thinking substance and material substance constitutes Descartes's famous dualism, which achieved widespread acceptance among intellectuals during the Enlightenment period (roughly 1640–1800). By positing an absolute dissimilarity between mind and matter, Descartes instigated a pervasive intellectual legacy in which material objects came to be regarded as being utterly physical. In other words, all vestiges of pattern, purpose, and principle, other than the principle of pure mechanism, were purged from anything material. In the wake of Descartes, matter was presumed to be a purely neutral stuff that does not act unless it is acted upon. This principle was soon spelled out explicitly in Newton's laws of motion. Over the years, other aspects of Cartesian dualism have faded from philosophical fashion, but the conviction that material objects are explainable solely in terms of physical cause and effect still persists even now, not only among philosophers but as a common assumption in Western worldviews.

In this book, that conviction is held to be utterly and, in fact, ruinously mistaken.

This coupling of materialism with mechanistic causation (efficient causality) as a bottom-line explanatory principle in nature has not always been as pervasive as it is today. Prior to Descartes and Newton the world was widely perceived as being characterized by a kind of aliveness or animation. We have seen that Aristotle himself advanced a doctrine of causality in which material and efficient causes were integrated with formal and final causes. Greek stories and plays from antiquity routinely depict the futility of attempts to circumvent an inexorable fate, a fate that is determined not by the overarching efficacy of physical causation but by the power of destiny itself—somewhat like the Hindu notion of karma. Moreover, prior to the Enlightenment, nature itself was viewed as having moral and spiritual properties. From antiquity through the Middle Ages and into the Renaissance, popular thinking tended to see nature not strictly in terms of physical cause and effect but as an arena in which physical causes were in constant interaction with a wide variety of purposes and invisible realities. Historian Roland Bainton reports that in the time of Martin Luther (1483–1546),

> certain elements even of old German paganism were blended with Christian mythology in the beliefs of these untutored folk. For them the woods and winds and water were peopled by elves, gnomes, fairies, mermen and mermaids, sprites and witches. Sinister spirits would release floods and pestilence, and would seduce mankind to sin and melancholia. Luther's mother believed that they played such minor pranks as stealing eggs, milk, and butter. . . . The education in the schools brought no emancipation but rather reinforced the training of the home.[14]

The plays of Shakespeare (1564–1616) are replete with omens, witches, and soothsayers whose predictions are fulfilled not through supernatural intervention but as the outcome of ordinary events; this mingling of various types of causation is part of the fascination that Shakespeare's plays engender. Today, cultures that are not in the grip of scientific materialism are still able to discern the presence of the spiritual within nature. As Black Elk (1863–1950), an Oglala Lakota teacher, is believed to have said, "Perhaps you have noticed that even in the slightest breeze you can hear the voice of the cottonwood tree; this we understand is its prayer to the Great Spirit, for not only men, but all things and all beings pray to Him continually in different ways."

Aphorisms like this one are indeed encountered in the world of educated Westerners but mostly on calendars or in greeting cards, not in books about science, philosophy, economics, or business.

The materialistic view of the world is not all bad. The idea of matter as passive, neutral stuff that is changed only by external forces works quite well when we are making calculations pertaining to most macroscopic objects—moving automobiles, hydraulics, construction projects, weather forecasting, and so on. After Isaac Newton (1643–1727) invented calculus, he was able to make extraordinarily accurate predictions about the motions of the planets. In his great work *Opticks* (1704), Newton summarized his view of the natural world concisely: "God in the

14. Bainton, *Here I Stand*, 26–27.

beginning formed matter in solid, massy, hard, impenetrable, movable particles, of such sizes and figures, and with such other properties, and in such proportion to space, as most conduced to the end for which he formed them."[15]

Newton saw the creative wisdom of God in the mathematical predictability of the motions of material objects, but it did not take long for Newton's intellectual descendants to decide that God was scarcely necessary in a universe that runs essentially like a machine. Just over a century after Newton, the French mathematician-philosopher Pierre-Simon Laplace (1749–1827) wrote, "Given for one instant an intelligence which could comprehend all the forces by which nature is animated . . . it [that intelligence] would embrace in the same formula both the movements of the largest bodies in the universe and those of the lightest atom; to it nothing would be uncertain."[16]

Laplace had utter confidence in the supremacy of efficient causes in nature, and it is possible to view his stance on this subject as something of a pinnacle in scientific-materialist thinking. Contemporary scientists do not usually state their worldviews so baldly and unconditionally as Laplace did, but on the other hand the presupposition of mechanistic materialism certainly has not been supplanted by any well-defined alternatives. Laplace's worldview has been tweaked, adapted, or sidestepped in various ways, but it has not disappeared.

An example of updated Laplacianism can be found in the ideas of contemporary biologist Ursula Goodenough. Goodenough acknowledges the reality of the novel characteristics that arise in emergent wholes. Cells, for example, have traits that molecules do not have. However, Goodenough avers, all holistic phenomena must finally be attributed to the interaction of their constituent parts. "Life," in her view, "can be explained by its underlying chemistry, just as chemistry can be explained by its underlying physics."[17] Genes, which serve as instructions for making certain proteins, may seem to be purpose-driven (they seem to have an Aristotelian final cause), but in fact they operate as they do simply because they have the right physical shape (an Aristotelian formal cause). Goodenough writes, "The CATTAAGCATAG promoter sequence has a distinctive shape, and proteins will bind to it if they have the correct

15. Newton, *Opticks* 3, query 31 (375–76).

16. Bartlett, *Familiar Quotations*, 479, quoting the introduction to Laplace's *Théorie analytique des probabilités*.

17. Goodenough, *Sacred Depths*, 28.

pockets and protuberances to do so."[18] Reductionism is the final truth. The appearance of life on earth may be a mystery, but the components that lead to life, at least, are not. "The workings of life are not mysterious at all. They are obvious, explainable, and thermodynamically inevitable. And relentlessly mechanical. And bluntly deterministic."[19]

A second example of biological determinism can be found in the work of Chilean biologist-philosopher Humberto Maturana, who coined the term *autopoiesis* to refer to systems that are capable of producing and maintaining themselves by creating their own parts. Maturana says that he finds the idea of autopoiesis exciting for two reasons. "In the first place it solves the problem of identity which two thousand years of philosophy have succeeded only in further confounding."[20] Maturana links the phrase "problem of identity" with Kant's notion of *das Ding an sich*, "the thing in itself"—that is, objects as they really are, independent of human modes of observation and conceptualization. In other words, the "problem of identity" is the metaphysical problem of describing ultimate reality or identifying the fundamental nature of being. Maturana holds that his concept of autopoiesis solves Kant's problem. Maturana continues,

> The second reason why the concept of autopoiesis excites me so much is that it involves the destruction of teleology.[21] When this notion is fully worked out and debated, I suspect it will prove to be as important in the history of the philosophy of science as was David Hume's attack on causality. Hume considered that causation is a mental construct projected onto changing events which have, as we would say today, associated probabilities of mutual occurrence. I myself have for a long time been convinced that purpose is a mental construct imported by the observer to explain what is really an equilibrial phenomenon of polystable systems. . . . I leave it to the reader to engender his own excitement in the discovery of a "purposelessness" that nonetheless

18. Goodenough, *Sacred Depths*, 53.

19. Goodenough, *Sacred Depths*, 46–47.

20. Maturana and Varela, *Autopoiesis and Cognition*, 66.

21. *Telos* in Greek means "end," "aim," "goal," or "outcome." *Telos* is the word used by Aristotle to identify his idea of a final cause. *Teleology* explains actions or events by citing the purpose they are intended to serve rather than the physical forces by which they arise. If a person walks for the sake of her health, the goal of good health is the final cause of her walking. An explanation based on *efficient causes* (which is the only type of explanation allowed by reductionistic physicalism) would explain walking in terms of nerve impulses and muscular contractions—and these biological explanations could be reduced to the principles of chemistry and physics.

> makes good sense to a human being—just because he is allowed to keep his identity, which alone *is* his purpose. It is enough.[22]

In response to this it is tempting to quip that Maturana writes for the purpose of denying the existence of purposes. Maturana attempts to explain away purposes as "constructs" that are "imported by the observer to explain" certain phenomena, while failing to see that this importing and explaining are themselves patently purposive. I suggest that, at this point, Goodenough and Maturana—eminent scientists though they are—join many other scientists in illustrating a very unscientific tendency: the tendency to *interpret their data in ways that are based on a foregone conclusion*. The foregone conclusion is antiteleological materialism. Goodenough, for example, presupposes that the physical shape of the gene is sufficient to explain the seeming purpose that the gene serves, thus arriving at the conclusion that the gene's functioning is "relentlessly mechanical" and "bluntly deterministic." It should be noted that this opinion is not actually the only way the gene's behavior could be interpreted. Rather, the mechanistic-deterministic outlook is an unfounded assumption that then guides the way the gene's functioning is explained. In the study of logic this fallacy is known as question begging.

Normally the tacit acceptance of a foregone conclusion is anathema in science, but, strangely, it is allowed here. The reader is invited to consider that the results of the scientific observations and measurements themselves do not prove the truth of materialism. For example, the brain can be studied at the cellular level by using sophisticated machines, but this method does not prove that the functioning of the brain as a whole can be explained solely in terms of the physical interactions of cellular components. To report scientific findings as if they substantiated a mechanistic or materialist outlook is, methodologically, scarcely any different from the tactic of a religious fundamentalist who presupposes the preeminent authority of the Bible. Indeed, Maturana's "excitement" about the "destruction of teleology" rather resembles the fervor of an evangelist defending a previously espoused faith stance.

A third type of reductionism is related to the doctrine of "the survival of the fittest," a phrase often attributed to Charles Darwin (1809–1882) but actually coined by Herbert Spencer (1820–1903). I watch a good many documentaries about nature and have noticed that many of them explain virtually all animal behaviors as results of

22. Maturana and Varela, *Autopoiesis and Cognition*, 67.

natural selection or "survival value." Sometimes the word *instinct* is used but with the assumption that instincts exist because they contribute to survival. (I examine the nature of instinct more fully in chapter 12.) In these documentaries a tacit desire to sound scientific seems to prohibit any explanation of animal behavior that cites conscious goals or individual enjoyment. For example, birds are typically said to sing or build their nests in a certain way in order to attract a mate. It is obvious, of course, that mating is necessary if the genetic material of a specific animal is to be reproduced, but alluding to this fact as if it were the ultimate explanation of all animal conduct is reductionistic. Philosopher Marjorie G. Grene writes,

> The authoritative position of "science" in our culture stems largely from the successes of the exact sciences. Others, be they biologists, sociologists, psychologists, hope to copy the rigour, the power of prediction of the sciences of inanimate nature, and so become more "scientific" in their turn. Descartes began it, with his beast-machines, his belief that nothing existed but thinking mind and inert-extended matter, that the whole range of living creatures to whom we are so evidently akin were but automata, simulating the feelings, drives, perceptions which we alone possessed. What a deadly vision of the world, and how dearly we have paid for its influence on our minds! A world of men (=minds) and machines—and then, if you like, animals reinstated in a sentimental niche to be our toys. Even evolutionary biology, founded on the kinship of men and animals, has gloried in reducing all alike to the single status of machinery-for-staying-alive, or strictly speaking, machinery-for-not-being-annihilated. The mysterious multifariousness of living nature, formerly a massive obstacle to mechanistic thinking, was transformed by Darwin's genius into an indefinite aggregate of random changes mechanically maintained. Life itself became machinery, made by no one for no purpose except barely to survive.[23]

Several years ago my wife and I were in a wooded picnic area where we happened to hear a mockingbird singing in a nearby tree. The bird methodically and very beautifully copied the songs of all the other birds that were singing. Although the singing may have attracted a potential

23. Grene, *Portrait of Aristotle*, 228. In fairness, I doubt that it is accurate to blame Charles Darwin for the use of the phrase "survival of the fittest" to explain life in terms of a mechanistic algorithm. It is significant that Darwin himself was explaining the origin of *species*—the separation of life into well-defined groups—not the origin of life itself.

mate,[24] it seemed undeniable that she sang as she did simply because she enjoyed it, as ornithologist Richard Prum has suggested.[25] On another occasion our family was living on the outskirts of a town in the mountains. I was up late one night when I looked out the window and saw four or five coyotes chasing one another's tails in a circle under the light of a full moon. What a captivating scene! Probably this behavior, too, could be explained away by someone as contributing eventually to survival, but I think it is undeniable that the coyotes were engaged in the activity simply because they found it enjoyable.

Persons who espouse an ardent belief in materialist worldviews might say that I am projecting my own feelings and values onto nature, but it would be just as logical to argue that scientific reductionism explains away the animals' apparent enjoyment because of the materialists' prior commitment to a reductionist ideology. Materialists claim that their commitment is an intellectual one, but I submit that it is also an emotional one, as is usually revealed when this conversation is pursued at any length with parties who disagree. For example, as Grene notes, many biologists reduce life to algorithms based on survival and genetics. This type of reductionism often turns out to be motivated by a resolute attempt to avoid teleology, failing to acknowledge that if organisms were not motivated by a teleological desire to survive, there would be no survival-of-the-fittest algorithm in the first place. Living creatures could not survive unless they were able to respond purposefully to their environment, seeking what they enjoy and avoiding what brings pain or fear. "Scientific-minded" people explain away this purposiveness by appealing to genetics and natural selection. This type of loyalty to preconceived dogmas is often lampooned by these same intellectuals when they are discussing religion, but it is strangely excused and even expected in science.

For instance, science writer Ethan Siegel assures us that "the Universe really is 100% reductionist in nature," and that "the whole isn't greater than the sum of its parts; that's a flaw in our thinking. Non-reductionism requires magic, not merely science."[26] In Siegel's view, reality can be explained entirely in terms of its two basic components: physical particles and the laws of nature. Any other explanatory factors, including

24. Robinson, "Mockingbirds."

25. See Prum, *Evolution of Beauty*. I owe my awareness of this book to Barrett, "Pragmatist Inquiry," 97, 100.

26. Siegel, "Reductionist in Nature."

various forms of emergentism and holism, are guilty of a fallacious "God of the gaps argument,"[27] which Siegel equates with the claim "that something magical, divine, or otherwise non-physical comes into play."[28] Siegel is perhaps unaware that the phrase "God of the gaps" was originally coined not by scientists criticizing religion but by theologians (yes, *theologians*) who were criticizing the superficiality of the idea that belief in God can serve to explain phenomena that science cannot account for (and which are therefore "gaps" in human knowledge).[29] Siegel warns his opponents not to argue against a "straw man" version of reductionism by oversimplifying it, and yet creating an oversimplified straw man is precisely what Siegel does when he lumps all things nonphysical together with the "magical," the "divine," and the "God of the gaps," thus peremptorily dismissing the possibility that nature itself might include factors that are not solely physical. Siegel is evidently untroubled by the fact that his reduction of the universe to particles and laws of nature fails to explain how laws, which are patently nonphysical, can have any governing influence upon particles. The statement that the particles themselves are simply law abiding is not an explanation but merely a restatement of the question. Siegel's reductionism actually reintroduces the problem of Cartesian dualism, which fails to explain how material substance and thinking substance can interact.

Siegel is one of many scholars who have replaced the older ideas of materialism with *physicalism*, which is usually considered to include not only matter as historically understood but energy and the patterns of interaction or "laws of nature" that describe the interactions between matter and energy. While some articulations of physicalism continue to disavow teleology altogether, others endorse intentionality and consciousness as *emergent phenomena* found in living organisms that have reached a sufficient level of complexity. This "emergent complexity" explanation of purposiveness may seem to be a matter of common sense, but there are critical problems with it. The first is that it is an unsolvable problem to pinpoint the stage of complexity at which sentience emerges. Is an atom sentient? If not an atom, then perhaps a cell? Or a sunflower? Or a centipede? What criteria and evidence can be cited as we address this question? The second problem is that it is equally difficult to explain

27. Siegel, "Reductionist in Nature."
28. Siegel, "Reductionist in Nature."
29. Wikipedia, "God of the Gaps."

just what part of the organism it is that actually experiences sentience.[30] Do the eyes, or the heart, or the nervous system experience sentience? Or only the brain? If so, how can this restriction of sentience be explained? (I address these questions in chapters 10 through 12.) Perhaps sensing these difficulties, not all scientists are as openly mechanistic, reductionistic, or antiteleological as Goodenough, Maturana, and Siegel. Many seem simply to dodge the question of purposiveness or teleology, adopting a sort of "don't ask, don't tell" approach. Nevertheless, whether in biology or in physicalist philosophy, lurking elements of reductionism and mechanism tend to reappear.

For example, in a wonderful video, biomedical animator Drew Berry presents computer-aided graphical animations of the processes entailed by human cell division, focusing on the replication of DNA and the transport of various bits of cellular material from their existing position to new locations as cells maintain and reproduce themselves.[31] These processes are constantly being carried out throughout our bodies and are essential to the preservation of life. The processes are staggering in their sheer numbers, astonishing in the seemingly well-designed efficiency of each process, and fascinating in the ways in which the varying processes work in concert in order to achieve the overall functioning of the cell. Even without explanatory commentary, watching the animations evokes an amazed sense that there is an unfathomable network of collaboration in which all the processes must participate in order to support life.

Of particular interest are the tiny units—essentially, very large molecules—that work to "unzip" the chromosomes and carry various packages of cellular material to the right location. Some of these servant units actually appear to have tiny legs; the upper portion of the unit attaches itself to the material requiring transport, and the "legs" move frontward in alternation, appearing to walk to a preestablished delivery point. Watching the video is like watching an animated cartoon in which little stick figures are busily marching about and carrying objects from place to place. Berry refers to these delivery agents as "astonishing molecular machines" or "DNA replication machines," adding, "you have billions of this machine right now working away inside you."[32] Berry explains that the "machine"

30. For an expanded examination of this point, see Conner, "Beyond Emergentism," 217–20.

31. See Berry, "Animations of Unseeable Biology."

32. Berry, "Animations of Unseeable Biology," 4:03.

> is able to feel when the cell is ready, when the chromosome is correctly positioned. It's turning green here because it feels that everything is just right. And you'll see, there's this one little last bit that's still remaining red. And it's walked away down the microtubules. That is the signal broadcasting system sending out the stop signal. And it's walked away—I mean, it's that mechanical. It's molecular clockwork.[33]

Is there a tension between Berry's ready use of the word *feel* and his assertion that the process itself is actually "mechanical"? This is a tension that chronically reoccurs in biology.

In a different, more detailed video, biologist Ron Vale refers to the delivery agents as "molecular motor proteins."[34] The proteins are essential to life, and yet they are described as machinelike. In still another video, the "motor protein" is said to "strut like a drunken sailor."[35] Despite being referred to as tiny "motors," the transport agents do appear to behave intentionally, and they certainly do serve a purpose. They are said to be "mechanical" and yet simultaneously likened to an intoxicated sailor. We might stop to consider that such a sailor, though probably disoriented, is certainly not much like a machine.

By drawing attention to this incongruity between purposiveness and mechanism, I do not mean to find fault with the video animations or with any of the statements made by the animations' presenters. The lecturers in fact show no evidence of wanting to grind any antiteleological axes; rather, they appear to be as amazed as their audiences by the single-minded effectiveness of the "little machines," the "motor proteins," which, even if they are assumed to have no internal purpose of their own, resolutely march back and forth in order to meet the needs of cellular functioning. The question raised by these videos, and by our previous comments about sentience as something that is alleged to "emerge" from biological complexity, is this: *At what level of complexity is it allowable to talk about purposes?* According to Goodenough and Maturana, purposes are never actually involved at all, for "the workings of life" are

33. Berry, "Animations of Unseeable Biology," 7:21–7:51. The colors red and green are added by the software to identify the transitional segments.

34. Ron Vale is a professor in the Department of Cellular and Molecular Pharmacology at the University of San Francisco. The video is available via Vale, "Molecular Motor Proteins." See also Veritasium, "Your Body's Molecular Machines," and Berry and Uno, "DNA Animation."

35. For the "drunken sailor" animation, see Harvard Medical School, "Molecular Motor."

"thermodynamically inevitable" and "relentlessly mechanical."[36] In this book, such reductionistic mechanism is deemed untenable—for reasons that will become clear in our consideration of relativity theory and quantum mechanics.

In the meantime, let us inquire further about the claim that terms such as *life* and *purpose* are attributable to entities only when those entities rise to a certain level of size and biological development. Many viewers of these videos might say that the motor proteins are simply too small to be described as purposive. As an illustration, imagine that you ask your doctor whether some part of your body is alive—your liver, your skin, or your bones. The doctor will answer immediately that yes, they are alive; if they weren't, your physical health would be in peril. But next let us consider a part of your body that is smaller and less complex, such as the electrolytes in your bloodstream. Ions such as sodium and chloride are essential to your bodily health, but hardly anyone would claim that, taken individually, those ions are alive or purposive. The ions are too small and simple to be regarded as alive. (I do not mean to conflate the terms *alive* and *purposive*, but in the present context we may associate the two.) Now, regarding the tiny "motor proteins," would a biologist say they are alive? Or that they experience any sense of purpose—even a very primitive one? The tiny "motors" are essential to our health, and, unlike sodium and chloride ions, they do seem to behave as purposive agents, unzipping DNA strands and carrying cellular materials from one location to another. They *appear* to be goal-directed, but there is a problem: an almost universally accepted belief that matter is inert and insentient makes it unacceptable to ascribe purposeful motivation to molecules, even if they are relatively large ones.

There seems to be a way out of this problem. Maybe the motor proteins do function with a purpose, but it is a purpose that is *imparted to them* by some larger entity, such as the cell as a whole or even the entire bodily organism. By way of analogy, consider a vending machine. The machine accepts coins or bills and in return delivers snack foods or beverages, even dispensing correct change when the user inserts more money than is needed. The machine appears to be self-animated and to serve purposes, but the purposes have been implanted by an external creator—the machine's manufacturer. Is it possible that the motor proteins, too, have an externally derived purpose, bestowed perhaps by the cell or

36. Goodenough, *Sacred Depths*, 46–47.

by the entire organism? At first this suggestion may seem promising, but there is an unavoidable problem. The problem is that the cell, or even the entire organism, is not an external agent like the manufacturer of the vending machine. We might say that the cell had to come first in order for the organism to emerge. Clearly, then, the purposiveness of the cell cannot have been derived from the purposiveness of the organism. When we apply the analogy of the vending machine to the cell with its motor proteins, what we are saying is essentially that *the vending machine designed and manufactured itself*. That is, the cell could not be the cell that it is unless it were already making use of the services of the motor proteins that are part of its functioning. What this suggests is either that the motor proteins must have been operable prior to the creation of the cell as it now exists, or perhaps that the cell and the motor proteins coevolved together. But in either case, the difficulty is that we still have not answered the question about whether and how purposiveness can be attributed to the motor proteins—for precisely when, where, and how, in the process of coevolving, did purposiveness or intentionality arise?

The materialist contention that such "motors" are the product of chance occurrence or random variations is untenable. Two astronomers, Fred Hoyle and N. C. Wickramasinghe, made rough calculations regarding the possibility of creating a single functioning enzyme by means of random molecular interactions. They started with the set of twenty amino acids that are contained in enzymes. A bacterial enzyme requires two hundred amino acids altogether. (Each of the twenty can be used more than once). Stuart Kauffman describes the calculation of the odds:

> The answer is obtained by multiplying the probability for each correct amino acid in sequence, 1 in 20, together 200 times, yielding 1 in 20^{200}, a vastly low probability. But since more than one amino acid sequence might be able to function to catalyze a given reaction, the authors concede a probability of 1 in 10^{20}. But now the coup de grâce: to duplicate a bacterium it would not suffice to create a single enzyme. Instead, it would be necessary to assemble about 2,000 functioning enzymes. The odds against this would be 1 in $10^{20 \times 2{,}000}$, or 1 in $10^{40{,}000}$. These exponential notations are easy to state, but difficult to take to heart. The total number of hydrogen atoms in the universe is something like 10^{60}. So $10^{40{,}000}$ is vast beyond vast, unimaginably hyper-astronomical.[37]

37. Kauffman, *At Home in the Universe*, 44.

Kauffman concludes that the likelihood of life emerging *strictly by chance* from nonliving matter is "comparable to the chances that a tornado sweeping through a junkyard might assemble a Boeing 747 from the materials therein."[38]

The problem of the emergence of purposiveness in living organisms is related to an issue in academic philosophy known as "the hard problem of consciousness." This is the problem of explaining why and how living creatures—notably, human beings—are able to experience personal self-awareness. We assume that conscious personal experience is made possible by the biological functioning of our brains. The human brain contains about 86 billion interconnected neurons and an estimated 100 trillion synapses, meaning that "an average cell body is in synaptic contact with 1,000 other neurons."[39] The "problem" exists because it is not at all clear how the transmission of this astonishing multitude of signals between these myriad neurons can result in the consolidated, unitary phenomenon that we experience as consciousness or self-awareness. Another way to phrase the problem is to ask how thoughts and feelings can arise from a matrix of nonsentient matter. This question relates to the mind–body problem that is associated especially with the philosophy of Descartes and which is studied today especially in the branch of philosophy known as "the philosophy of mind." The phrase *hard problem* was first coined by philosopher David Chalmers in his paper "Facing Up to the Problem of Consciousness" in 1995.[40] Chalmers distinguished between the relatively "easy" (in his opinion) problems related to the biological-neurological functioning of the brain as it receives and processes signals from the body and then sends signals back. The "hard" problem is to account for the distinctive quality of self-awareness, which we experience as a single, integrated process and not as a vast agglomeration of firing neurons.[41] With regard to our cells' army of motor proteins, there is surely no need

38. Kauffman, *At Home in the Universe*, 45.

39. Goodenough, *Sacred Depths*, 95.

40. Chalmers, "Facing Up to the Problem." Chalmers was born in 1966 in Sydney, Australia, and has continued his academic career in the United States.

41. See Wikipedia, "Hard Problem of Consciousness." Wikipedia does not claim to be an authoritative resource for academic purposes, but typically its articles are very helpful. For example, the article on the hard problem cites over one hundred references. In this book I avoid relying on Wikipedia to document any points that might be uncertain or disputed. On the other hand, in an interdisciplinary work such as this one, Wikipedia is an accessible way of verifying basic facts such as dates, word meanings, biographical data, and so on.

to ascribe conscious experience, but it seems obvious that if we were able to solve the hard problem of consciousness, we would most likely also discover an avenue of approach to the question of purposiveness even at the cellular level.

The position taken in this book is that the conundrum of whether and how to assign purposiveness to motor proteins, and the larger "hard problem" of human consciousness itself, are the result of an obdurate loyalty among most scientists and philosophers to the Aristotelian–Cartesian–Newtonian notion of matter—matter viewed as enduring, neutral, inert stuff that takes up space and has mass—in other words, Newton's "solid, massy, hard, impenetrable, movable particles."[42] Matter so defined can have no thoughts, no feelings, no motives, and no self-caused action and is therefore held to be static unless acted upon by an external force. This is why Goodenough feels led to characterize even biological processes as "thermodynamically inevitable."

The ironic thing to recognize is that this confident belief in materialism or physicalism is not supported by any solid evidence. (The pun is unintentional.) The objects that surround us *look* solid, and in the experiences of ordinary life these objects take up space. But these appearances of solidity fail the tests of science. A thousand years ago most people believed that the earth was flat. After all, it *looks* flat. But evidence opposing this belief slowly accumulated, culminating in the voyages of Columbus (1451–1506) and Magellan (1480–1521). Similarly, in our own time evidence against physicalism has been mounting, and yet many scientists and philosophers still cling to the idea that material substance is at the base of reality. In chapters 4 through 8 we will consider evidence from special relativity and quantum mechanics in support of a worldview that is not materialistic but processive and organismic.

In the meantime, the hard problem really is *hard*. Confronted by a stark contradiction between a physical-mechanical conception of matter and our own experience of self-awareness and purposiveness, we appear to have three options. The first is to acquiesce to strict physicalism, which means that we must view self-awareness and intentionality as mere epiphenomena—as ancillary neurological by-products that have no actual effects on our behavior. Purposiveness is thus illusory, what Maturana

42. Newton, *Opticks* 3, query 31 (375–76).

refers to as "a mental construct imported by the observer to explain what is really an equilibrial phenomenon of polystable systems."[43]

The second option is to postulate that self-awareness and intentionality do exist but that they emerge as realities from physical components only when those components achieve some minimal level of organismic complexity. This option may at first seem plausible, but it entails several difficulties, including the following four: (1) Keeping in mind the virtual impossibility revealed by the Hoyle–Wickramasinghe calculation, there is the difficulty of explaining how the minimal level of biological complexity was achieved in the first place. (2) There is the difficulty of explaining why the inert matter of which nature is composed tends to organize itself into repeating patterns and self-perpetuating types. In other words, we must ask why nature seems to follow mathematical "laws," the presumed source of nature's repeating patterns. I address this problem more thoroughly in the next chapter. (3) There is the difficulty of explaining exactly what it is that the physical components do—that is, just how the atoms and molecules collide or fit together—to cause consciousness or sentience to spring forth. Atoms and molecules interact in countless ways; which of these ways triggers sentience? (4) There is the difficulty of explaining where, precisely, among the relevant molecules, the experience of sentience is located, for in a physicalist ontology everything must be located somewhere. It is unfeasible to attribute sentience to specific molecules because the ascription of sentience to physical objects has already been ruled out. It will then be alleged that sentience characterizes the entire biological organism qua organism, as a manifestation of emergent holism. But the term *holism* is disallowed by many physicalists. And even if we accept the idea of holism, there is a problem as to how the word may be applied. Consider an example. The mineral corundum is very hard. However, corundum is composed of aluminum and oxygen; aluminum is rather soft, and oxygen is a gas. Hardness is said to "emerge" in the corundum molecules as a whole. The difficulty with this example is that sentience is not an externally observable property like hardness; sentience is an internal experience of individual selfhood that is had by organisms. Which occurs first: the sentience, which causes the molecules to become an organism, or the organism qua organism, which causes sentience to appear? This amounts to saying that the organism has

43. Maturana and Varela, *Autopoiesis and Cognition*, 67.

experience because it is an organism, which may be true enough, but it is a circular statement, not an explanation.

The third option is simply to relinquish pure physicalism altogether and give serious consideration instead to a worldview that ascribes something like sentience, or at least an ability to evaluate and utilize information, to nature "all the way to the ground floor." This is the idea that the ability to process information, even information in forms that are conceptual rather than physical, exists even in the most minute subatomic entities. If we regard this ability to process information as primordial, then intuitively it makes sense to think that (a) entities could spontaneously use information to enhance their own complexity and (2) that this enhanced complexity could make it possible to employ information in increasingly creative ways. (The claim that nature experiences human-like feelings or consciousness is often referred to as *panpsychism*. This book does not endorse panpsychism, but, rather, *panexperientialism*. I explore this distinction in chapter 12.)

The idea of particles making use of nonphysical (e.g., mathematical) information is often dismissed peremptorily by persons who deem it self-evident that physical material cannot make use of conceptual (nonphysical) data. Nevertheless our observation of the spontaneous complexification that occurs in nature does suggest that the creativity of nature relies on more than purely physical interactions. We will encounter evidence of this that is even more compelling when we examine relativity and, especially, quantum physics. In the meantime, the position taken in this book is that the prevalent doctrine of reductionistic materialism is actually little but a form of "inculcated ideological imprisonment,"[44] a worn-out shibboleth required for acceptance by the orthodox society of Western scientists and philosophers.

Up to a point, of course, Goodenough is absolutely correct. Life processes do follow the rules of thermodynamics. My argument is not that life somehow violates the principles of chemistry and physics. I am saying that nature has, despite those principles, devised ways of collecting and utilizing energy that would otherwise have been radiated away in all directions—with the result that self-creation and self-preservation can and do take place. The position now being advanced is that self-initiation

44. Credit for this phrase goes to Adam Gopnik, an American critic and essayist, in Gopnik, "Real Thing," 14. Gopnik, it should be noted, applies this phrase not to any debate about materialistic science but to art criticism, and particularly to the work of certain nineteenth- and twentieth-century American painters.

and creativity are results of the fact that *nature traffics in information.* Self-initiation and creativity can occur because entities are able to make use of information in such a way as to instantiate patterns of self-organization—and, in fact, not merely to repeat but to *improve upon* those patterns so that they tend to become increasingly complex. In order to clarify this claim, it is to the topic of *information*—what it is, and how it is utilized—that we now turn.

3

Information About Information

IN TODAY'S WORLD THE word *information* is used in several ways. Some of these simply differ in emphasis, but others are actually in conflict. The goal of this chapter is to clarify the issues and ambiguities that are typically involved. These issues and ambiguities are addressed more conclusively in chapter 9, after we have had a chance to consider contributions made by relativity theory and quantum mechanics. At present, however, it is necessary to establish several preliminary concepts and assumptions.

When discussing the idea of "information" it is important to notice that the noun is derived from a root word, the verb *inform*. Breaking the word down further, we arrive at "in-form," meaning "to impart form" or "to give form to." Considering the verb in this way provides a reminder that "information" normally presupposes an intentional or teleological context, for "to inform" or "to be informed" cannot help but suggest a setting in which there is at least one active agent who is able to recognize data qua data and then either convey that data or make use of it. These are goal-oriented activities. Indeed, some persons assume that "information" refers chiefly to ways in which human beings record data—in writing, on film, or in electronic memory. It is possible to define "information" solely in terms of the concepts utilized by intelligent creatures; this might include not only human beings but animals who are capable of forms of reasoning. Indeed, the argument being developed in this book is that even the smallest parts of nature—electrons and photons, for example—cannot be understood apart from their capacity to receive and transmit information. But this claim requires explanation.

As a contrasting outlook it is sometimes argued that information consists simply of objective data or uninterpreted facts. Information may simply be written on a piece of paper or saved on a microchip. Information such as the density of iron, the chemical formula for glucose, and the distance from the earth to the sun would appear to be simply facts, with no interpreter or perceiver needed. However, further thought reminds us that facts such as these are expressed as they are because of the role of some active observer. One way of getting at the involvement of this observer is to recognize that the information is necessarily expressed by employing some system of units that have been invented by human beings. The density of iron is 7.9 grams per cubic centimeter, the formula for glucose is $C_6H_{12}O_6$ (where C, H, and O are symbols for carbon, hydrogen, and oxygen atoms), and the distance from the earth to the sun is about 93 million miles (150 million kilometers). Grams, centimeters, miles, and chemical symbols are all units or concepts invented by people so that, at least in these cases, "information" seems unavoidably to be linked to human thinking.

A naysayer may object that yes, of course, the symbols or units are human inventions, but the facts themselves are just facts. The facts would be the same even if no human beings existed to be aware of them. But, we might reply, is it really possible that *facts* may be "just facts"?

As an illustration, let us consider an inactive volcano located above a large underground reservoir of magma. The volcano has not erupted in years, but pressure is building inside the magma chamber. If pressure continues to build, the time will come when the expansive force of the molten magma surpasses the restraining weight and the structural rigidity of the overlying rocks. When that happens, the magma will break through and the volcano will erupt. If scientists are able to obtain reliable data about the temperature and pressure of the magma and about the thickness and structural resistance of the surrounding rock, they can make calculations and then publicize a timely warning about the probability of an eruption. But, as our naysayer will remind us, the volcano will erupt whenever the physical conditions require it to, *regardless of any information obtained by human beings.*

I.

The claim that information exists in nature apart from human awareness is an idea that we must consider seriously. The analysis of relativity and quantum mechanics in chapters 4 through 9 will bring clarity on this point. The present chapter presents some critical preliminary ideas.

As a first step in the discussion, it is necessary to respond to the naysayer. We remind the naysayer that our original inquiry was not about the workings of physical mechanisms themselves but about *the nature of information*. What, exactly, *is* information?[1] The naysayer has attempted

1. At this point it is necessary to take note of the interesting overlap between *information theory* and the study of *thermodynamics* in chemistry. The mathematical equations used in both fields display a noticeable similarity, and both chemistry and information theory employ the notion of entropy. The word *entropy* was coined in 1854 by German chemist Rudolph Clausius as a concept in the mathematical analysis of the unavoidable loss of heat in mechanical engines and in all spontaneously occurring physical or chemical processes. For example, if a hot brick is placed next to a cold brick, heat will flow from the hot brick into the cold brick until both reach the same temperature, a state known as "thermal equilibrium." *Disorder* is then said to have increased in the bricks because it is no longer possible to distinguish one brick from the other by measuring its temperature. To clarify this point, imagine the difficulty of shelving several hundred books if none of the books has been marked with any classification information such as the Library of Congress system. There would be no standardized means of organizing the books. The point is that there is a connection between having *less information* and having *greater disorder*. In the case of the two bricks, at thermal equilibrium both *the amount of order* and *the amount of information* have *decreased* because the temperature of the bricks can be stated by taking only one temperature measurement instead of two. Also, the two bricks can no longer be differentiated on the basis of temperature. When the two bricks cannot be told apart, the system is said to be less organized.

In 1948, electrical engineer Claude Shannon at Bell Labs published a paper, "A Mathematical Theory of Communication," in which, for reasons that are not entirely clear, he used the term *entropy* to refer to the amount of uncertainty that was present in transmitted information. Static in a phone line, for example, is disorderly and makes it much more difficult to sort out or make use of the data that are being sent. *Uncertainty* in communication theory thus parallels *disorder* in physical chemistry (see Gray and Haight, *Principles of Chemistry*, 66–69, and Wikipedia, "History of Entropy"). I refer to Shannon's idea of entropy now because our naysayer is liable to cite the mathematical similarities between thermodynamics and information theory in order to press the claim that information can be treated purely as a set of mathematical relations and physical conditions. A full response to this argument is beyond the scope of this book, but a condensed reply is that in both physical chemistry and information theory, the symbols, equations, and technical terms such as *entropy* have all been invented by human beings in order to further human goals—such as, in chemistry, the minimization of wasted energy, and in communications, the improved efficiency of information transmission. The mathematical formulae may look objective and impersonal, but it is nevertheless the case that they are couched in a purposive context—a context from

at the beginning of the debate to stipulate that "information," properly speaking, should refer *only to physical conditions themselves* and not to the ways in which those conditions may be described, measured, symbolized, recorded, or communicated. When the idea of "information" has thus been defined solely in terms of preexisting physical conditions—a definition that frankly seems rather arbitrary and not very practical—then the whole of nature can be reduced to a series of forces, pressures, masses, and collisions, of which an erupting volcano is a good example. It is not hard to understand why this "physical-conditions-only" definition might be insisted on by persons who embrace an a priori commitment to physicalism or scientific materialism. We recall Ursula Goodenough's statement that "the CATTAAGCATAG promoter sequence has a distinctive shape, and proteins will bind to it if they have the correct pockets and protuberances to do so."[2] The pockets and the protuberances themselves are alleged to be the only type of information that nature really needs. In this case, "information" is simply a matter of having the correct physical form, like puzzle pieces. The information in this case is simply that the puzzle pieces either fit or they don't. This definition of information can be illustrated by the children's game called "musical chairs." As music plays, the children circle about, and a chair is removed so that the number of chairs is one less than the number of children. When the music stops the children run to find a seat, but one child gains the information as a sheer physical fact that she has nowhere to sit. For physicalists, the type of information discovered by this child becomes an epistemological paradigm.

As an example of some of the central issues pertaining to the contrast between *physical* information and *interpreted* information, imagine that you are in Arizona visiting the Grand Canyon. As you gaze out across the vastness of the canyon, you can see numerous layers of rock. Most of us know that the process of depositing these layers took millions of years and that, as we look deeper into the canyon, we are, as it were, looking farther back into the past. For many visitors these simple facts sum up the extent of their information about the canyon—but for a geologist familiar with the area there is much more information contained in the rocks: information about the ages of various layers, the names of constitutive minerals and their chemical composition, the properties of the rocks such as hardness and resistance to erosion, the regional

which they cannot be isolated.

2. Goodenough, *Sacred Depths*, 53.

extensiveness of various strata, and the names of some of the deposits, such as "the Kaibab Formation." Clearly there is a sense in which the information is objectively present in the rocks whether or not human beings know about it—but just as clearly, in another sense, the mind of the geologist holds and understands a great deal more information than the mind of a casual tourist. Perhaps, then, it is best simply to agree that there are two definitions of information—one pertaining strictly to physical conditions, and the other pertaining to the knowledge recognized by creatures with minds.

In an attempt to settle this issue or at least to clarify the inquiry as to what information really is, some scientists have proposed that the reality of information can be verified if it can be shown that information actually has *mass*.[3] From the standpoint of materialism or physicalism, verifying that information has mass would presumably enhance information's ontological status as something that really does exist. The ascription of mass to information may seem a little less strange if one is aware of the fact that energy itself adds to the mass of an object. For example, though it may seem surprising, it is true that a spring that is either compressed or stretched weighs a tiny bit more than a relaxed spring. This is due to an increase in the potential energy contained in the spring.[4] This agrees with Einstein's explanation of mass–energy equivalence, which is presented below in chapter 4.

It will be objected, however, that energy is not the same thing as information. One proponent of the claim that information has mass, physicist Melvin Vopson, bases this claim on his proposed "mass-energy-information equivalence principle," stating that information "transcends into mass or energy depending on its physical state."[5] It is not clear what the word *transcends* means in this context. Vopson goes on to warn us that all of the data being generated by human beings may eventually "redistribute" earth's matter from physical atoms to digital

3. Puiu, "Fifth State of Matter." See also Landauer's article, which states that "information is inevitably tied to a physical representation and therefore to restrictions and possibilities related to the laws of physics and the parts available in the universe" (Landauer, "Physical Nature of Information," abstract). Note that though Landauer does acknowledge the representational—and therefore the teleological—nature of information, he verifies its ontological status physically and not functionally or teleologically.

4. Bobsmith76, "A compressed spring has more mass," Physics Forums, Mar. 8, 2012, https://www.physicsforums.com/threads/a-compressed-spring-has-more-mass.585199/.

5. News Staff, "Newly Proposed Experiment."

information—which Vopson posits as a "fifth state of matter," alongside liquid, solid, gas, and plasma.[6] But even if it turns out that information does have mass, the claim that all the data now being generated by human beings would somehow add to the mass of the earth is not reasonable. As Vopson himself notes, nature itself is already permeated with information. For example, every electron in every atom has its own set of quantum numbers. Considering the immense number of atoms that constitute planet Earth, this entails an unimaginably large quantity of information. (The relevance of quantum numbers to atomic orbitals is discussed below in chapter 7.) The creation of information by human beings may more reasonably be understood not, as Vopson suggests, as the creation of some new state of matter but rather as a mere rearrangement or reconfiguration of matter that already exists (whether in the form of a computer chip or simply the quantum information in an atomic system). The only difference related to humanly created information is that the rearrangement was made on purpose in order to provide a configuration that can be deciphered by sentient creatures: us. But obviously the information contained in the layers of rock in the Grand Canyon does not somehow become more massive simply because of the presence of a geologist.

There are in fact several manifest problems with defining information solely in terms of physical conditions. In the first place, though science is often confidently touted as humanity's most reliable and comprehensive source of information, science itself cannot be understood—nor would science even be possible—if the definition of information were restricted exclusively to physical conditions. Science is not merely a vast listing of facts. Science is a human endeavor that inherently entails observation, measurement, experimentation, mathematical analysis, the reporting of results, and acts of assessment and interpretation by a community of scholars. Every single one of these activities is purposive and—dare we utter the word?—teleological. Every paper or book that is published, every experiment that is conducted, and, even, every table of logarithms or chemical properties exists because the results were presumed at the time to be of some value or usefulness. Of course science is objective in the sense that scientists are not permitted to alter their data in order to suit their personal preferences or expectations. But science itself is always motivated and conducted within a framework of values,

6. American Institute of Physics, "Digital Content."

whether those values are stated openly or assumed tacitly. And though it is expected that those values will not be allowed to modify the details of the data or the conclusions that are reported, the values held by researchers and their financial underwriters do influence which investigations are undertaken, the methods by which the research is conducted, the ways the results are reported, and which aspects go unreported because they are deemed irrelevant, inconclusive, or, in some cases, counterproductive to the aims of those who have a vested interest in the research.

But the incompatibility between science as a purposive activity and the claim that information consists of "physical facts only" is not the most serious problem with the physicalist definition of information. A deeper problem is that nature herself seems to rely on information qua information—that is, on what we refer to when we use words and phrases such as *pure data, principles, patterns, general conditions*, or *mathematical laws*.

Consider the commonly used phrase *laws of nature*. The notion that there are laws of nature was not always a part of everyday parlance. We have already noted that pre-Enlightenment worldviews regarded nature as being populated with all sorts of causes and influences that today would be considered mythological or unscientific. But today, nature is said to follow "laws" partly because nature does indeed follow regular patterns that can be described mathematically but also—and this reason remains largely unacknowledged—because *the belief that nature is law abiding can be construed as supportive of the dogma of mechanistic physicalism*. It is to be expected that a world that operates ultimately by means of efficient causes *should* exhibit lawlike patterns of action.

But the role played by "laws of nature" is not as simple and certain as it may seem. For one thing, the so-called laws are not nearly as unbending and absolute as many people think. In Matthew David Segall's apt phrase, the "laws of nature" are really more like "cosmic habits."[7] In most contexts the laws seems lawlike because they are being applied to very large statistical samples. For example, even in the head of a small pin there are about 10^{19} (ten quintillion, or one followed by nineteen zeroes) atoms of iron.[8] Statistically, that's a very large sample! With that sample size we get a high degree of predictability concerning the properties of

7. Segall, *Physics of the World-Soul*, 109.

8. Author's calculation, assuming that a small pinhead may be approximated as a disc 1.4 mm in diameter and 0.6 mm thick and that a steel pin is composed mostly of iron.

the pinhead. Predicting the behavior of a single atom, however, is a different matter.

As an analogy, imagine that you are in an orbiting space station looking down at some highly populated area such as southern California. Within your field of vision there are about twenty-four million human beings. Now imagine that it is 2:00 a.m., local time. It is safe to guess that most of the people are asleep or at least in bed. But if we start selecting individuals or specific households at random, we will soon find persons who are awake and engaged in some sort of activity, even at 2:00 a.m. My point is that individual behaviors are much more difficult to predict than statistical trends. Nature, in important ways, resembles this example—predictable when macroscopic objects are observed, but much less predictable at the atomic and subatomic levels. We will consider this uncertainty further in our discussion of quantum mechanics.

The question we must now ask is how patterns and regularities can appear at all, if nature is composed of nothing but physical particles and physical forces. In what ways does nature use information? And where do the data—presupposed by the "laws"—come from? Some patterns, of course, occur solely because of material interactions. The sand on some beaches is fine and powdery; on others, it is very coarse—and some beaches are composed of pebbles or even small rocks. How does this sorting out, this typology of beaches, arise? The differences between various beaches are determined by factors such as the force of the waves, the depth of the tides, the age of the beach, and the properties (such as hardness) of the minerals that are available. However, nature displays other patterns that are holistic, patterns that appear to bring about increasing complexity and novel functionality. The motor proteins that we considered in chapter 2 are an example. What information or pattern was used to produce those proteins?

This leaves us once more with questions about the functioning of the laws of nature, wondering whether we might now add the question of whether those laws also include hitherto unrecognized "laws of self-organization and complexity."[9] We will postpone further consideration of this question, but for now, the valid insight contained in the common notion that there are laws of nature is simply the recognition that numerous patterns or regularities are maintained in nature from place to place and time to time. Scientific evidence suggests that these patterns

9. Phrasing borrowed from Stuart Kauffman's *At Home in the Universe: The Search for the Laws of Self-Organization and Complexity*.

or requirements are consistent throughout the universe. Another way of phrasing this is to say that the laws refer to *relationships that are constantly being copied and repeated.* Indeed, the very essence of nature depends on this repetition of patterns. When Galileo dropped weights from the leaning Tower of Pisa (or rolled metal balls down an inclined plane; there are conflicting stories), it always took the same time for the same weight to fall the same distance. This regularity shows that nature somehow takes mathematical account of things like distances and weights. This "taking account" is what we really mean when we say that nature traffics in information. The constant *repetition* of these physical states and behaviors suggests that they are based on more than the mere fortuitous, moment-to-moment arrangement of physical shapes. But then, we must ask, where does the uniformity, the repeatability, come from?

II.

In one of his lectures, "The Relation of Mathematics to Physics," physicist Richard Feynman was describing Newton's law of gravitation, which Newton applied with great success to predict the motions of the planets. In his lecture, Feynman said that gravitation can be described using ordinary language but added that mathematical equations provide a more direct approach. The equation looks like this:

$$F = (G \times m_1 \times m_2) / r^2$$

where F is the force of attraction between two bodies, m_1 and m_2 are the masses (or "weights," colloquially) of the bodies, and r is the distance between the two bodies. G is the "gravitational constant," a numerical value originally measured empirically, which has certain values depending on which units are used (grams or pounds, feet or meters, etc.). In his lecture Feynman presented this equation and then wondered how to explain the lawlike quality of the mathematical formula.

> Nevertheless it is kind of mathematical, and we wonder how this can be a fundamental law. What does the planet do? Does it look at the sun, see how far away it is, and decide to calculate on its internal adding machine the inverse of the square of the distance, which tells it how much to move? This is certainly no explanation of the machinery of gravitation! You might want

> to look further, and various people have tried to look further. Newton was originally asked about his theory—"But it doesn't mean anything—it doesn't tell us anything." He said, "It tells you *how* it moves. That should be enough. I have told you how it moves, not why." But people often are unsatisfied without a mechanism.[10]
>
> If this were the only law of this character it would be interesting and rather annoying. But what turns out to be true is that the more we investigate, the more laws we find, and the deeper we penetrate nature, the more this disease persists. Every one of our laws is a purely mathematical statement in rather complex and abstruse mathematics. Newton's statement of the law of gravitation is relatively simple mathematics. Its gets more and more abstruse and more and more difficult as we go on. Why? I have not the slightest idea. It is only my purpose here to tell you about this fact.[11]

As Feynman says, the deeper question is *how* or *why* nature is mathematical. Does the mathematics itself somehow exercise a physical influence? Do the equations themselves somehow have the power to act as physical causes? Aren't principles merely *descriptive* rather than efficacious? The math describes what happens; it does not explain *why* it happens. Thus Feynman paraphrases Newton's response to questioners: "It [the equation] tells you *how* it moves. That should be enough. I have told you how it moves, not why."

But Feynman's phrase describing the laws of nature as "purely mathematical statements" is thought provoking. I have portrayed one type of information as facts associated with specific physical conditions—the density of iron, erupting volcanoes, layers of rock, and a shortage of chairs in a children's game. In comparison with this embodied information, mathematical information appears to be exactly the opposite, for mathematics is not *intrinsically* tied to any specific physical conditions. Mathematical information, in and of itself, seems to be completely generalizable, purely abstract. As a thought experiment, we can imagine a metaphysical system that is truly monistic. In such a system, by definition there is only one reality—perhaps the singularity that is sometimes thought to have produced the Big Bang. But in a pluralistic universe, as soon as there are even two actual things, the *possibility* of addition and

10. Feynman, *Character of Physical Law*, 31.

11. Feynman, *Character of Physical Law*, 33.

all the other mathematical relationships springs into existence, expanding into larger numbers without end. These mathematical relationships include not only addition, subtraction, multiplication, and division but algebra, trigonometry, systems of geometry, matrices, and so on. If it be objected that these mathematical relations do not actually exist, for they are only abstractions, the reply must be that this begs the question, for the subject now being considered is not specific physical existents but *information.* As an example, consider that in algebra the derivation of the quadratic formula relies entirely on symbols, and the formula is valid regardless of any concrete application—but no one would claim that the quadratic formula does not convey information. Information is, in fact, its very essence.[12]

At this point, then, we have two types of information: information that is embodied in physical objects or events, and information such as mathematical information that is purely conceptual. In the remainder of this book I argue that there is a third type of information, which can be viewed as a hybrid combination of the concrete type and the conceptual type. But before considering the nature and the sources of the hybrid type, we must first consider another objection.

Our naysayer may say that all of this is silly, for the role of mathematics in nature can be easily explained—to wit, "The gravitational attraction between planets and other objects is mathematically consistent because all of the objects in the universe are constituted of atoms—or of the particles that make up atoms, along with some others. The particles of each type are all identical; they have the same mass, and therefore they are all attracted to one other by the same gravitational force. The gravitational attraction between large objects such as planets is merely the result of combined attraction of the individual particles of which they are made. It's simple arithmetic."

This explanation may seem logical, but it does not really explain anything. There is a circular quality to the reasoning. The naysayer is saying essentially that the laws of nature are mathematically uniform because all the types of particles that populate nature are mathematically uniform (the same mass, charge, and spin). However, we can imagine at least hypothetically that the universe might have been formed in such a way that the force of gravity took on varying values in each particle or from one part of the universe to another, but empirically this is not the

12. The ontological status of mathematical entities is a much more complicated question than can be examined here. See Horsten, "Philosophy of Mathematics."

case. The naysayer's account does not explain why there is uniformity; it simply shifts the source of the uniformity from macroscopic objects to minute particles. By reducing the behaviors of large objects to the sum of the behaviors of their parts, the naysayer has introduced the new problem of explaining how and why all the specific particles (protons, electrons, neutrons, and others) got to be identical in the first place. Was there some pattern or principle that acted upon all of them from the moment of their creation? And this question takes us back to the original question of how a *principle* or a *pattern* can act as a *cause*. In the words of science historian James Gleick, "A law is not a cause; yet it is more than merely a description."[13] It is the meaning of this phrase, "more than merely a description," that we must explore in greater depth.

The question of whether mathematics or patterns can, in and of themselves, influence nature is not a new one. In antiquity Plato (ca. 425—ca. 347 BCE) theorized that evidence of the influence of patterns or forms is detectable both in the natural world and in the processes of human reasoning. It made sense to Plato to think that the forms must exist prior to physical things in order for the natural world to have been patterned upon them. In other words, for Plato the forms are nature's chief form of information. But Plato's student, Aristotle, rejected Plato's theory of forms because he believed that Plato had failed to provide any explanation of the way the eternal forms are connected to or influential in the temporal world, thus "depriving the sensible world of most of its reality and meaning."[14] Going further, Aristotle denied the ability of patterns or numbers to act on their own as causes. "But if there are to be ideas [*eidos*, sometimes translated "forms"] or numbers, they explain nothing; or if not nothing, at least not movement. And how can magnitude or continuum consist of nonextended parts? For numbers will not, either as movers or as forms, produce a continuum."[15] Aristotle is describing the same issue that we are considering right now: how forms, ideas, or numbers can influence the physical world—that is, what kind of information they are.

The dispute about the functioning of patterns or ideas in nature was reformulated during the Middle Ages in the debate between realists and

13. Gleick, introduction to Feynman, *Character of Physical Law*, ix.

14. Copleston, *Greece and Rome*, 375.

15. Aristotle, *Metaphysics* 12.10, 1075b25–30 (269). By *continuum* Aristotle seems to have meant almost the same thing we mean today when we speak of "the space-time continuum." See also 1075b15–21.

nominalists. Usually this debate is viewed as a question about metaphysics or, more precisely, about the ontological status of *categories* or *classes*. But the debate is relevant to our present discussion because virtually all information, in order to be intelligible, depends on the use of classes or categories. Therefore an analysis of the nature of classes or categories has direct relevance to a discussion about the reality of information.

The nominalist-realist debate was expressed in terms of "particulars" and "universals," with *particulars* referring to specific objects (a single oak tree) and *universals* referring to a class of objects (*oak*, as a type of tree). Realists such as John Duns Scotus (ca. 1265–1308) held that universals have a real, substantial existence, regardless of human awareness of them. This stance resembles the views of Plato. In opposition, nominalists like William of Ockham (1285–1347) viewed universals as mere names, words that are employed as a matter of convenience in order to designate the things human beings wish to refer to. According to nominalism, universals have no independent existence. The things that seem to us to be separated into types by nature itself (for example, oak, elm, and pine trees) are actually particular objects that people classify together by inventing systems of names. An intermediate position known as *conceptualism* was advanced by philosophers like Peter Abelard (ca. 1079–1142), who explained universals as names that refer to immaterial entities called concepts. Concepts are real at least insofar as they exist in the mind and allow us to deal with the world and serve to guide our behavior. We form a concept of a meal plan that we later implement for a family dinner or a imagine a design for a piece of furniture that we are about to build. Scholars carried forward this quarrel between nominalists, realists, and conceptualists into the coming years, and it is not hard to see how that quarrel landed three centuries later in the lap of Descartes, who framed his own dualism as a legacy of its oppositions.

This brief foray into the history of philosophy may seem to have steered us off course, but it adds to our awareness of current options if we understand how our forebears wrestled with questions about the ontological status of information and about how information may be operative not only in the minds of human beings but in the natural world. Today, instead of "universals" we might use words such as *class*, *category*, *type*, *pattern*, *abstraction*, *idea*, *concept*, *form*—or *information*. Ockham maintained that *universals* are mere words or sounds. As a parallel, today some persons would maintain that *information* consists of mere ones and zeroes in a computer's memory or of molecular records stored physically in the connections between the brain's neurons. The other side of the debate—the side of Plato and the realists—is not very common among today's scholars. In this age of physicalism and "hard science," Plato's stock has gone down. But Plato's insights have not disappeared entirely from the intellectual marketplace. They contain too much truth for that to happen.

The third option, conceptualism, may seem most sensible in our own time. Universals are said to have no independent existence, but they do alter the physical world when they exist as mental concepts. This assertion seems obvious enough, but today, as we have noted, the claim that concepts exist in the mind elicits the "hard problem of consciousness," in which the challenge is not to defend the mere existence of concepts but to explain the existence of the mentality or consciousness that makes use of them.

In contemporary philosophy of mind, one approach to this problem is known as *neutral monism*. The words *neutral monism* appear to have been chosen as a direct response to Descartes's dualism: *monism* is intended to indicate that reality is not divided into a dualism between thinking substance and material substance but that there is, rather, *one* ("monistic") type of reality. As for the word *neutral*, it seems intended to indicate that neither mind nor matter is given a favored position, for both matter and mind are considered to be equally present in whatever it is that is deemed to be ultimately real. Unfortunately the label *neutral monism* is an ill-chosen one. The word *monism* insinuates that reality has a unitary, static character—that there exists a single substance in which all things are unified. *Neutral* similarly prompts notions of an immobile, disengaged state of affairs, like a vehicle's transmission that has been taken out of gear.

Beyond the hard problem of consciousness, there is another and perhaps even more difficult problem. It is that, even if we do manage to explain the presence of concepts in the human mind, we still face the problem of Plato and Aristotle and of the realists and nominalists: namely, how is it that concepts, patterns, and types can function in nature itself—if in fact they do function, as they certainly seem to. Considering all this, it appears that what we now need is an understanding of information (1) that avoids the static nature of substances and emphasizes that information is inherently a process of ongoing give and take; (2) that acknowledges that information is utilized throughout nature—information understood not only in the form of physical conditions but in forms that have a conceptual aspect; and (3) that is therefore able to affirm that, in some very rudimentary way, the physical and the conceptual are not mutually exclusive.

III.

In the chapters ahead I argue that one consequence of these three requirements is that human consciousness must be discarded as one of the two contrasting realities constituting the mind-body problem. The other pole of the problem—matter defined as inert physical stuff—must also be discarded. The position taken in this book is that the mind-body problem is, even today, still stated in terms that are implicitly Cartesian and that these dualistic presuppositions make the problem impenetrable.

The "hard problem" assumes at the outset that the essence of human experience is found in consciousness, which is populated with "qualia"—perceptions, thoughts, feelings, and so on. Thus, today's philosophers, who mostly want no part of Cartesian dualism, nevertheless persist in granting paradigmatic status to *human* forms of mental awareness. And yet these same philosophers would surely acknowledge that the human mind is an outcome of ages of biological development, gradually generated over eons from the much simpler perceptual functionings of our earliest forebears on the tree of evolution. The stance taken in this book is that human consciousness may more appropriately be understood not as a paradigm of mental operation but as a more complex elaboration of much simpler modes of responsiveness.

When the "hard problem" employs human experience as an initial premise, it implicitly views human awareness as something like Descartes's thinking substance—a clearly defined type of reality that can unequivocally be said to be present or not present in either-or fashion. Epistemology has thus unthinkingly taken it for granted that *human* knowledge is the definitive example of knowing. Philosophers who are steeped in traditional epistemology and its problems naturally tend to disregard more primitive forms of experience. When animal knowledge is mentioned, our first thought is usually to reverse engineer our own experience into the species in question, quickly arriving at the obvious but mistaken thought that it is silly to attribute feelings such as joy, pain, and yearning to oysters, seeds, and, most implausibly of all, to electrons. Of course what I am claiming is *not* that oysters, seeds, and electrons do experience joy, pain, and yearning but that joy, pain, and yearning are complex emotions that are not suitable as starting points in epistemology. Rather, they are emergent from a prior history of simpler forms, and the simpler forms should be the starting point for our analysis.

When we do speak of *human* experience, it is essential to recognize that in real life consciousness typically is not much like the distinct, logical, focused process of Descartes's *cogito, ergo sum*. More often, consciousness is a tenuous, jumpy, come-and-go affair, which might be better illustrated not by epistemology's typical example of clear, single-minded thinking but by the ordinary experiences of daily life. Consider a person who is driving a car through traffic while chatting with a sibling about their mother, whom they are on their way to visit. The driver is, mostly, engrossed in the conversation but grants to road conditions a level of awareness sufficient to avoid an accident. Moreover, though not thinking about it directly, the driver and the passenger are both under the influence of vague feelings of anxiety concerning a surgical procedure that their mother has just scheduled. Rather than dwelling on Descartes's clear and distinct ideas, their consciousness flits about with varying foci and fluctuating emotional intensity—and yet this is precisely the way consciousness meets our needs in ordinary life.

Examples like this one reveal that all of us constantly have important types of *experience* that are only partly conscious and, indeed, of which we may actually not be conscious at all. Our bodies depend upon an unremitting ebb and flow of vital data, most of which never emerges into consciousness. *Experience*, therefore—and *not consciousness—should be regarded as the primary locus of information*. The word *experience* is broader than *consciousness*. *Experience* is applicable not only to the nonconscious aspects of human life but to the lives of animals and, in some way, to plants.[16] Instead of being equated with experience, consciousness should be viewed as an evolved activity by which certain organisms with more complex brains focus their attention upon varying aspects of their environment. In other words, consciousness is a survival mechanism that helps organisms find food, avoid threats, socialize with others, and otherwise cope with an ever-changing environment. In this

16. Currently there is research in the area of "plant intelligence." Unfortunately this subject is frequently greeted with skepticism, mostly because, with Descartes, most people continue to equate intelligence with conscious reasoning. The following videos may be of interest: WIRED, "Blob Solves Mazes," and TED, "Secret Lives of Plants." I discuss plant intelligence more fully in chapter 12.

regard, consciousness is an evolved biological ability like seeing, hearing, and tasting. In contrast to experience, consciousness should not be deemed a primary epistemological category—or, for that matter, as a primary metaphysical category, either.

I stated above that we need an understanding of information (1) that avoids static, substance-oriented categories and emphasizes instead the processive character of information; (2) that acknowledges that information—including abstract information—is utilized throughout nature; and (3) that is therefore able to affirm that the physical and the conceptual are not mutually exclusive. One way of detecting the purposive character of physical information is to consider the way that experience transmutes physical data.

In epistemological discussions it is a frequent practice to cite a common human experience such as seeing a color—let us say, red. The perception of redness counts as an example of a quale (pronounced "kwah-lay," singular of *qualia*) in the hard problem of consciousness. Redness also counts as information in our present discussion. The complication appears when we ask what *redness* really is or where the experience of redness really originates. The heart of this problem is that redness, physically, originates in nothing more than radiant energy of a particular wavelength—in this case, about seven hundred nanometers. (A nanometer is one billionth of a meter.) There is thus a physical component of redness and a numerical component of redness. But when we actually *see* a color we do not see vibrations; we see the quality of the color itself. Physical information and pure, mathematical information are involved, but ultimately the color *red* cannot be understood simply as an objective attribute of certain objects, as strict physicalism maintains. In both humans and animals, redness is, rather, an experience *constructed* by various bodily processes and by consciousness itself.

We encounter this same problem if we analyze our experience of sound. Human beings and many animal species experience sounds as various tones, but these tones are caused by alternating compressions and rarefactions in the air or other medium that exerts pressure on our eardrums. This is the point of the old riddle, "If a tree falls in the forest, does it make a sound?" There is an obvious distinction between the atmospheric vibrations that are caused by a falling tree and our conscious experience of the sound itself. Thus the *constructed* nature of seeing and hearing reveals the need to convert pure physical data into forms—which might also be called *symbols*—that are useful in guiding the creature's responses to the environment.

Pondering this difference between purely physical data (e.g., photons striking the retina) and symbolic forms (the perception of a flower or a mountain peak) leads again to the mind–body problem and to epistemology's pernicious gap between subject and object or between inner experience and outer causation—a gap that drove Kant to give up on metaphysics altogether. Our analysis leads to the realization that the information we receive in the form of sight and sound unavoidably entails a process of neurological-mental construction—that is, to a type of subjective interpretation. This act of interpretation points to the goal-oriented or teleological aspect of the ways in which information is often utilized in nature. These issues are addressed more fully in chapter 13.

IV.

These considerations lead us to recognize three types of information. I now present these three types not as an established, formal typology but rather as a preliminary, analytical tool.

The first type is information may be viewed sheerly as a description of physical conditions. The sentence "It rained yesterday" conveys information that is mainly physical. The words *rain* and *yesterday* are relatively simple to define in physical terms. Even on planets where there are no life-forms and therefore no language and no conscious experiences, it is possible to have rainfall and an alternation between day and night. Information of the first type may be considered to be present in, or at least the result of, physical facts themselves.

Information of the second type may be referred to as *conceptual.* I am using the word *conceptual* partly as a reference to the conceptualist position found in medieval philosophy. Conceptual information may be viewed as a combination of physical information and pure information. This combination can be illustrated in the example of a shot attempted by a basketball player. The idea of a basketball shot originates in physical conditions, but for the player who is about to shoot, the shot is still only a concept. Conceptual information may be recorded in writing, in a photograph, as patterns on a data chip, or as neurological traces in the mind. The physical traces are essentially information of the first type, but the traces themselves represent information of the second type insofar as they are able to be *considered* or *interpreted* by an experiencing subject.

The word *concept* itself refers to something that is held and employed mentally, and all mental activity is at some level purposive or goal oriented. This purposiveness pertains even to thoughts that are seemingly pointless. I am now referring not only to the Freudian claim that there are no accidental or purposeless impulses but more broadly to the affirmation that, no matter how dysfunctional they may actually be, all mental processes originate in the organism's natural urge to preserve its own health and satisfaction. Human thought relies on conceptual information almost constantly, even when thinking about concrete objects or processes.[17]

The third type consists of information that is purely abstract or nonphysical. The easiest examples to imagine are mathematical, but there are other forms of purely ideational information as well. As mathematical examples, the principles of algebra and calculus seem to convey information that has no intrinsic dependence on specific objects or situations. In relativity and quantum mechanics we encounter empirical evidence that mathematical information is utilized in the basic processes by which nature organizes itself. In fact it seems virtually indisputable that nature's use of mathematical information is ubiquitous. In chapter 9 I argue that the systematic, cumulative use of this third type of information has led slowly but effectively to the formation in the physical world of macroscopic objects and, eventually, to the development of living organisms.

However, this claim prompts two questions that have troubled philosophy and science since the time of Plato and even earlier—two questions that so far have not been decisively answered. The first question concerns the *ontological status* of numbers, mathematical relationships, and other pure abstractions. Mathematicians themselves routinely distinguish between *real* numbers and *imaginary* numbers, with the latter type referring to the square roots of negative numbers.[18] But do numbers—even "real" numbers—really exist? If so, *how* and *where* do

17. I discuss this notion of conceptual or combined information more fully in chapter 10.

18. Readers may want to be reminded that a negative number multiplied by another negative number yields a positive result so that arithmetically there is no number that, when multiplied by itself, produces a negative value. Imaginary numbers are represented as multiples of i, the square root of -1. Imaginary numbers themselves can never be applied to actual objects or concrete situations, but they are nevertheless significant in mathematics and even in physics.

they exist? Persons who are devout nominalists or strict physicalists are prone to deny that numbers have any intrinsic, independent existence. For example, philosopher Richard Rorty (1931–2007) argued that the number 17 has no essence, no existence in itself,[19] and physicist–computer scientist Rolf Landauer (1927–1999) asserted that "information is not a disembodied abstract entity; it is always tied to a physical representation."[20] This is not the place to present an argument against the views of Rorty and Landauer, but I do disagree with them and would affirm that numbers, mathematical relationships, and other pure possibilities are not tied necessarily to localized physical embodiment. Numbers and math do *exist* in some sense, even if their existence entails a need to give them an ontological category of their own.

One approach to this issue is to suggest that the ontological status of numbers resembles the ontological status of *potentia* (*dýnamis*) in Aristotle's philosophy. Aristotle posited a basic metaphysical distinction between actuality and potentiality, which resembles the distinction in modern physical science between kinetic energy and potential energy. Aristotle anticipated Newton's first law of motion when he observed that all objects, whether at rest or in motion, tend to stay in their current state unless acted upon by an external influence. But Aristotle noted that, despite this natural inertia, things throughout nature are typically subject to change in one way or another and therefore can exist in a state of *possibility* or *potentiality*. According to F. E. Peters,

> All of these usages [by previous Greek philosophers] pertain to *dynamis* as a "power," but in the *Metaphysics* Aristotle develops another sense of *dynamis*, i.e., potentiality, and he distinguishes the two in *Meta.* 1045b–1046a; potentiality cannot be defined, but only illustrated (*ibid.* 1048a–b), e.g., the waker is potentially the sleeper; the passage from potency to actuality (*energeia*) is either through art or by an innate principle (*ibid. I1049b–1050a*), hence the necessity of the first mover (see *kinoun*) always in a state of *energeia* (*ibid.* 1050b).[21]

Aristotle suggests that in virtually all areas of life there is at least an implicit need to distinguish between realizable possibilities and unrealizable imagination. The former are relevant to actual processes in a way in

19. Rorty, *Philosophy and Social Hope*, 52–71. I thank Demian Wheeler for drawing my attention to this article.

20. Landauer, "Physical Nature of Information," 188.

21. Peters, *Greek Philosophical Terms*, s.v. "dýnamis" (43).

which the latter are not. With this distinction in mind, we observe that actual possibilities do exist in an ontological state that could be identified as *the possible.* This stands in contrast to the ontological state of the *impossible* or the *imaginary.* Persons possessed by a nominalist temperament may reject the application of the term *ontological* to either category, but in fact the process of distinguishing between the possible and the imaginary is a constant aspect of our whole mental life. It would be inappropriate to insist that unrealized possibilities must exist in some specific place or that they must be preserved physically, whether in a physical arrangement of particles, in a computer's memory, in the human brain, or in the "mind of God." This same comment can be made pertaining to the existence of mathematical entities.

The second question pertains to physical influence. This question asks how it is possible for pure information, such as mathematical relationships, to affect physical reality. In many instances numbers are merely descriptors of existing conditions. When someone says, "There are seven crows in the cornfield," this is simply a report regarding current circumstances. The difficulty comes when mathematics or pure information appears to influence the outcome of events that have not yet taken place. Earlier in this chapter we pondered this puzzle in association with Richard Feynman's remarks about Newton's law of gravitation. Feynman said, "What does the planet do? Does it look at the sun, see how far away it is, and decide to calculate on its internal adding machine the inverse of the square of the distance, which tells it how much to move?" Feynman added, "This is certainly no explanation of the machinery of gravitation!"[22]

There does seem to be a sense in which nature utilizes mathematics or pure information not merely to *describe* but to *influence* outcomes. But how can a natural process "be aware," as it were, of information that is not already in some way present in its physical environment? In human experience we guide our behavior by *imagining* a result that has not yet been actualized; but is there any means by which nature itself can envisage potential consequences? In an effort to answer this question, some philosophers and a few scientists have endorsed the idea of panpsychism, defined by philosopher Anthony Flew as "the theory that . . . the world is rendered more comprehensible on the assumption that every object is

22. Feynman, *Character of Physical Law*, 31.

invested with a soul or mind."[23] In other words, panpsychism is the view that mind or a mind-like aspect is a pervasive characteristic of nature and of all natural objects. As a companion idea, the phrase *world soul* is sometimes used to associate a specific name or identity with this panpsychism; some writers regard the "World Soul" as an actual reality, while others acknowledge it only as a metaphor.[24]

In my view, though there is a grain of truth in these concepts, panpsychism and the notion of a world soul are untenable beliefs. In chapter 10 I elaborate on my reasons for rejecting these doctrines. At present we must simply take note that when considering panpsychism we must distinguish the claim that nature does make use of pure information from the claim, which I believe is excessive, that the universe has a mind or a mind-like capacity or that nature as a whole manifests any overall goal or singleness of purpose. In this chapter it would be premature to try to settle this matter, for this book attempts to base its analysis of nature's use of information on an interpretation of the theory of special relativity and on quantum mechanics, both of which we have yet to consider.

23. Flew, *Dictionary of Philosophy*, s.v. "Panpsychism."

24. For example see Segall, *Physics of the World-Soul*, and Fidler, *Restoring the Soul.*

4

Special Relativity: Experiments and Principles

IN THIS CHAPTER AND chapter 5 we consider the special theory of relativity. Chapters 6 and 7 will present the basic principles of quantum mechanics. Attempting to summarize these two subjects is, obviously, a challenging task. Relativity theory is usually introduced to students only after they have completed foundational courses in physics, algebra, analytic geometry, and calculus. Quantum mechanics is held to be at least as daunting as relativity theory, if not in fact worse. Physicist Richard Feynman put it drolly:

> There was a time when the newspapers said that only twelve men understood the theory of relativity. . . . There might have been a time when only one man did, because he was the only guy who caught on, before he wrote his paper. But after people read the paper a lot of people understood the theory of relativity in some way or other, certainly more than twelve. On the other hand, I think I can safely say that nobody understands quantum mechanics.[1]

Fortunately our assignment is not as demanding as it may seem, for we need not grasp relativity and quantum physics as thoroughly as would be expected of professional scientists. Our goal is more modest. We need to understand relativity and quantum theory only well enough to assess their potential contributions to a radically new outlook on the basic

1. Feynman, *Character of Physical Law*, 123.

characteristics of nature—that is, to perceive how relativity and quantum theory can contribute to the formulation of a new philosophical cosmology or metaphysics. In pursuit of this goal we can bypass the complicated mathematical methods that are integral to the development of contemporary physics. It will be sufficient to understand the outcomes of various experiments and the general conclusions that those experiments have yielded.

But before focusing on actual experiments, it is important to be clear about the meaning of the word *theory* as it is applied to special relativity and quantum mechanics. In some contexts, theory refers to ideas that are hypothetical, conjectural, or speculative—that is, to one proposed explanation among several others. Using *theory* in this sense, creationists have attempted to cast doubt on Darwin's discoveries about evolution by claiming that Darwin's ideas are "only a theory." Scientists, however, often use *theory* in a very different way, to designate an organized, coherent system of ideas that explains and interprets a series of empirical observations. *Theory* thus does not refer to something that is still debatable but rather to that aspect of a scientific topic that is more abstract, ideational, and explanatory, in contrast to reports of specific experimental results. This latter definition of theory is what scientists are using when they speak of the theory of special relativity, quantum theory, or, for that matter, of the theory of evolution. The basic truth of these theories is firmly established.[2]

With this in mind, I will now turn to a consideration of the Michelson–Morley experiment and how it contributed to the special theory of relativity. The purpose and design of the apparatus used by Michelson and Morley are not difficult to understand once one understands the assumptions that inspired the experiment. Those assumptions were derived from considerations pertaining to the wave nature of light or radiant energy.

I.

From April through July of 1887, at what is now Case Western Reserve University in Cleveland, physicists Albert A. Michelson and Edward W. Morley conducted an experiment that was destined to revolutionize

2. For a fuller discussion of the nature of scientific theories, see Barbour, *Issues in Science and Religion*, 162–71.

physics. Their experiment led to basic alterations in the methods of calculation by which we analyze physical motion, it deepened our ideas about measurement and perspective, and it transformed our conceptualization of space and time themselves, leaving them transformed into what we now refer to as the space-time continuum. As the two scientists began their work they had not the slightest expectation of producing any of these results. Their intention was to merely to measure the fluctuations in the speed of light caused by light's transmission through the medium that was then referred to as "the luminiferous ether." To do this, they employed an instrument designed by Michelson and now known as the Michelson interferometer.

To understand the ether, the interferometer, and the results of the experiment itself, it is necessary to know something about the wavelike character of light. In particular, we must understand what an interference pattern is.

> The wave theory of light became fully established in the early nineteenth century, when an English physician, Thomas Young, and a French engineer, Augustin Fresnel, gave simple and fairly complete explanations of interference and diffraction on the basis of wave theory. Today, a phenomenon is believed to be a wave or, at least, to have wave characteristics when and only when interference effects can be produced.[3]

Prior to the work of Young and Fresnel, there had been a long-standing controversy over the question of whether light consists of waves or particles. The Dutch scientist Christiaan Huygens (1629–1695) had championed the wave theory, but Huygens's ideas were abandoned in favor of the corpuscular theory of Isaac Newton, first set forth in his great work *Opticks* in 1704. Newton presented a variety of observations and arguments in favor of particles. One of the most straightforward is that light is reflected off shiny surfaces at the same angle at which the light has struck that surface. If you shine a flashlight at a mirror from an angle—say, 40°—the light will reflect off of the mirror in the opposite direction but also at 40°. This behavior is very similar to the behavior of a tennis ball bouncing off of a tennis court or a cue ball bouncing off of the side cushion on a pool table. Light appears to reflect or bounce off of shiny surfaces in the same way that spherical objects bounce off of flat surfaces.

3. Sears and Zemansky, *University Physics*, 898.

For this and other reasons Newton contended that light is composed of infinitesimal ball-like particles that he called *corpuscles.*

The trouble with the corpuscular theory is that it cannot explain interference or diffraction, two effects that are consistently associated with the transmission of light. In order to understand interference, it is instructive to think about what a wave really is. Waves are not found only in air and water. It turns out that waves are a distinctive phenomenon that arises throughout nature. Waves can be analyzed mathematically. The equations are beyond the scope of this book. However, a general understanding of waves is essential to our examination of relativity and quantum mechanics. This is particularly true as we seek to understand the nature of interference.

In chapter 3 we mentioned sound waves, noting that although we hear sounds such as musical notes as unwavering pitches, the sound waves themselves actually consist of alternating rarefactions and compressions in the air. The vibrating source (any vibrating object) causes air molecules themselves to press on the molecules next to them and then bounce back, creating a chain reaction of rising and falling pressure that tends to spread out from the source of the sound in all directions through the atmosphere. Sound waves are typically described in terms of *frequency*, which designates the number of waves being transmitted per second. On a piano, the note A above middle C is usually tuned to a frequency of 440 cycles (vibrations) per second.

Vibrations caused by alternating fluctuations in pressure are not peculiar to the air. When we look at water we can *see* waves—or at least, we can see the effects of the waves on the water's surface. Water waves appear as the familiar alternation of swells and low points, known as peaks and troughs, that we observe in the ocean. However, as with the air, water waves are actually caused by a chain reaction of molecular collisions, spreading outward from the source of the disturbance.

Imagine that you have tossed a small rock into the still waters of a pond. As the rock enters the water it creates waves radiating outward in concentric circles. The *wavelength* may be understood as the distance between one peak and the next, or between one trough and the next. The wavelength is the distance required for the wave to make one complete up-and-down cycle. Now imagine that you have tossed two rocks into the water simultaneously and that the rocks enter the water at locations ten feet apart. As the waves spread out from each location, the waves caused by the first rock will run into the waves caused by the second rock. Where the peak of one set of waves encounters the peak of the other set, the pressures causing each wave to rise up will join forces, and the water will rise even higher above the normal level of the pond. Similarly, where two troughs coincide, the water level will drop even lower. But if the peak of one set of waves coincides in position with the trough or low point of the second set, the peak and the trough will tend to neutralize one another. The two sets of circular waves will create a curved crisscross pattern on the surface of the pond, and the waves are then said to *interfere* with one another. The phenomenon of interference pertains not only to water waves but also to invisible sound waves in the air and even to light waves and all types of electromagnetic radiation. Understanding the concept of interference is essential to the study of radiant energy (including light) and also to the study of quantum mechanics.

Consider what happens when light of a certain wavelength ("monochromatic" light) is projected from two sources onto a flat board. Imagine also that two narrow slits have been cut in the board and that behind the board there is a white screen so that light passing through the two slits shines on the screen. As the light passes through the slits, light waves radiate outward from each of the two openings and intersect with one another. The intersecting waves create the same type of interference pattern that occurred when we dropped the two rocks into the pond. The pattern appears as regular stripes of light and dark areas on the white screen. The alternating light and dark regions would not occur if light consisted of particles. If light consisted of particles, they would pass through the two slits and produce two distinct areas of illumination on the whiteboard, resembling the pattern that would be formed if we sprayed paint from two nozzles onto a piece of canvas. The alternating light and dark lines that do appear constitute more than merely an interesting pattern. Interference patterns can be created and studied using an instrument called an *interferometer*. Physicists, being the clever folks that they are, can measure the distance between the two slits, the distance from the two slits to the whiteboard, and the distance between the lines on the whiteboard, and then calculate the wavelength of the monochromatic light being used.[4]

4. Sears and Zemansky, *University Physics*, 898–99.

II.

Michelson and Morley knew how to use an interferometer and certain mathematical calculations to determine a light beam's wavelength. As noted already, what they proposed to do was to measure the velocity of the luminiferous ether relative to the surface of the earth. Their experiment was based on several assumptions. Examining these assumptions is now necessary in order to understand the outcome of the experiment.

First, in the late nineteenth century it was believed by virtually all scientists that, since light consists of waves, it must be transmitted by a medium. We have already described the essential role of air or water in the transmission of sound waves. Common sense seemed to suggest that light waves, also, must be carried by a medium that would transmit the vibrations that constitute light's frequency. This medium was called "the luminiferous ether." It was presumed that, since we can see light from distant stars, the ether must be present throughout the universe.

Second, Michelson and Morley assumed that the earth was moving through the ether at a high rate of speed. The surface of the earth seems stable to us, but astronomy reminds us that we are actually in constant motion. In the first place, the earth is rotating on its axis—spinning like a very large merry-go-round. A second source of motion is that the earth is orbiting the sun. The orbit is almost circular, but it is actually an ellipse. The distance traversed through the earth's orbit every year is roughly 584,000,000 miles. One potential issue that had occurred to scientists was that the ether might be carried along with the surface of the earth (the earth's continents and oceans), just as, more or less, the atmosphere is. However, the possibility that the ether moves along with the surface of the earth seemed unlikely. Scientists recognized that if the velocity of the ether were influenced by local topography, this would distort the appearance of the sun, moon, stars, and planets, and even of distant objects on the earth's surface. But we do not observe such a distortion. Therefore, it was reasonable to believe that the ether was not influenced by the earth's motion.

Third, Michelson and Morley assumed that any variations in the velocity of the ether would produce a corresponding change in the velocity of light, *as measured*. Such changes are observed pertaining to sound waves; we refer to this as the Doppler effect. You have probably noticed that if you are standing near the roadside and a noisy vehicle passes by—say, a motorcycle, a train engine, or a driver honking his horn—the noise has a higher pitch as the vehicle approaches and a decreasing pitch as the vehicle reaches you and then recedes into the distance. Note that we are referring not simply to the *volume* but to the pitch or frequency of the sound. The motion of the car toward you effectively causes the sound waves to reach your ears at a higher rate, which causes the pitch that you hear to be higher. But as the car draws even with you, the pitch will drop, and as the car travels away the pitch will fall even more. It is as if motion toward or away from the sound source either compresses or stretches the wavelengths, which causes a corresponding increase or reduction in frequency.

The Michelson interferometer is assembled according to a pattern resembling a plus sign, which can be imagined as arms pointing west, north, east, and south.[5] As an illustration, visualize four identical plastic pipes, each about five feet long, that have been attached to one another by a single connecting joint so that they resemble a large plus sign. Now imagine that this plus sign of pipes has been suspended in a river so that water can flow through the pipes and from one pipe to another in any direction. Suppose that one pipe is pointed directly upstream and that the opposite pipe is, obviously, pointed downstream. Therefore the other two pipes will be oriented so that one is pointed at the near bank and the other is pointed at the far bank. It is easy to see that water will flow through the pipes pointed upstream and downstream at virtually the same speed that water is already flowing in the rest of the river. In contrast, the water in the pipes that are pointed at the banks of the river will hardly flow at all, for those pipes are oriented in a direction perpendicular to the river's flow.

Now imagine, however, that the assembly of pipes has been rotated 45 degrees so that the four arms of the plus sign are oriented diagonally to the river's flow. In other words, if you are standing on the bank, the plus sign now looks more like a large X. Now, flowing water will enter the two pipes that are pointed upstream and flow in a direction diagonal to the river's flow until the water in the pipes is emitted at the downstream end of the structure. In this case, since all four of the pipes are oriented diagonally in the river, *all the water in the pipes will flow at the same speed.*

The construction of the interferometer is similar to this design of the pipes in the river. Of course light does not flow like water in a pipe. Instead of pipes, Michelson used an arrangement of mirrors and

5. Additional information may easily be found by researching "Michelson interferometer." A schematic diagram is available at HyperPhysics, "Michelson Interferometer."

lenses. The mirrors were arranged so that some light would flow straight through the apparatus and some light would flow across the apparatus. The clever thing about this is that because the two beams travel for a significant distance[6] perpendicularly to one another, *any existing ether wind should influence one beam more than the other*, just as rotating the assembly of pipes in the river would cause a contrast in the velocity of the water flowing in the pipes.

Michelson and Morley could not measure the actual speed of the light as it moved through the interferometer. Instead, they designed the interferometer so that the two beams of light were reunited or overlapping and let them shine together on a white screen. The two beams created an interference pattern on the screen. Physically turning the interferometer would cause one of the light beams to change its direction in the ether more than in the other direction, and this should therefore noticeably alter the appearance of the interference pattern. Michelson and Morley anticipated that any motion of the ether relative to their interferometer should be detectable as visible change in the number of fringes—the interference pattern—that their interferometer displayed.

The experimenters had mathematical formulae in mind that would have enabled them to calculate the velocity of the ether "breeze" itself. In an effort to obtain accurate results, they had gone to a good deal of trouble and expense. In the actual experiment, in order to minimize any disturbances caused by intrusive vibrations and temperature changes, the interferometer was assembled in a room in the basement of a stone dormitory. Vibrations were further reduced by placing the apparatus on top of a block of sandstone about five feet square and a foot thick and then floating the stone in a large tub of mercury. The stone block would have weighed about 3,800 pounds, but it would easily have floated because mercury is over twice as dense as sandstone.

With everything in place, the two scientists began to take measurements. They rotated the interferometer and made measurements in many different directions. They worked for several weeks, rotating the device and cross-checking their results. But no matter what they did, and contrary to all their expectations, no measurable variations were detected. As one physics textbook puts it, "The fact that no shift at all was observed was most baffling, and to this day the Michelson–Morley experiment is

6. Michelson and Morley adjusted the mirrors so that the light would reflect back and forth several times before the interference pattern was created. This lengthened the path that the light beam traveled.

the most significant 'negative result' ever obtained."[7] How could it be possible that the ether wind had no effect on light waves? Concerned scientists all over the world were perplexed.

After two or three years without any new proposals, two physicists advanced the odd idea that the negative outcome of the experiment could be understood *if the actual length of one of the interferometer's arms were somehow shortened.*

> At first it did not occur to physicists to give up the notion of the ether. Lorentz and, independently, Fitzgerald realized that the negative result of the Michelson–Morley experiment could be explained if a linear object like a rod or a portion of the interferometer moving parallel to itself through the ether with speed v were to undergo a shortening by a factor
>
> $$\sqrt{1 - v^2/c^2}$$
>
> *Merely by virtue of its motion through the ether*, a length . . . of the interferometer should be contracted.[8]

As a way of analyzing this suggestion, first let us consider the mathematical formula being used. In the expression

$$1 - v^2/c^2$$

the letter v represents the velocity of the interferometer moving in the direction of the supposed ether wind, and c represents the speed of light. The meaning of the math becomes clear if we focus on the fraction v^2/c^2, which can also be written $v^2 \div c^2$. The speed of light is immensely greater than the velocity of any of the moving objects that we observe in our daily experience. For familiar examples of motion, the numerator (v) of the fraction would be very much smaller than the denominator (c), so that even objects with comparatively high speeds would result in a very low value for the fraction itself. This low value, when subtracted from 1, causes almost no change in the value of the mathematical expression.

Let us now make use of another example. Though not a precise parallel, consider that airline flights traveling westward across the United States usually require more time than the return flights to the east. This discrepancy is caused by the jet stream, a persistent high altitude wind

7. Sears and Zemansky, *University Physics*, 909.

8. Sears and Zemansky, *University Physics*, 909–10; emphasis original.

that blows mainly from west to east. The distance by land from, say, New York City to San Francisco is obviously constant. However, aircraft in flight do not travel in contact with the ground but with the air. Since the jet stream is blowing from west to east, a plane flying against the wind from New York to San Francisco is required, in effect, to fly *farther through more air* than it would have to fly if it were flying from San Francisco to New York. This explains why, though the jet itself is traveling *through the air* at about the same speed in each direction, the flight westward takes longer than the flight eastward.

But now consider a very strange idea: if, during the trip westward, we could somehow *actually shrink the physical distance* from New York to San Francisco, then the flight times that were required in each direction could be adjusted so as to come out equal. But this idea is completely implausible, isn't it? Undeterred by this seeming inconceivability, Lorentz and Fitzgerald imagined that if Michelson and Morley's interferometer were somehow shortened only slightly in the direction of the ether wind, this shortening could compensate for the fact that there appeared to be no wind. But, though this notion was successful mathematically, there are several reasons why it was unsuccessful as an explanation. In the first place, the idea of physical shortening is ad hoc; it attempts to explain one phenomenon only. In the second place, it is not supported by any other empirical evidence; in fact it runs counter to other observations. In the third place, it is not supported by any general theory or set of assumptions.[9]

This left scientists with no prospective line of attack concerning the problem of light and its mode of transmission. For eighteen years physicists were perplexed by the negative results of the Michelson–Morley experiment. And then, finally, in 1905 Albert Einstein (1879–1955) formulated the theory of special relativity.[10]

III.

Einstein was born in the southern German town of Ulm in 1879 to a family of nonpracticing Ashkenazi Jews. His mother was Pauline Koch, and his father was Hermann Einstein, a salesman and engineer. When

9. Sears and Zemansky, *University Physics*, 910.

10. Einstein, "Zur Elektrodynamik." This was translated into English as "On the Electrodynamics of Moving Bodies."

Albert was still a baby his family moved to Munich, where his father and his uncle Jakob started a company that manufactured electrical equipment. In school Albert excelled in math and physics. When he was twelve he decided that the world could be explained as a "mathematical structure," and when he was thirteen he read Kant's *Critique of Pure Reason*, declaring afterwards that Kant was his favorite philosopher. At seventeen, he entered the Federal Polytechnic School in Zurich, receiving a federal teaching diploma in 1900. He received a PhD from the University of Zurich in 1905. Einstein was married twice—first to Mileva Marić, in 1903, and then to Elsa Löwenthal, whom he married in 1919. He was unsuccessful in his search for a teaching position, so he accepted a job as an assistant examiner in the Swiss Patent Office in Bern. Einstein worked there in obscurity until the publication in 1905 of four highly original papers that almost immediately brought him worldwide fame.

Though Einstein was aware of the Michelson–Morley experiment, in his paper he never mentions the experiment by name.[11] In fact, according to Michael Polanyi, the special theory of relativity began to emerge when, as a school boy, Einstein wondered about "the curious consequences that would occur if an observer pursued and kept pace with a light signal."[12] This thought experiment was not directly relevant to the experiment of Michelson and Morley. Thus, Lorentz and Fitzgerald had derived their mathematics from a geometrical analysis of the interferometer, but Einstein used a completely different tactic. He ignored Lorentz and Fitzgerald's ad hoc hypothesis that physical objects might undergo an actual reduction in length and made instead a much more daring assumption: that *space and time themselves are not a single, objective reality for all observers.*

> The negative result of all attempts to discover the ether wind suggested to Einstein the bold principle that an observer cannot determine motion with respect to any absolutely established system of coordinates or frame of reference. Observations of events occurring in a system moving with respect to the observer depend on the relative motion of system and observer, not on the motion of the system with respect to the hypothetical "ether." . . . Since [this] principle denies the existence of an absolutely preferred system, the speed of light, which within any

11. Isaacson, *Einstein*, 117.

12. Polanyi, *Personal Knowledge*, 10.

> system is always measured to have the same numerical value, must have the same value in any other system.[13]

In other words, Einstein held that the characteristics of space and time vary, depending upon the conditions in which measurements are being made. This idea was so novel and extraordinary that it had not occurred to anyone else. How did Einstein come up with it?

> Einstein's discovery of special relativity involved an intuition based on a decade of intellectual as well as personal experiences. The most important and obvious, I think, was his deep understanding and knowledge of theoretical physics. He was also helped by his ability to visualize thought experiments, which had been encouraged by his education in Aarau. Also, there was his grounding in philosophy: from Hume and Mach he had developed a skepticism about things that could not be observed. And this skepticism was enhanced by his innate rebellious tendency to question authority.
>
> Also part of the mix . . . was the technological backdrop of his life: helping his uncle Jakob to refine the moving coils and magnets in a generator; working in a patent office that was being flooded with applications for new methods of coordinating clocks; having a boss who encouraged him to apply his skepticism; living near the clock tower and train station and just above the telegraphy office in Bern just as Europe was using electrical signals to synchronize clocks within time zones.[14]

The special theory of relativity is not usually introduced by referring to its implications for our ideas about space and time. Starting with a discussion of space and time would most likely be too philosophical to suit physicists. Instead, relativity is typically explained as a change in our way of interpreting measurements pertaining to objects that are moving relative to the position of an observer, with special reference to the speed of light, which is held to be constant regardless of the motion of the measurer. "The underlying principle in the theory of special relativity may be stated as follows: there is no preferred frame of reference for

13. Oldenberg and Holladay, *Atomic and Nuclear Physics*, 26–27. The authors explain Einstein's thinking as a response to Michelson–Morley, but Polanyi calls this explanation a "historical fiction" that "can be found in every textbook of physics" (Polanyi, *Personal Knowledge*, 9). Whether or not Einstein himself was directly motivated by Michelson–Morley, the negative results of their experiment certainly prepared physicists for Einstein's conclusions.

14. Isaacson, *Einstein*, 113.

the measurement of the speed of light; the speed is the same for every observer, without regard to the magnitude or direction of his motion."[15]

The meaning of this sentence may at first seem plain enough, but it is important to understand that the theory of special relativity goes far beyond the simple statement that "all motion is relative"—an obvious fact that has been recognized since the motion of objects was first studied mathematically.[16] The saying that "all motion is relative" merely expresses the insight that the *measured* speed of a moving object will be reduced for a measurer traveling in the same direction with the object and increased for a measurer moving away from that object. For example, imagine that you are sailing through the air on a hang glider at a speed of 25 miles per hour (relative to the ground) and a hawk flies past you at a speed of 45 miles per hour (relative to the ground). If you are carrying a radar gun and you measure the speed of the hawk, your radar gun will show a speed of 20 miles per hour because the hawk is flying 20 miles per hour faster than you are. This example illustrates the meaning of "relative motion."

But now imagine that you are in a spaceship traveling parallel to a beam of light. Suppose you are traveling at great speed, say, 90 percent of the speed of light. Classical Newtonian physics predicts that as that light passes you, if you measure its velocity, that velocity will be only 10 percent faster than yours, just as the eagle's speed was only 20 miles per hour faster than your hang glider. Amazingly, however, in the spaceship example, classical physics would be wrong. The special theory of relativity predicts, correctly, that no matter how fast you are moving, any beam of light whose speed you measure will be moving at, well, *the speed of light*, because the speed of light is always the same for all observers. This implies that, contrary to common sense, space is not extended uniformly in all directions throughout the universe, and so that time does not pass at a uniform rate throughout the universe.

Mathematically it turns out that in order for the speed of light to be constant regardless of one's frame of reference, *time must actually pass more slowly* for observers moving at high speeds, and *length must actually contract* in the direction of travel. This does not mean that if you are on a spaceship moving at high speed, things seem to move in slow motion or that objects seem to be squeezed together in the direction of travel. None of the arms on your spaceship's interferometer will actually

15. Sears and Zemansky, *University Physics*, 910.

16. Sears and Zemansky, *University Physics*, 81.

shrink. On the spaceship you would still move around at the same rate and your clock would keep time in the same way. *It is time itself that dilates.* The changeable, relativistic qualities of space and time are referred to as "space contraction" and "time dilation." Here is equation for time dilation (if you find this math difficult to understand, have no fear; simply skip this and the next paragraph and resume reading at the paragraph that begins, "As we study the equation"):

$$t_m = t_o \times \sqrt{1 - v^2/c^2}$$

where t_m represents time as it is measured by a clock in motion and t_o represents time measured by a clock that is stationary. It is noteworthy that this equation contains the same expression, $1 - v^2/c^2$, that was derived by Lorentz and Fitzgerald, despite the fact that Einstein's process of reasoning was quite different from that of Lorentz and Fitzgerald.

If we ponder the equation we can tell that when the value of v (velocity) is small, then the fraction v^2/c^2 will also be small, and $1 - v^2/c^2$ will therefore be very close to 1. Any number multiplied by 1 equals that same number, so at the comparably low velocities that are familiar to us on the surface of the earth, t_m (the time that passes for the moving object) will be virtually equal to the "ordinary" time described by Newton's equations. But as the value of v draws very close to c (the speed of light), then the value of $1 - v^2/c^2$ will come closer to $1 - 1$, thus becoming smaller and smaller. This means that less and less time passes from the perspective of a "stationary" observer. I have used scare quotes here to remind the reader that the word *stationary* itself is a relative term. Being earthbound as we are, it is understandable that we ordinarily regard the earth's surface as a fixed point from which objective measurements can be made. But astronomy reminds us that, even without taking the theory of relativity into account, in the universe as a whole there is no anchor, no stable framework, no determinate starting point. Measurements of distance must always and only be made by accepting at least one point of origin *arbitrarily.*

As we study the equation for time dilation, one very remarkable implication is that, at the actual speed of light (when $v = c$), the calculated time must be multiplied by zero (actually, the square root of $1 - 1$). Multiplying any quantity by zero yields zero as a result. This means that, at the speed of light, *time does not pass at all.*

> Einstein's theory of [special] relativity . . . lays down some very curious effects which start to make themselves felt when an object is moving at a very high speed. In particular, the traveller's time scale is slowed down. If he could move at, say, 99 per cent of the velocity of light, he could travel to Alpha Centauri and back in a period which would seem to him to be about ten years; but on his return to Earth he would find that he had been away for much longer than ten years because his time scale, relative to that on Earth, has been changed. Worse, one's mass increases at tremendous speeds. . . . [A] body travelling at the full 186,000 miles per second would find that its mass had become infinite and its time scale had come to a complete stop, which is another way of saying that it can't be done.[17]

Again, the fact that Einstein's conclusions are referred to as a "theory" does not mean that special relativity is merely one intellectual option among others. Special relativity has been verified in numerous ways.

> A highly accurate clock was flown round the Earth in a jet aircraft, travelling at its maximum possible speed. This is very sluggish when compared with the velocity of light, but it was enough to show that when the experiment was over the travelling clock had "lost" a tiny amount of time compared with a similar clock which had been kept in a laboratory on the ground.[18]

Numerous other tests and observations have confirmed the accuracy of the special theory.

> The time dilation effect can . . . be checked experimentally. The Earth is under constant bombardment by cosmic rays, which are not really rays at all, but high-velocity atomic nuclei coming from all directions in space. As the nuclei hit the Earth's upper air, they smash into the air particles and are broken up; the air particles are broken too, and particles called mu-mesons [now known as muons] are created. They last for so short a time before decaying that they ought not to survive for long enough to reach the Earth's surface, yet they do, because their time scale, relative to ours, has been slowed down.[19]

When cosmic rays, usually consisting of high-velocity protons, interact with atomic nuclei in the upper atmosphere, the primary product

17. Moore, *Travellers*, 18.
18. Moore, *Travellers*, 20.
19. Moore, *Travellers*, 20.

is the muon. The muon is an unstable particle with a mean lifetime of 2.2 millionths of a second. Muons created by high-energy protons typically continue at almost the speed of light in about the same direction as the instigating proton. Due to its very brief lifespan, if the muon were moving at a lower speed it would travel for only a short distance before decaying, but the time dilation effect of special relativity (from the viewpoint of the earth) allows muons to continue all the way to the earth's surface.[20] In the earth's frame of reference the muons have a longer half-life due to their velocity. On the other hand, from the viewpoint (the frame of reference) of the muon, it is the length contraction effect of special relativity that permits this longer trajectory, since in the muon's frame its *lifetime* is unaffected, but the *length contraction* causes distances to the earth's surface to be far shorter than these distances are when measured in the earth's frame of reference. Both effects are equally valid ways of explaining the fast muon's curious survival over distances.[21] Again, it is important to emphasize that *neither the earth's frame of reference nor the muon's produces the "real truth."* Both measurements are true and "real," even though they appear to conflict. This may be regarded as one of the so-called paradoxes of relativity. We will suggest in the next chapter that the paradoxes and contradictions entailed by special relativity are not actually paradoxes and contradictions at all; it is our intuitive conviction that space and time are objective "receptacles"[22] that create the contradictions. Moreover, the "paradoxes" of relativity turn out to be a good deal less paradoxical if we begin with the pragmatic theory of truth's idea that truth is to some degree derived from usefulness rather than from "objectivity" and conclude with the metaphysical notion that reality is based not on objective material conditions but on experience. We consider these ideas more fully in chapters 8, 9, and 10.

In the meantime, the lesson now to be learned is that if, at the speed of light, time does not pass at all, then this timelessness must pertain to photons, which are particles of light or, perhaps more appropriately,

20. Muons are passing around us and through us every day. You can actually detect them in your own home. See CosmicWatch, "Cosmic Muons."

21. See Wikipedia, "Muon."

22. This use of the word *receptacle* originates in Plato's dialogue *Timaeus* and is generally understood to suggest "a pre-cosmic 'container' filled with chaotic proto-elements"—in other words, a universal space in which all specific objects exist and all specific events take place (Gorey, "Atomism and the Receptacle"). See also Zeyl and Sattler, "Plato's *Timaeus*," sec. 6 ("The Receptacle").

packets[23] of light. The name *photon* was coined by American chemist G. N. Lewis in a letter to the magazine *Nature* in 1926. It may occur to the reader that, just when we had settled on the belief that light consists of waves, we now find ourselves resorting once again to the notion of particles.

> The present standpoint of physicists, in the face of apparently contradictory experiments, is to accept the fact that light appears to be dualistic in nature. The phenomena of light propagation may best be explained by the electromagnetic wave theory, while the interaction of light with matter, in the processes of emission and absorption is a corpuscular phenomenon.[24]

This last sentence, which detects the wave nature of light in certain circumstances and the particle nature in other circumstances, is very important. We examine the philosophical implications of this idea more closely in our next chapter. For now, we must take note of the fact that even when light does resemble a particle, the particle in question is a very peculiar one. Existing nontemporally is not the photon's only weird attribute. Photons also are understood to be massless. That is, if they could be weighed on a scale, their weight would be zero. In a way, this follows from the statement made above that at relativistic speeds mass increases, and at the speed of light, by calculation, mass would be infinite. As noted already, this is impossible. But since a photon has *no mass*, it would not acquire an infinite mass at the speed of light since zero is zero no matter how much we multiply it.

The photon's zero mass is not merely theoretical. It is also observed empirically. Daily experience shows that, ordinarily, when a moving object collides with a stationary object, the moving object imparts at least some of its momentum to the stationary object. If you throw a rock and it strikes a bottle balanced on a fence post, the rock will have some visible impact on the bottle. But when light strikes an object, the object does not recoil. It doesn't even wobble a little.[25] When you open your kitchen curtains in the morning and sunlight suddenly illuminates the breakfast table, the light does not knock over the salt and pepper shakers. It will

23. The term *packets* is applied to photons by Sears and Zemansky, *University Physics*, 824.

24. Sears and Zemansky, *University Physics*, 824.

25. Technically photons do impart momentum but to a much lesser extent than material objects do.

warm them, but that's because of the radiant energy the photons transmit, not because of the collision of two masses.

Besides being timeless and massless, photons generally are also deemed to take up no space. The size of a photon proves to be a slippery subject, for nobody can hold a photon down and measure it with a caliper. Historically photons have been understood as "point particles," which means that, like a point in geometry, the photon has no physical size; it has a location but no dimensions. However, sometimes photons do appear to have a size, but physicists speak of this size in terms of "cross section," which has to do with the effective distance at which the photon interacts with objects nearby—and this varies depending on the circumstances.[26] So, photons are timeless, massless, and spaceless—or if they do have a size or shape, it varies according to their circumstances. Photons thus have none of the traditional characteristics of matter, except that what they *are* very good at is imparting energy. We investigate this further in our next chapter.

26. See Koberlein, "About the Size of It."

5

Special Relativity: Philosophical Implications

I HAVE SUMMARIZED THE special theory of relativity and some of its physical exemplifications. We are now ready to consider the theory's philosophical implications. Naturally it is possible to minimize philosophy and to study special relativity only with regard to its practical application in the physical sciences—primarily physics, astronomy, and sometimes chemistry. But as stated already, practical science is not our emphasis here. We are interested in discovering the relativity's meaning for philosophical cosmology and metaphysics—that is, its contributions to a fundamentally new description of nature as a whole. As we begin, we need to be clear about some limitations pertaining to our manner of proceeding and about the overall scope and purpose of our discussion.

A first observation is that the interpretations presented here are not meant to serve as a new, definitive, stand-alone interpretation of special relativity. The ideas articulated in this chapter are in no way intended as a response to the extensive body of scholarship that already exists concerning the philosophical implications of Einstein's work. I make no attempt to survey the conclusions of others, interesting and important though they may be. The reasons for this are twofold: first, my purpose is not to create a detailed metaphysical system but to articulate some general ideas that are suggested by the special theory of relativity and that represent a significant departure from the conventional notions that have been maintained in Western worldviews for several centuries; and second, these general ideas are intended specifically to show how a scientific worldview may legitimately be not merely compatible with the human experience of meaning but supportive of it. In other words, I admit at the outset that my analysis of special relativity is intended to support a specific line of reasoning. This does not mean that the analysis is distorted; it does mean that certain concepts are emphasized deliberately.

A second comment is that, rather than adopting the precise language of philosophy or the mathematical methods of science, I intentionally rely more heavily on metaphors and other figures of speech. There are three reasons for this. The first is a merely practical one—that, although the ideas now being presented should be of interest to both scientists and philosophers, this book is intended to be accessible to persons who are not specialists in either field. The proposals that are put forward are envisioned not as technical scientific innovations but as contributions to a new worldview that is meant to be comprehensible for any thoughtful person.

The second reason why technical jargon is not appropriate is the admission that, in discussions about relativity and, especially, about atomic theory, scientists themselves have been forced to rely on metaphors and figures of speech. We would do well to recognize that the phrase *frame of reference* itself is composed of two metaphors. The rather subtle term *complementarity* is an attempt to explain the coalescence of two seemingly diverse conceptualizations of certain physical phenomena. Scientific cosmology often employs the phrase *fabric of space-time* or *fabric of reality*, but of course the word *fabric* is entirely metaphorical—and the metaphor "fabric of space-time" is in my view not a very helpful one. Particles are routinely described as having *spin*, though the particles themselves are not really tiny spinning spheres; spin is detected by the particle's behavior in a magnetic field. Quarks are a type of elementary particle and a fundamental constituent of matter, but quarks are said to exist in six "flavors": *up*, *down*, *top*, *bottom*, *strange*, and *charm*. Indeed, the name *quark* was first given to the particle by physicist Murray Gell-Mann, who had encountered the word in James Joyce's 1939 book *Finnegans Wake*. The third and most substantive reason for employing figurative language has to do with an appropriate degree of intellectual humility. Language and thought have intrinsic limits. In today's world it is no longer feasible to embrace Descartes's resolute confidence in "clear and distinct ideas." We should allow ourselves to be forewarned by the fact that the histories of both science and philosophy are cluttered with discarded ideas that were once endorsed as indisputable truth.

Fortunately, the goal we need to pursue is a broad one that does not require validation from precise measurements, mathematical analysis, or meticulous logic. In this chapter we shift from a description of the results of specific experiments to a consideration of the characteristics of nature as a whole. This more wide-ranging subject matter requires us to eschew a façade of exactness or certitude that would be unwarranted or misleading. As we attempt to describe the general characteristics of nature, it is prudent to recognize that language is innately ill-suited to perfectibility of expression. Theologian Bernard Meland wrote,

> Language and reality, while related, do not coalesce. Yet, neither are they dissociated. The language of any people is a formulation out of their structure of existence; yet their structured existence participates as bodily event, and possibly much more, in the Creative Passage which connotes the whole of the ongoing reality.
>
> This, then, is not the old distinction between appearances and reality, but one of accounting for individuated, creatural difference along with designating identity both with the community of creatures and the ultimate Creative Passage in which reality as ultimacy inheres.
>
> . . . Our language, in other words, is inevitably, and of necessity, made up of fallible forms and symbols.[1]

Though many people speak of "literal language" or "literal meaning," it is a conspicuous irony that the notion of *literal meaning* is actually very difficult to interpret literally. All language is referential and symbolic. The sounds we utter acquire their meaning by association with experiences gained from our physical senses, by construction using combinations of other words, or as products of various mental processes that may or may not be consonant with the experiences of other persons. Naturally there are contexts when precision of statement is crucially important. We think, for example, of certain legal proceedings, of vital medical information, or of the principles that are mandatory for the safe construction of buildings, aircraft, and automobiles. Other contexts, however, call for language that is not and should not be so precise. Theologian Harvey Potthoff wrote, "It is dangerous to assume that for every idea or every insight there is an exact word or an appropriate symbol. There is an ultimate loneliness of the mind and spirit. And so we use the linguistic tools which are available, knowing they are inadequate. It is likely that we have

1. Meland, *Fallible Forms*, 22–23.

neither words nor syntax for the most important truths about the world and human life."[2]

As Potthoff suggests, it is this less exact, more figurative type of language that is appropriate to the very stuff of life. Consider words of friendship, affection, loyalty, and love; of literature, art, music, and the appreciation of beauty; of moments of intensity, deliverance, and redemption; of humility and repentance, and of courageous self-affirmation; of commitment to our highest ideals; of the most powerful expressions of moral suasion and social reform; of religious feelings and our encounter with ultimacy; of life's endless quest for higher hope and deeper meaning. In all of these circumstances, language that sounds literal, overly precise, or emotionless is conspicuously out of place and may even seem stuffy or comical. I am not suggesting that statements about the world as a whole must be *solely* metaphorical or figurative. Some statements are either true or false, helpful or unhelpful, and some metaphors, after all, are more apt than others. My point is simply that it is a mistake to pretend that any worldview or system of metaphysics can be composed entirely of statements that are mathematically precise, purely objective, or philosophically unassailable. This disclaimer needs to be stated explicitly, for over the centuries numerous scientists, philosophers, and theologians have demonstrated an obstinate preference for the precise types of expression that bolster the appearance of indisputable accuracy.

We these thoughts in mind, we now turn to four specific propositions about a new depiction of the world that is suggested by the special theory of relativity. But, as Whitehead said, "the exactness is a fake."[3]

I.

Space and time are not uniform and constant throughout the universe but may more accurately be understood as modes of relation. This statement may sound simple enough, but its ramifications are subtle and far-reaching. Though we may easily form the words, the idea that space and time are "modes of relation" remains counterintuitive and difficult to accept. In chapter 2 I noted that the perception of matter as inert, neutral stuff is first embedded in our worldview by our formative childhood experiences. The same comment can be made regarding our experience of the

2. Potthoff, *God and the Celebration*, 116.

3. Whitehead, *Essays in Science*, 96 ("Immortality").

constancy and uniformity of space and time. As soon as babies learn to reach out and touch objects, the unfluctuating sameness of space seems to be verified directly. Throughout our lives, our eyesight and hearing tell us that there is a consistency in the ways that objects are nearer or farther away. Though we do have personal experiences when time seems to drag by or fly by, the regularity of the ways objects move, the time required for the completion of familiar physical processes, the regular cycle of day and night, and our reliance on clocks and other time-keeping devices all seem to corroborate our feeling that time passes at a constant rate.

Our intuitive endorsement of this perception has been reinforced by centuries of philosophical and scientific literature. Plato's *Timaeus*, written in about 360 BCE, describes space and, by implication, time as a "receptacle" that must exist constantly and uniformly as something like a cosmic container within which all the action of the physical world takes place. Plato describes this receptacle as "the nurse of all generation,"[4] adding that space "is eternal, and admits not of destruction."[5] Plato's description soon acquired the weight of authority and seemed in any case intuitively obvious. However, the most influential proponent of the receptacle theory must, in fact, be not Plato but Isaac Newton, whose principles of mechanics presuppose that space and time are neutral absolutes that exist as a necessary precondition of all motion and physical change.[6] Interestingly, Newton was actually opposed at this point by one of his own contemporaries, Gottfried Wilhelm Leibniz (1646–1716), who deduced, rather ahead of his time, that, as I have said, space and time are "modes of relation."

> I have many demonstrations, to confute the fancy of those who take space to be a substance, or at least an absolute being. . . . But if space is nothing else, but that order or relation; and is nothing at all without bodies, but the possibility of placing them; then those two states, the one such as it now is, the other supposed to be the quite contrary way, would not at all differ from one another. Their difference therefore is only to be found in our chimerical supposition of the reality of space in itself. . . . The case is the same with respect to time.[7]

4. Plato, *Timaeus* 49b (Jowett).

5. Plato, *Timaeus* 52b (Jowett).

6. Time and space are treated as "absolutes" in Newton's *Mathematical Principles of Natural Philosophy*, first published in Latin in 1687.

7. Leibniz, "Third Reply," 26.

Newton himself did not deign to reply directly to Leibniz's arguments but instead allowed his friend and supporter Samuel Clarke (1675–1729), an English philosopher and clergyman, to respond. The Leibniz–Clarke correspondence took place in 1715 and 1716.[8] By then Leibniz and Newton were already prominent disputants, having engaged from 1699 through 1711 in a quarrel about which of them had first invented calculus. Interestingly, in calculus today it is Leibniz's system of notation and not Newton's that is more widely used.[9] At any rate, concerning the nature of space and time, up until 1905 scientists and philosophers sided with Newton. But with the publication of a single journal article, Einstein changed all that.

Let us ponder another thought experiment—a rather technical one—concerning the way in which special relativity treats space and time not as absolutes but as modes of relation. Imagine that an unstable atomic nucleus expels two electrons, each moving away from the originating nucleus at nine-tenths of the speed of light and in precisely opposite directions. From this point on we will need to make use of two equations, but readers who dislike equations need not feel excluded; after the math is finished, the results are interpreted in plain English.

Using the familiar symbol c to denote the speed of light, we say that one electron has a velocity of $0.9\ c$ and the other a velocity of $-0.9\ c$. One value—it doesn't matter which one—has a minus sign because the electrons are moving in opposite directions. In classical mechanics, the velocity of the electrons relative to one another would be calculated simply by subtracting −0.9 from 0.9, just as, in chapter 4, I subtracted the velocity of the hang glider from the velocity of the hawk. In arithmetic, subtracting a negative number produces the same result as adding a positive number, so the expected result would be that the two electrons are separating from one another at a velocity of 1.8 times the speed of light. This seems obvious—but it is unacceptable. Velocities exceeding the speed of light are forbidden by special relativity.

The correct solution requires us to use Einstein's relativistic equation for velocity. If v' equals the velocity of the two electrons relative to *one another*, this equation is

8. Alexander, *Leibniz–Clarke Correspondence*, ix.

9. Leibniz's notation was better suited to generalizing calculus to multiple variables, and in addition it highlighted the operator aspect of the derivative and integral. As a result, much of the notation that is used in calculus today is the invention of Leibniz (Tomforde, "History of Calculus").

$$v' = (v_1 - v_2) / [1 - (v_1 \times v_2)/c^2)]$$

Because $v^1 = -v^2$, we make the substitution that $v^2 = -v^1$. This yields

$$v' = 2v / [1 + (v^2/c^2)]$$

and $v' = 0.99475\ c$.

Stated in ordinary language, the two electrons, relative to one another, are separating at a velocity of 0.994 times the speed of light. This seems reasonable and is allowable, knowing what we know about the speed of light as an absolute limit on velocity.

Now, just for fun, let's replace the two high-speed electrons with two familiar terrestrial objects—say, two cars speeding past one other in opposite directions, each at 90 miles per hour relative to the surface of the highway. First, so we can compare the velocity of the cars with the speed of light (which is expressed in miles per second), we convert miles per hour (90) to miles per second (0.025). Then, calculating their rate of separation by using the same relativistic formula that we used for the electrons, a calculator reveals that, after they pass each other, the cars are separating at a speed (v') of 180 miles per hour, which is what we would have assumed intuitively. Relativistic equations make virtually no difference at the comparatively low speeds of which automobiles are capable. We see that at ordinary velocities the relativistic equation produces virtually the same result as is predicted by Newtonian physics.[10]

The larger lesson of this thought experiment is that *there is no single, correct frame of reference*—no system of time or space that can be relied upon as a neutral or objective standpoint for making measurements everywhere. If, when we first considered the thought experiment, it seemed that the two electrons must be moving apart at 1.8 c, that's because we assumed uncritically that the position of the original emitting nucleus could serve an objective frame of reference. The mistaken intuition behind this assumption is that space and time extend uniformly from the standpoint of the emitting nucleus into the whole universe. The counterclaim that space and time are modes of relationship maintains that space and time (or the space-time continuum) are a means of creating or describing the separateness between two objects or events (what Leibniz

10. I am very grateful to chemist John Michael Nicovich for guiding me through the mathematics involved in this thought experiment.

called an "order of relation") and that these descriptions vary depending on the measurer's physical setting. When considering the two electrons, the only way to determine their velocity is *to relate them directly to one another*. The position of the originating nucleus becomes irrelevant. This fact gets at what we mean when we say that space and time are modes of relation rather than independently existing absolutes. This relativistic or relational understanding of space and time does not mean that they are merely subjective or imaginary or illusory. Space and time are objective realities—as our calculations have demonstrated—but their objective properties are described by relativistic mathematics rather than by Newtonian mechanics.

II.

A second lesson from special relativity is that *the phrase "frame of reference," which is indispensable in any interpretation of special relativity, strongly suggests that something like perspective or viewpoint is inherent within nature.* This in turn implies that *nature itself is radically relational.* It is as though the myriad relations that compose nature insist on being measured somehow, even by physical objects, and the measurements vary depending on the standpoint from which they are made. This idea, in turn, summons up the image of an active agent making deliberate observations. The word *reference*, after all, is derived from the verb *refer*, and *frame* hints at the existence of a previously constructed template or system of demarcation. Ordinarily verbs such as *refer* and *frame* would involve purposive action.

However, at this point we need to be careful. It must be made explicit that the existence of a frame of reference does not depend on conscious agency. The dilation of time indicated by the previous chapter's orbiting atomic clock and the high-velocity muon's extended life demonstrate that "frame of reference" pertains to actual physical facts, whether or not some observer is making measurements. In chapter 3 I quoted Richard Feynman's rhetorical question, "What does the planet do? Does it look at the sun, see how far away it is, and decide to calculate on its internal adding machine the inverse of the square of the distance, which tells it how much to move?"[11] Clearly nature does not need to "look at things" or use a calculator. "Frame of reference" does not imply that perspective

11. Feynman, *Character of Physical Law*, 31.

or viewpoint is somehow the actual determinant of physical outcomes. However it does imply that *all physical processes are relational.* Let us expand upon this.

In human life, the phrase *frame of reference* may make us think of the familiar experience of encountering objects or situations from contrasting orientations. For example, sometimes we scrutinize a painting closely in order to perceive the artist's techniques, or we stand at a distance in order to absorb the entire work. We may hear a certain piece of music by attending a concert or, in contrast, by listening to a recording. Variations in "frame of reference" apply not only to the objects that we inspect but to personal interactions. After a funeral it is common for family members to share memories and opinions about the person who has died. Many memories will prompt general agreement, but almost always there are discrepancies regarding details, along with differing assessments of the person's traits and achievements. In human experience, "frame of reference" and "perspective" are swayed by the constant influence of various purposes and motives, whether conscious or unconscious.

This is not the case when "frame of reference" is used in explanations of special relativity. In human life, purposes and motives are complex psychological phenomena. In physics, the phrase *frame of reference* is a way of allowing for the fact that measurements of both space and time are altered when the relation between two objects is characterized by a difference in velocities. But whether or not two objects are moving in relation to one another, there is a sense in which nature itself takes their motion into account as the gravitational attraction between them is established. It is as if each object's velocity is "measured" implicitly.

This brings us to the deeper insight that the phrase *frame of reference* implies a kind of radical relationality that exists throughout nature. The occupants of the universe constantly "refer" to one another. For example, our own Milky Way Galaxy contains an estimated one hundred to four hundred billion stars occupying a disc-shaped area approximately one hundred and seventy thousand to two hundred thousand light-years across.[12] The "barred spiral" configuration of the Milky Way's stars is the result of the aggregate gravitational force of the stars operating over vast distances for billions of years. Galactic gravitation may be imperceptible

12. One light-year is the distance light would travel in one year, which equals 5,869,713,600,000 (5.8 trillion) miles. The estimated diameter of the Milky Way is about 1,000,000,000,000,000,000 (one quintillion) miles.

to us, but its power is constant and efficacious, operating over eons of time.

The universe's tendency to form galaxies is an obvious example of cosmic connectedness, but interstellar gravity is not the best example of special relativity. A better example is the red shift of the light that we observe from galaxies that are receding from ours. From the earth's frame of reference, light from distant galaxies is converted to a redder color (towards a longer wavelength) because those galaxies are rapidly moving away from ours. From the frame of reference of those galaxies, *our* galaxy's light would be red-shifted. The relativistic masses of objects moving at velocities near the speed of light are increased because of that velocity. But in space there is no stable center point, no geometric origin, no overriding frame of reference. Therefore there is no impartial way to determine which mass is moving. An observer on the object that we deem to be moving would detect that *we* are moving and calculate that *our* mass has been increased. In a collision, the amount of energy released would be the same regardless of which mass was considered to be in motion. Relativistic mass is thus not a settled amount for an object[13] but a property that varies whenever there is a contrast between the velocity of that object and of another object,[14] and *the word* velocity *has meaning only when at least two objects are involved in a comparison*. In Newtonian physics, mass was held to be conserved; the mass of an object remained unaffected by its relationship with any other object. In special relativity, the strict conservation of mass is no longer an accepted principle.

The bottom line of all this is that the phrase *frame of reference* ultimately follows from the fact that the real things of which the universe is made up are what they are not because they are composed of a static, internal essence but because of the ways in which they function or interact. Superficially, the relationships that permeate nature may seem to be purely mechanistic, but the deeper insight is that these relationships depend upon the constant transmission of information related to velocity, mass, energy, frequency, electrostatic charge, spin, gravitational attraction—and possibly other factors. As we recognized in chapter 3, the attempt to force all of this information into the preconceived mold of pure physicalism is very unconvincing and, in fact, simply implausible.

13. Physicists do use the terms *rest mass* or *invariant mass*, but in our current context these phrases would suggest the static depiction of matter maintained in classical physics, which would lead to errors in many contemporary applications.

14. See Wikipedia, "Mass–Energy Equivalence."

The universal applicability of the phrase *frame of reference* points to the radical relationality of the universe and to the indispensable role of information as that relationality's necessary vehicle. In chapters 6 and 7 we will see that this principle of cosmic relationality is equally applicable to the phenomena of particle physics.

III.

Third, *the relationality of nature suggests in turn that nature must be fundamentally dynamic and processive rather than static.* It seems intuitively obvious that entities that are naturally interactive are also naturally in process. We see this illustrated amply in astronomy, in the recurring formation, evolution, and deaths of stars and entire galaxies, and in biology, in the ceaseless mutation and adaptation of organisms, including even viruses, and in the continual workings of the metabolic and regenerative processes that sustain life. In physics, an intrinsic dynamism is evident in the creation of heavy elements by the immense heat and pressure of stars and in nature's tendency to produce holistic structures such as atoms and molecules. Internal dynamism is also evident in the instability of certain nuclei and of particles such as neutrons and some quarks. In fact, quarks—one of the basic building blocks of matter—are virtually never found in isolation but survive in relation with one another in the larger hadrons that their interaction creates.

However, illustrations from astronomy, biology, and physics are not the best examples, for they all depend on particular circumstances. Arguably, the most general example of the inherent dynamism of nature is found in the equivalence between matter and energy. This equivalence is quantified in Einstein's formula

$$E = mc^2$$

which has been called "the world's most famous equation."[15] People all over the world are now familiar with the idea that matter can be converted into energy. There is a general (if vague) awareness of matter–energy equivalence not only because of nuclear weapons but because of nuclear reactors as a source of electricity. The first human-made self-sustaining nuclear reactor was created under Stagg Field at the University of Chicago in December 1942, and the first atomic bomb was detonated at the

15. As stated by David Bodanis's book titled *E=mc²: A Biography of the World's Most Famous Equation.*

Trinity Site in south-central New Mexico on July 16, 1945. These events were kept secret until after the destruction of Hiroshima, three weeks after the Trinity test. The first nuclear reactor used as a public utility was introduced in the Soviet Union in 1954. Since 1990, about one-fifth of the electricity produced in the United States is generated by nuclear power.[16] A single uranium fuel pellet weighing about one-third of an ounce creates as much energy as one ton of coal, 149 gallons of oil, or 17,000 cubic feet of natural gas.[17] People today are also familiar with nuclear medicine, which employs radioactive isotopes, and with nuclear magnetic resonance spectroscopy, which is now used routinely in hospitals. Familiar facts such as these may contribute to the public's casual awareness of the connection between matter and energy, but they have not led very many people to wonder what the matter–energy equivalence means or how it should alter our understanding of nature itself.

In chapter 2 I argued that the age-old depiction of matter as a static substance that takes up space is no longer needed—that this depiction has been rendered anachronistic by the discoveries of relativity and quantum mechanics. Now, as we consider Einstein's discovery of the matter–energy equivalence, we are in a position to affirm the fundamentally processive nature of physical reality.

The proposition I am now advancing is that matter may more helpfully be viewed as being composed of *structured patterns of events* in which the structures are based partly on previously occurring structures and partly on the contributions of information qua information. This dynamic, process-oriented concept of matter as structured events is suggested by the matter–energy equivalence which was discovered as one of the implications of the theory of special relativity. There is a radical aspect to what is being suggested here. It is that we dispense altogether with the intuitive idea that physical nature consists of self-identical stuff or substance that takes up space and replace that idea with the affirmation that process itself—vibration, if you will—is the basis of physical nature. In other words, the wave, and its field, are in a significant way more fundamental than the particle, though there is an obvious sense in which waves culminate in particles when they interact with their environment. I will introduce additional evidence for this idea when we

16. See Fernández, "Nuclear Share in Electricity."

17. See NEI, "Nuclear Fuel."

discuss quantum mechanics. But in the meantime, I confess that, for many people, this is an idea that takes some getting used to.

The vibrations or patterns of events that we now have in mind may be organized not only in space but also in time so that the persistence of a single enduring particle may be understood as a *serially ordered* chain of events. For example, an electron may be regarded as a series of "electronic" events, each derived from its predecessor and each in turn contributing defining conditions and characteristics to its successor. As an illustration of how enduring identity can be maintained by means of process, we can recognize that this is how we explain the identity of a human being—not simply as a material body of organized cells but as a history of events and memories that are more or less integrated into an accumulating experiential unity. An amnesiac with no memory has effectively lost her identity.

The suggestion now being made is that not only persons but material objects may better be understood not in terms of unchanging substances but as the result of continuous histories. By attributing the stability of enduring objects to accumulating histories of patterned events, I am discarding the theories of fixed substance that have dogged the steps of Western thought and replacing them with the affirmation that nature, down to its very core, is dynamic, relational, and processive. This is suggested by the interchangeability of matter and energy. The events that compose the world are held to be *atomistic*, from the Greek word *atomos*, which does not mean "small" but "indivisible" or "cannot be cut."[18] This processive depiction of reality is *atomistic* because—unlike ordinary daily events such as dinner conversations, football games, and committee meetings—the elemental events or occasions now being described are not prolonged or extended in time. Instead, they spontaneously arrive at a conclusive determination. This must be true because *in order to be able to influence other events, an event itself must become something definite and concrete*. These events are not the scheduled events of human experience, which of course involve many interactions and can be protracted indefinitely, but the metaphysical events that, I am arguing, constitute

18. *Webster's Third New International Dictionary* (1966), s.v. "atom." The Greek philosopher Democritus (ca. 460—ca. 370 BCE) observed that if we ponder any solid object we can imagine cutting it in half, over and over again. However, Democritus reasoned, there must be some limit to this process of cutting since different objects have differing properties, and these differing properties indicate that each substance must be composed of specific types of *atoms* whose peculiar characteristics determine the properties of the object as a whole.

reality. These events are discrete, limited by their own definiteness. I examine the reasons for this discreteness more fully in chapters 7 and 8, but at present let us consider an illustration from ordinary experience.

Imagine that you and another person have agreed to go out for dinner, but you haven't chosen a restaurant. As you drive along you discuss various dining establishments and narrow your choices to two possibilities: Sarah's Soirée and Ed's Edibles. But, as sometimes happens, the more you talk about the two locations, the less willing either of you feels to sway the verdict. At length, as both of you are feeling less and less decisive, you come to a T-intersection in the road. Sarah's is located to the left and Ed's is to the right. The inescapable facticity of the layout of the local streets now compels you to make a decision. Of course, you can pull off onto the shoulder and continue your deliberations, but as the philosopher William James (1842–1910) trenchantly observed, the very failure to decide amounts to a decision—in the negative, so that, as long as you make no decision, you never arrive at either Sarah's or Ed's, or at any other restaurant, for that matter. The point of the illustration is that it is your confrontation with sheer, physical circumstances (an intersection in the road) that forces you to do something definite.

This example is much too anthropomorphic and psychological to be applied directly to photons, light waves, or other particles, but it nevertheless clarifies a fundamental principle: *that indeterminate processes are forced to terminate in a definite outcome when they encounter their environment or if they are to have any impact on that environment.* This statement will be of great importance when we consider quantum mechanics. In that discussion, I will consider this type of forced termination in connection with what has been referred to as "the collapse of the wave function." For now, the important thing to see is that this commonplace occurrence pertaining to human decision-making illustrates something profound about the ontological status of photons and other subatomic processes.

As another example, the notion that reality is composed not of substances but of atomistic processes is a helpful way to approach the riddle of wave–particle dualism. As I concluded the discussion of the properties of light, I noted that light sometimes behaves like a wave and sometimes like a particle. In other words, the results of experiments involving light seem dualistic and contradictory. But if we ponder this contradiction it becomes evident that the seeming incompatibility of wave and particle is actually based on the old materialist presupposition that *the physical world necessarily consists of static material.* Aristotle held that the "material cause" must be a stable substance that does not change unless it is acted upon by an external influence. Though Kant argued that the "thing in itself" is unknowable, he nevertheless assumed that *das Ding an sich* must be some sort of stable reality, an objective thing that, as it were, exists stably with its characteristic attributes, waiting to be examined. Whether or not scientists are familiar with Aristotle and Kant, it is this intellectual predisposition to think in terms of static types of reality that leads them to assume that particles must simply be particles and waves must simply be waves and that these two kinds of reality are not merely contrasting but *dualistic* and *contradictory*. This assumption rests on a lingering substance metaphysics that lurks in the background.

If we imagine that both waves and particles are patterns of events, then it is not difficult to imagine that there could be conditions in which one pattern could flow into the other, depending on the current pattern's encounter with other patterns—for both patterns are the result not of fixed substance but of serially ordered events. When a photon is detected by a measuring instrument, it is detected as single, particulate unit. However, the probability of detecting the photon is calculated by equations that pertain to waves. This suggests that when electromagnetic waves interact with matter, they may be said to *behave like* particles because of the quantified way they impart energy to the matter that is disturbed. In other words, the processive dynamism of the events themselves allows for a fluid interchangeability of waves and particles in the course of experimental observations.

I must admit that I personally have not found the wave–particle relation described in just this way in any other literature, either philosophical or scientific, and that the description of wave–particle conversion through dynamic process is probably more nearly in the philosophical category than the scientific. In any case, the key to solving the wave–particle conflict is to jettison the image of particles as enduring physical objects and adopt instead the idea of serially ordered, relational processes whose very nature is to be responsive to their environment. Admittedly this suggestion invites a more detailed explanation, but we need not delve into that now, for our present goal is simply to describe the processive, dynamic worldview that is suggested by relativity's matter–energy equivalence.

IV.

The fact that time does not pass for photons or light waves suggests that there is such a thing as nontemporal process—that is, that there is a dynamic type of reality that somehow exists apart from the passage of time. For most people the notion of nontemporal process is very difficult to imagine. We must nevertheless do our best with it, for the idea is entailed by the phenomenon of relativistic time dilation and also, though perhaps less obviously, by our hypothesis that space and time are modes of relationship.

Let us take refuge once more in a specific illustration. The Andromeda Galaxy is approximately 2.5 million light-years from the earth. Though it appears from the earth only as a dim, fuzzy point of light in the night sky, Andromeda is visible even without binoculars. So, if you look in just the right direction on a dark night, photons from the Andromeda Galaxy will strike your retinas, which will send signals to your brain. It takes only five or ten photons arriving in less than a tenth of a second for your occipital lobe to register the arrival of these photons as a conscious experience. As for the photons themselves, we must say that they do not experience time because (1) this is what is calculated by Einstein's equation for relativistic time dilation, and (2) because, as an aspect of special relativity, we have posited that space and time are not cosmic absolutes but ways of describing physical relationships. As long as nothing that the photon can interact with gets in its way, it has no external reality to respond to. In other words, *time does not pass for photons because, until they act upon something in their environment, they are ontologically solitary*. So the photon travels along in splendid isolation, blissfully unaware, as it were, of the fact that from the standpoint of an external frame of reference it is in motion. Astronomers make good use of this situation; they can observe areas of the universe as they existed millions or even billions of years ago, for the light that has traveled from those areas to the earth has not changed at all. The ordinary notion of physical alteration is, in fact, inapplicable to photons, for they are timeless. It has been said that a photon's moment of birth is frozen into the moment of its death. Meanwhile, from *your* perspective as a stargazer, if you do see Andromeda, you get to experience photons that have not changed in *two-and-a-half million years*.

But, one may ask, light doesn't travel all over the universe as *particles*, does it? Isn't light transmitted as waves? And if so, how is it conceivable that a wave, which is defined by vibration and wavelength, can be *nontemporal*? Isn't the process of vibration intrinsically time consuming? How can there be "nontemporal vibration"? The response that this contradiction is merely one of the many "paradoxes of relativity" is more nearly a dodge than an answer. In search of a better answer, one possibility is to return again to our discussion about the nature of information. Suppose we imagine that light waves and photons are not primarily *physical* objects but *information*. Of course, identifying photons with information is only a partial truth, but it is nevertheless somewhat helpful to think of a photon as a *packet* of information. The information in this case would be quasi-mathematical in nature—essentially, information about wavelength, frequency, and energy.

To illustrate the notion of a "packet of information," we can refer to the protocols of telecommunications. In the old days of the telegraph, data were sent not as ones and zeroes but as dots and dashes. In Morse code there was no combination of dots and dashes to represent the period at the end of each sentence. Therefore, in order to terminate each sentence, the telegraph operator was required to send the word *stop*. Today it is still impractical to send digital data as an unpunctuated or "unparsed" stream of bytes. Data are instead sent in "packets" that start with a header indicating the electronic address of the receiving device. Depending on the stated protocol, packets can be defined in advance as a specific number of bytes or by including specific combinations of bytes that identify the packet's beginning and its end. This allows for error checking and also can allow the protocol to reassemble packets that have arrived out of order. It is interesting to reflect on the fact that, though the bytes themselves may be viewed simply as ones and zeroes, the signals thus sent represent not only numerical values but are routinely converted into written language, pictures, and sounds. The point in any case is that in the quasi-numerical data in the telecommunications example we can discern *a nontemporal aspect*. Data are expected, in a way, to be non-time-dependent. As an example of the time-transcendent nature of information, the value of pi, 3.1416 . . . , is always *that* value, regardless of circumstances and regardless of the passage of time. I submit that this nontemporal quality can in fact be attributed to all information *when viewed simply as* pure, conceptual *information*. In telecommunications, though packets of data are transmitted as electrical pulses, the data, considered *as data*, have a permanent, nontemporal quality. During transmission the data are expected to remain unchanged, and the data themselves may be sent by a variety of physical technologies without significantly altering the end result that the data produce.

Thinking of a photon as something like a packet of information is at least a little helpful in understanding the nontemporality of the transmission of radiant energy. Some—perhaps most—information is time sensitive. In sentences such as "Barbara has written a book," or "The milk is about to turn sour," or "The Devonian period lasted for sixty million years," the information being conveyed is true only within certain time parameters. Photons (or light waves) are not time dependent in this way because the information they contain always stays the same from the moment of their creation to the moment of their termination. Macroscopic objects can slow down or speed up; this does not pertain to electromagnetic waves. Each wavelength or frequency of the electromagnetic spectrum carries its own specific amount of energy. This amount is expressed as a direct proportionality between frequency and energy, calculated by the formula

$$E = h\,\nu,$$

where E represents energy, ν (the Greek letter *nu*) equals frequency, and h is Planck's constant, a specific value that appears frequently in the equations of physics. Basically this formula indicates that higher frequencies of light carry higher levels of energy. Energy throughout the universe is believed to be conserved, which is another way of saying that specific frequencies (or wavelengths) of radiation retain their value virtually unalterably—until they "collide" with an entity to which they can impart their energy. I use scare quotes around *collide* because the notion of a photon colliding with an object is actually just one more metaphor in physics' ever-growing collection of figures of speech.

But, metaphorical or not, collisions with photons certainly do impart physical energy; the entire universe illustrates this constantly. Radio waves cause small electrical currents in antennas; infrared waves impart higher kinetic energy to atoms, which is what we refer to as heat; and ultraviolet light excites the electrons in atoms so that materials seem to glow in the dark, sometimes causing ionization. X-rays and gamma rays—the most energetic waves—knock electrons completely away from their atoms. The resulting ions react with neighboring atoms in unpredictable ways, which is why X-rays and gamma rays sometimes cause cancer (and why the much less energetic radio waves used by cell phones do not). Frequencies of visible light resonate with the frequencies of the electrons of the materials upon which they are incident. When resonance occurs between a photon and an electron, the photon is absorbed by the electron, which is then said to be energized or "excited." The electron soon returns to its original energy level, emitting a photon that carries a specific frequency (or specific energy) of the color of light that we may then see. In all of these processes the role of mathematics is absolutely indispensable—not only in the ways that we human beings describe the processes but in the ways in which the processes *actually occur*. This points to the informational aspect of the processes themselves. In other words, natural processes are actually responsive to quantitative information.

We began this section by affirming the reality of nontemporal process and are now relying on the inherent presence of mathematical ("pure") information in the transmission of radiant energy to serve as an illustration involving that nontemporal quality. Mathematical information changes if the object it is describing changes, but mathematical values, in themselves, simply are what they are. This information itself, even if nontemporal, describes vibration that imparts energy, and this seems to entail a sort of dynamism or tension—thus the phrase "nontemporal *process*." Perhaps this tension is in a very primordial, elemental way somehow a little like the mental tension being maintained by two persons who cannot decide where to go for dinner. There is no definitive outcome, and yet there is a kind of dynamism or vibratory quality within the uncertainty itself.

In fact, though I have been speaking of the transmission of radiant energy (light, etc.) as if it were an expanding throng of photons, the image of photons streaming through space is fundamentally flawed. It would be more accurate to think of each photon as the particle that occurs (or perhaps the packet of information that is communicated) when electromagnetic waves interact with individual particles such as electrons, protons, and neutrons. Radiant energy, in and of itself, consists of waves that can be (and frequently are) transmitted even through empty space. As the Michelson–Morley experiment proved, no medium of transmission is necessary. This implies that it is vibration, and not substance, that has ontological primacy. This seemingly bizarre fact amounts to another verification of our previous assertion that nature is fundamentally dynamic and processive rather than static. The waves themselves are the reality, and require no underlying substance. I will provide further evidence for this seemingly strange state of affairs in our consideration of quantum mechanics.

In summary, these are the ideas we have gained from the special theory of relativity. (1) Space and time are not uniform and constant throughout the universe but may more accurately be understood as modes of relation. (2) The phrase *frame of reference*, which is indispensable in any interpretation of special relativity, strongly suggests that something like perspective or viewpoint is inherent within nature. This in turn implies that nature itself is radically relational. (3) This relationality of nature implies that nature must be fundamentally dynamic and processive rather than static or substance based. And (4), the fact that time does not pass for photons or light waves indicates that there is such a thing as nontemporal process—that is, a dynamic reality that is somehow real apart from the passage of time. With these thoughts we conclude our interpretation of special relativity and turn our attention to quantum mechanics.

6

Quantum Mechanics: Experiments

No one would dispute that quantum mechanics is a complex and difficult subject. The reasons for its difficulty may be summarized as threefold. The first is simply that the mathematical methods that are required by quantum mechanics are idiosyncratic and complicated. Some types of math have been invented chiefly to deal with the subject of quantum mechanics itself. But even not considering the math, the concepts of quantum theory are demanding. As Richard Feynman noted, there is much about quantum theory that students are not taught until their third or fourth year of graduate school.[1]

The second source of difficulty is that quantum mechanics entails a picture of nature that diverges radically from orthodox mechanistic materialism and even from what seems to be ordinary common sense. In quantum mechanics, in certain circumstances things are said to exist in two differing states simultaneously; this is called *superposition*. Moreover, things that were once held to be certain are said to be uncertain, while, conversely, other things—primarily the integral, quantized description of atomic processes—are held to be certain. Contrary to what many people think, quantum uncertainty is not caused by the inherent impreciseness of experimental methods. Rather, uncertainty—or, as some authorities prefer, *indeterminacy*—is built into the prerequisites of quantum theory itself. Taken as a whole, the picture of nature presented by quantum physics just seems implausible. Richard Feynman, a celebrated authority in the field, said,

1. Feynman, *QED*, 9.

> Finally, there is this possibility: after I tell you something [about the quantum depiction of nature], you just can't believe it. You can't accept it. You don't like it. A little screen comes down and you don't listen anymore. I'm going to describe to you how Nature is—and if you don't like it, that's going to get in the way of your understanding it. It's a problem that physicists have learned to deal with: They've learned to realize that whether they like a theory or they don't like a theory is *not* the essential question. Rather, it is whether or not the theory gives predictions that agree with experiment. It is not a question of whether a theory is philosophically delightful, or easy to understand, or perfectly reasonable from the point of view of common sense. The theory of quantum electrodynamics describes nature as absurd from the point of view of common sense. And it agrees fully with experiment. So I hope you can accept Nature as She is—absurd.[2]

The third difficulty is that even among scientists there is no single, widely accepted means of working our way out of this absurdity. Quantum theory has been an established part of science for over a hundred years, and yet there are still broad disagreements concerning its interpretation. Explaining quantum mechanics often seems more like deciphering a poem or interpreting a painting than explicating a scientific theory. As Feynman says, many physicists have simply taught themselves to ignore questions of interpretation, focusing their attention instead on mathematical techniques and experimental results. Feynman commented, "While I am describing to you *how* Nature works, you won't understand *why* Nature works that way. But you see, nobody understand that."[3] Unable to answer "why" questions, physicists restrict their attention to testable results.

Despite the attitude, which has been expressed not only by Feynman but by many others, that nobody really understands quantum mechanics, in chapter 7 I will present certain understandings or interpretive conclusions about quantum mechanics that I believe are, if not inescapable, at least plausible, and even convincing. As a way of giving those conclusions a firm footing, in the present chapter, instead of attempting to summarize the abstract principles of quantum mechanics, I focus on the actual experiments that eventually forced physicists to embrace those principles. My goal is to emphasize that quantum mechanics truly is an empirical description of nature itself and not merely a set of abstract

2. Feynman, *QED*, 10.
3. Feynman, *QED*, 10.

ideas invented by a group of very clever theoreticians. This emphasis on experiments means that the current chapter is, at some points, rather detailed. To some readers, these details will be very interesting. On the other hand, if you find that some of the details are elusive or even simply off-putting, I hope you will nevertheless plow through to the end. The goal of the chapter is to provide compelling evidence for the amazing worldview that quantum theory entails.

It may be said that quantum mechanics made its initial appearance in Berlin on December 14, 1900, when German physicist Max Planck (1858–1947) presented certain findings pertaining to black-body radiation to the German Physical Society (*Deutsche Physikalische Gesellschaft*). For several decades physicists had been baffled by the characteristics of light emitted from black bodies; experimental results seemed enigmatic and, in fact, unintelligible. As a theoretical physicist Planck had worked for five or six years with various mathematical descriptions of laboratory results pertaining to black-body radiation, but the equations he developed always seemed to fail when applied to certain conditions. At last Planck found an approach that succeeded. But lamentably, in Planck's opinion, the new mathematics required an assumption about the nature of light that seemed so arbitrary and implausible that Planck, a methodical scientist with a conservative disposition, was reluctant to announce it. In fact, Planck later described the publication of his findings as "an act of despair. . . . I was ready to sacrifice any of my previous convictions about physics."[4]

Despite Planck's personal misgivings, the ideas that grew out of his work were destined to transform our understanding of radiant energy, atoms, chemical bonding, and numerous other physical phenomena. Planck had discovered a direct proportionality between the frequency of light and its energy; the mathematical ratio between the two is a numerical value that was soon given the name "Planck's constant."[5] His pioneering research earned Planck the 1918 Nobel Prize in Physics "in recognition of the services he rendered to the advancement of Physics by his discovery of energy quanta."[6] A year after Planck's death, the Ger-

4. As quoted in Wikipedia, "Max Planck." Concerning Planck's hesitancy about his conclusions, see *Encyclopædia Britannica*, 15th ed. (1974), s.v. "Planck, Max."

5. The value of Planck's constant is 6.6256×10^{-24} erg-seconds, and it is symbolized as h or, in some calculations, as $\hbar$ ("h-bar"). The letter h is thought to have been derived from the German noun *Hilfsgrösse* ("auxiliary variable").

6. The Nobel Prize, "Nobel Prize in Physics 1918."

man scientific organization known as the Kaiser Wilhelm Society was renamed the Max Planck Society. The society now includes eighty-three institutes devoted to the study of a wide assortment of subjects. But, ironically, despite all these honors, in much the same way that Hendrik Lorentz had never fully accepted the assumptions of special relativity, Planck never fully accepted the principles of quantum mechanics.

> Planck . . . frequently spoke out against the new ideas to which he, of course, had personally given the incentive, but which had now become much more revolutionary. Thus, Planck, to the end of his life, belonged to the group of older physicists who recognized the successes of the younger generation but who refused to see anything conclusive in their theoretical constructions.[7]

I. BLACK-BODY RADIATION

I have said that Planck's radical new ideas were prompted by his study of black-body radiation. In physics the phrase *black body* refers to an object that absorbs all of the energy that impinges upon it. Generally this absorbed energy causes the temperature of the black body to rise, and at some point the material will begin to radiate its energy back into its environment. For example, we have all seen heated pieces of metal that became red-hot or even white-hot. This color of the glowing material is scientifically significant. In the context of an experiment, the advantage of a black body is that the light that is given off reveals something about the innate properties of the heated material itself. This is different from what happens when we simply shine a light on an object. Shining a light on an object tells us what color of light the object reflects, but for a black body there is no reflected radiation—thus, its blackness—so that the light frequencies that are radiated are due entirely to emissions produced by the heated material itself. Moreover, ordinary light sources—whether a table lamp, a flashlight, or even sunlight—all create their own peculiar wavelengths that subtly alter the appearance of the surfaces they illuminate. This problem does not pertain to black-body radiation because the light it emits did not originate in its environment but was caused by its own heat.

Solid bodies that are hot enough to glow generally emit a jumble of frequencies known as "indistinct spectra" (a mixture of wavelengths or

7. *Encyclopædia Britannica*, 15th ed. (1974), s.v. "Planck, Max."

colors). The situation is different with gases. When gases are heated and the light they emit is passed through a spectroscope, incandescent gases produce unique patterns of parallel ribbons of light. In other words, with gases, only isolated parallel lines or bands appear. *Bands* are emitted by *molecules* of gas, and *lines* are emitted by *atoms*.[8] This difference suggests that the bands or lines that occur are determined by the molecular or atomic structure of the gas being heated.

Prior to the development of quantum mechanics, this result was perplexing; it was contrary to what was predicted by current understandings of the relationship between light and energy. In the nineteenth century, Scottish mathematician-scientist James Clerk Maxwell (1831–1879) had formulated what is now considered the classical theory of electromagnetic radiation. Maxwell's work yielded the first explanation of electricity, magnetism, and light as three differing forms of the same physical reality. Maxwell's equations are still used today by electrical engineers; his principles are fundamental for making a radio or designing an electric motor. But Maxwell's theories cannot account for the characteristics of black-body radiation.

The black body that Planck and other physicists employed produced a variety of wavelengths of light. The black body itself was typically constructed as a closed box with graphite walls and a small hole on one side. Graphite was often used to cover the interior because graphite absorbs about 95 percent of the radiant energy that strikes it. Heating elements were incorporated so that the interior could be maintained at a stabilized high temperature. Because there was only one small opening, the box trapped almost all of its own heat so that its behavior approximated that of a perfect absorber. A diffraction grating was placed just outside the hole. (A diffraction grating is a flat surface covered with finely spaced grooves or slits; the grooves or slits separate light into the colors of which it composed, much as a prism does.) As the interior of the box was heated and began to glow, the grating separated the radiation emerging from the hole into specific wavelengths. A photo-sensitive instrument was used to detect the intensity of the various wavelengths.

The trouble encountered by Planck and his predecessors was that the intensity of the diverse wavelengths did not match the results expected by classical physics. Maxwell's equations predicted that as temperature increased, the energy of the light emitted should increase proportionally

8. Maron and Prutton, *Physical Chemistry*, 613.

without any maximum limit. The wavelengths that were actually emitted through the hole did increase slowly, shifting from red to yellow to white, then to a slightly blue color, and finally fading into the ultraviolet range, which is not visible to the human eye. Blue wavelengths are indeed more energetic than red ones, but the change was not as mathematically proportional as it should have been, and the wavelengths simply refused to progress beyond the near-ultraviolet range. This observed upper boundary on black-body emissions seemed totally arbitrary and was completely unexpected. The persistent upper limit came to be known, with only slight facetiousness, as "the ultraviolet catastrophe," not because it caused any physical disaster but because it had disastrous consequences for the physicists' existing theories and expectations.

As a theoretical physicist Planck was proficient at manipulating mathematical formulae. He had to be; we might stop to appreciate the fact that Planck and other nineteenth-century scientists possessed no electronic calculators. The equations they developed are complicated even by today's standards, employing exponential and trigonometric functions, logarithms, the calculation of various roots, and constant, tedious long division. All of this math had to be done by laborious manual effort. (Perhaps some of this was done by graduate students!) Every time a new equation was proposed, all the calculations had to be started afresh. Clearly this did not deter Planck, for he experimented with various equations long enough to recognize something odd—that if the energy that was emitted by a black body were *quantized*—not simply as a mathematical method but presumably as an empirical description of the emitted radiation itself—then the experimental measurements could be predicted successfully, and the ultraviolet catastrophe would be eliminated. Planck had no explanation as to *why* this quantum interpretation worked, and as noted already, he intensely disliked the whole idea—but it did work, and so he did report it.

At this point we must pause to consider what "quantization" means and why physicists were so averse to it. Imagine that you are a pitcher for a fast-pitch softball team. Your pitches are truly fast, but your coach wants to know precisely how fast they are, so a radar gun is brought in and your pitches are measured. Sure enough, your pitches range from 50 to 70 miles per hour, which in your particular softball league is splendid. However, as you and your coach review the figures, you both notice something strange: the measured velocities are all multiples of five. Pitches are either 55, 60, 65, or 70 miles per hour. There are no "in between fives" velocities such as 59 or 67. The coach and the operator of the radar device ask you to throw a few more pitches; surely the present results are a fluke and the strangeness in the data will disappear. So you throw another dozen pitches, but the pattern persists. The velocity of every pitch is a multiple of 5 miles per hour. The coach looks suspiciously at the radar gun and at its operator, who shrugs helplessly. Another operator and another radar gun are brought in, but the measured velocities continue to occur in multiples of five. Totally bewildered, the coach summons all the other pitchers and has the velocities of their pitches tested, and, surprisingly, the results are the same: multiples of five only. By now the coach is so perplexed and distracted that he loses his focus on the game and is forced to retire and move to an island resort in the Caribbean.

This illustration is lighthearted, but the reaction of the German physicists to Planck's presentation was not. Planck was too highly respected to ignore, but no one had any idea why emitted radiation should be analyzable only in terms of whole numbers (integers). Furthermore, the results could not be blamed on Planck or his individual laboratory. The black-body anomaly occurred everywhere, no matter who was doing the testing or which black-body device was being used.

We now know that the necessity of using only whole numbers is caused by the structure of atoms themselves. Increased heat in any material consists of its atoms or molecules colliding with one another at elevated rates of speed. These collisions cause some of the electrons in some of the atoms to transition to a higher energy level—an "excited state"—than the level they normally occupy, known as their "ground state." For reasons that we will soon consider, the possible energy levels of electrons in atoms are discrete, not infinitely variable. This discreteness is where the quantized behavior of the light emitted by heated bodies actually originates. Electrons in an atom are less stable in an energized state than they would normally be, and therefore they soon "fall back" to their ground state. (The phrase *fall back* is a metaphor.) It turns out that each electron in an atom has its own, preestablished state of energy. There is no gradual variation. Since energy is conserved throughout the universe, when the electron reverts to its ground state, it is required to emit exactly the same amount of energy that it had gained from any heat-produced collision. This is why the spectrum of light that is emitted is observed as lines or bands and not as a random mixture of all frequencies. This emitted energy is carried away by what we now call a photon, a quanta of radiant energy. Pertaining to black-body radiation, this energy is observed as glowing light coming through the opening in the device. Planck's studies were published, and scientists became familiar with his conclusions, but, as had been the case with the results of the Michelson–Morley experiment, no one could explain the results.

II. THE PHOTOELECTRIC EFFECT

There is a familiar story that Benjamin Franklin discovered electricity by flying a kite during a thunderstorm. This allegedly took place in the summer of 1752. Actually electricity was known long before Franklin, but electricity was thought to consist of positive and negative fluids of uncertain composition. The primary result of Franklin's work was the determination that lightning is indeed a form of electricity and that the destructive power of lightning—which before Franklin had been an unpreventable cause of many fires—could be diverted by the use of lightning rods that were grounded in the earth. Electricity does not readily flow through air, which normally serves as an insulator or a *dielectric*, but in a lightning bolt the insulating effect of air is overcome by lightning's immense electromotive force, typically about three hundred million volts.

The vacuum pump had been invented during the 1650s. Pumping all the air out of a sealed glass tube allowed scientists to experiment with electrical phenomena in the absence of air; the removal of air allows the electricity to jump off of any wire and flow through space much more easily. In 1869 German physicists observed that electrical current in a vacuum tube will readily flow from a charged negative electrode to a positive terminal. The electricity itself is invisible, but if a small amount of gas is left inside the tube, the gas molecules will glow when they are struck by the moving electrons. This glow made the flow of electricity visible. In 1869 no one knew either about the structure of the atom or about the nature of electrons, but the scientists were able to measure the strength of the current that flowed through the tube.

Then followed the seemingly peculiar discovery that in a vacuum tube electrical current will also flow from a piece of metal if the metal is illuminated by light. This phenomenon is known as the *photoelectric effect*, and it was investigated by physicists J. J. Thomson, Wilhelm Hallwachs, and Philipp Lenard during the years 1886 through 1902. The photoelectric effect does not require that the metal be heated or that it be subjected to a high voltage; the light itself is the chief cause of the current that flows. However, the light must be of at least *a certain minimum frequency*. This was unexpected. Another unexplainable fact was that different metals (iron, copper, etc.) required differing minimum frequencies of light for any electrons to be discharged. Intensifying the brightness of the light yielded more electrons, but higher intensity did not increase the electrons' velocity. On the other hand, shifting the color of the light further into the ultraviolet range caused the electrons to have higher energies, but without augmenting the number produced.

> It is found that with a given material as emitter, the wavelength of the light must be shorter than a critical value, different for different surfaces, in order that any photoelectrons at all may be emitted. This critical wavelength, or the corresponding frequency, is called the *threshold frequency* of the particular surface. The threshold frequency for most metals is in the ultraviolet (critical wavelength 200 to 300 nm), but for potassium and cesium oxide it lies in the visible spectrum (400 to 700 nm).[9]

Readers may want to be reminded that wavelength and frequency are inversely proportional to one another (the shorter the wavelength, the higher the frequency) and that the amount of energy transmitted by a light wave is directly proportional to its frequency (higher frequencies transmit greater energy).

From the standpoint of classical physics, the strangest thing about the photoelectric effect is that there is a minimum frequency of light needed to initiate the effect and that the velocity of the escaping electrons also depends on the frequency and not on the intensity of the light (the number of photons being emitted). "When a photon collides with an electron at or just within the surface of a metal, it may transfer its energy to the electron. This transfer is an 'all-or-none' process, the electron getting all of the photon's energy or none at all. The photon then simply drops out of existence."[10]

At first it may be hard to see what is so odd about the photoelectric effect, but let us consider an illustration. In many restaurants, plates of food coming out of the kitchen are placed on a counter where they are retrieved by waiters and then taken to patrons. Often there are heat lamps positioned over the counter to keep the food warm. It is easy to imagine that if the food is not warm enough, more heat lamps can be added over

9. Sears and Zemansky, *University Physics*, 957.

10. Sears and Zemansky, *University Physics*, 958.

the counter to maintain a higher temperature. In other words, having more lamps (more intense light) produces warmer plates of food. Heat in a material, we may recall, actually consists of molecules moving and colliding with one another at greater speed.

In the photoelectric effect, adding more lamps does not cause the emitted electrons to move at higher speed. Higher velocities are caused not by more intense light but by *shortening the wavelength of the light being applied.* Since ultraviolet light has a shorter wavelength (higher frequency) than visible light, ultraviolet light causes the electrons that are emitted from the surface of the illuminated metal to travel at a higher velocity; in other words, the electrons absorb more energy. And if the light is below a certain threshold frequency, *no electrons will be emitted at all.*

Another aspect of the photoelectric effect that classical physics could not explain was the "all-or-nothing" way that light imparts energy to the emitted electrons. With ordinary objects in everyday experience, all collisions produce some effect, no matter how minor. For example, suppose you are hammering a nail into the wall in order to hang a picture. If the picture is large you may have to use a larger nail, which will require the nail to be driven with more force. Additional force may also be needed if the wall is made of a harder substance—say, hardwood paneling instead of plaster. We are accustomed to altering the way we swing a hammer based on the assumption that higher hammer speed will drive a nail more quickly. For those of us who are not well practiced with the use of hammers, another consideration is that not all blows are equally effective. Some blows do not strike the head of the nail squarely, often resulting in a nail that is bent or in an aggravating blow to the thumb.

In the photoelectric effect, there are no glancing blows. Either *all* of the light's energy is absorbed or none at all. As a metaphor we can remember Cinderella's glass slipper, which would fit only the foot of Cinderella. Similarly, the energy of the incoming photon must precisely match the energy of the moving electron in the metal. Physicists were nonplussed by this. Then, again, it was Albert Einstein who published an explanation.

In chapter 4 I mentioned Einstein's paper on special relativity, which appeared in September 1905. Amazingly, however, during that year Einstein published four important papers altogether, and his paper on the photoelectric effect was actually the first.[11] In that paper Einstein

11. Einstein, "Erzeugung und Verwandlung." This was translated into English as "On a Heuristic Viewpoint Concerning the Production and Transformation of Light."

proposed that light consists of a stream of particles whose energies are related to their frequencies according to the formula developed by Max Planck. (I presented Planck's formula near the end of chapter 5.) As a particle, light conveys energy that is directly proportional to the frequency of its wave. Einstein explained that when light particles, which are now called photons, encounter a metal surface, if a photon's energy is sufficient to remove an electron, a collision with that electron produces the photoelectric effect. The energy necessary has a specific threshold because of the definite, quantized amount of energy with which the electron is bound to its atom. We now refer to this definite amount of energy as an electron's *ionization energy*. The precise magnitude of this energy varies depending on which metal is being used and which electron in the atom is involved, and the energy has a specific value because of the quantized nature of atomic structure. I describe that structure in the paragraphs below.

Initially many physicists felt that Einstein's proposals created more questions than answers. This was primarily because scientists had settled comfortably into the belief that light consists of waves rather than particles. However, after years of laboratory measurements, in 1915 the eminent experimental physicist Robert Andrews Millikan (1868–1953) confirmed the accuracy of Einstein's predictions,[12] and after sixteen years of waiting, in 1921 Einstein won the Nobel Prize. His analysis of the photoelectric effect was cited specifically. Ironically, Millikan had undertaken his experiments precisely because he wanted to disprove Einstein's theory, for Millikan openly disagreed with the whole idea that light could consist of particles.

We have reviewed black-body radiation and the photoelectric effect in order to understand the origins of quantum theory, but quantum mechanics achieves what may be its most important application in the fields of atomic models and the nature of chemical bonding. We now turn our attention to the structure of the atom itself.

The journal received this paper on March 18, 1905, and published it on June 9.

12. Holton, "Millikan's Measurement."

III. RUTHERFORD'S ALPHA SCATTERING EXPERIMENT

In chapter 5 I noted that the Greek philosopher Democritus (ca. 460—ca. 370 BCE) introduced the term *atom* to designate the smallest possible fragment of any material. Democritus said that if we consider any solid object, we can imagine cutting it in half over and over but that there must be some endpoint to this cutting. Democritus reasoned that since different objects have differing properties, these differing properties must indicate that each substance is composed of specific types of atoms whose peculiar characteristics determine the properties of the object as a whole. Thus the word *atomos* actually means not "smallest" but "cannot be cut."[13]

For 2,200 years little real improvement was made to Democritus's concept. John Dalton (1766–1844) proposed that the chemical elements have different types of atoms and that chemical compounds are composed of combinations of atoms, but, like Democritus, Dalton held that atoms cannot be cut. Next, based on experiments with cathode ray tubes, during the late nineteenth century J. J. Thomson (1856–1940) discovered the electron, theorizing that electrons were one of the constituents of all atoms. This implied that atoms cannot be regarded as simple objects having no parts. In an attempt to clarify this, William Thomson (1824–1907, also known as Lord Kelvin) suggested that atoms are diminutive balls of positively charged material in which electrons are embedded. This depiction became known as the "plum pudding" model. Since individual atoms themselves normally have no net electrical charge, it was assumed that in each atom the charge of the positive "pudding" must balance the total negative charges of the electrons. However the plum pudding model soon had to be abandoned on both theoretical and experimental grounds. It is tempting to say that the pudding model was found to be unpalatable.

In 1896, in his laboratory at the University of Würzburg in Bavaria, Wilhelm Roentgen (1845–1923) discovered X-rays. Roentgen was experimenting with a cathode ray tube, a glass tube containing no air but containing wires that can carry electricity. The wire was actually emitting electrons when Roentgen noticed that a chemically treated screen nine feet away was strangely glowing. The cathode ray tube was producing x-rays, which traveled through the glass of the tube and also through most

13. See *Webster's Third New International Dictionary* (1966), s.v. "atom."

other solid materials that Roentgen then inserted between the tube and the screen. This discovery was entirely unintentional.

During that same period, French physicist Henri Becquerel (1852–1908) was studying phosphorescence, the emission of light of one color caused by a material's prior exposure to light of another color. (Some of us can remember wrist watches whose numerals and hands were coated with phosphorescent paint; exposing the watch to light during the day would enable the user to read the time in the dark at night.) As part of his research Becquerel was using pitchblende, an ore of uranium that also contains traces of radium (though radium itself had not yet been discovered). After hearing about Roentgen's discovery, Becquerel wondered whether there might be a connection between phosphorescence and the newly detected X-rays. Becquerel's method was to wrap photographic plates in black paper to prevent exposure to light, place coins and other pieces of metal on the plates, cover the setup with pitchblende, and expose the ore to the sun to see whether sunlight would stimulate phosphorescence in the ore. If the phosphorescence included X-rays, they would travel through the black paper and create an image on the plates. Becquerel found that the pitchblende did indeed create a photographic shadow of the metal objects, and he interpreted this as evidence of a link between phosphorescence and X-rays. He was wrong.

Then, as luck would have it, during the last week of February 1896, the weather in Paris turned cloudy and rainy. In frustration Becquerel placed his arrangement of photographic plates and pitchblende in a drawer in his laboratory. On March 1, he got out the plates and developed them, thinking that he would see only a very faint image. Instead, the shadows were even more vivid than they would have been if the pitchblende had been left in sunlight. The next day, Becquerel reported at the Academy of Sciences that uranium salts emitted radiation without any exposure to light. As a result of an unforeseen interruption by the weather, Becquerel had discovered radioactivity. His discovery, like Roentgen's, was purely accidental.

Shortly after Becquerel's discovery, two other physicists working in Paris, Marie Curie (1867–1934) and her husband, Pierre (1859–1906), undertook efforts to purify pitchblende. Working for over three years, Marie and her assistant, Andre Debierne, laboriously refined several tons of ore. First they discovered polonium, named after Marie's native Poland, and in 1902 they finally isolated one-tenth of a gram of pure radium chloride. (This is not much radium—about one-thirtieth of the weight of a penny.) The Curies and Becquerel were jointly awarded the 1903 Nobel Prize in Physics for their investigation of radioactivity. On a rainy day in 1906 Pierre Curie was killed in a Paris street when he slipped and fell under a horse-drawn wagon. A month later Marie was appointed to succeed her husband as chair of the physics department at the Sorbonne, the first woman to hold a position on the regular faculty. In 1910 she isolated pure metallic radium and received the 1911 Nobel Prize in Chemistry—then the only woman to receive a Nobel Prize, and, at that time, the only person, male or female, to receive it twice. In 1934 Marie Curie died from leukemia caused by four decades of exposure to radioactive substances.

During the early 1900s New Zealand–born British physicist Ernest Rutherford (1871–1937) and, working independently, French physicist Paul Villard (1860–1934) discovered that radioactivity consists of three types of radiation. Rutherford named them alpha rays (the least penetrating), beta rays (moderately penetrating), and gamma rays (the most penetrating). Further research revealed that alpha rays and beta rays are really not "rays" at all but particles. Beta particles are simply electrons, and alpha particles are actually doubly ionized helium nuclei (two protons and two neutrons, held together by what we now call the nuclear "strong force"). Rutherford could not have guessed the composition of the alpha particle at that time, for no one knew what ions were or even that atoms had a nucleus. It was Rutherford himself who was about to make those discoveries.

Almost everyone today is familiar with the symbolic image of the atom—a round dot representing the nucleus with three identical ellipses surrounding the dot and representing the orbits of the electrons. This icon has the advantage of being easily recognized, but otherwise it is thoroughly misleading. In a more accurate depiction, the nucleus would have to be so small that it would be nearly invisible, and the paths of the electrons would (1) not be identical and (2) would have to be shown in gray scale as cloudy areas where the electrons are more or less likely to be found, for the electrons in an atom do not travel in predictable, planetlike orbits. An atom with three electrons would be the element lithium, apparently chosen as a model for no other reason than that it makes a nice looking diagram.

Rutherford and his assistants, Hans Geiger and Ernest Marsden, did not know any of this when they began work in 1906. They were using radioactive radon gas and discovered that the radon emitted radiation that traveled through the glass and caused a scintillation—a tiny flash of light—when the radiation impinged on a zinc sulfide screen. The individual flashes of light at varying points on the screen suggested that the radiation was composed not of rays but of particles. Further investigation showed that the particles had virtually the same mass as helium atoms but that, unlike helium atoms, the particles were deflected by electric or magnetic fields. This showed that alpha particles, as Rutherford called them, possess an electrical charge of their own—a positive charge, since alpha particles are repelled by other positively charged objects. (Identical charges repel one another; opposite charges attract.) Rutherford concluded that alpha particles are helium atoms that are somehow lacking their electrons. Since alpha particles can travel through a sealed glass tube but helium atoms cannot, Rutherford also theorized (correctly) that atoms without their electrons (in other words, a nucleus by itself) are a great deal smaller than normal atoms.

Rutherford was then curious to see what would happen if alpha particles were used to bombard a material heavier and denser than glass. He chose gold metal foil, partly because gold is among the densest of all materials (the density of gold is 19.3 grams per cubic centimeter, while the density of lead is 11.3 and the density of iron is 7.9) but also because gold is exceedingly malleable; it can be hammered into exceptionally thin sheets. In 1910 Rutherford placed a small piece of radium in a metal capsule with walls thick enough to confine the radium's natural radioactivity. A small hole was then drilled into the capsule, allowing a narrow column of alpha particles to stream forth. A piece of extremely thin gold foil was placed in the path of the stream, and a detector screen was positioned behind the foil. Whenever an alpha particle struck the screen, a little flash of light could be observed. The screen could be moved to different locations behind the gold foil, or off to the side, or even around to the front of the foil near the emitting capsule.

Alpha particles can be thought of as the heavy artillery of atomic physics. They are actually not a single particle but are composed of two protons and two neutrons held together by what is now called "the nuclear strong force"—the strongest force known in the universe, but one which is effective only over extraordinarily small distances. To refer to the masses of single particles, atoms, and molecules, scientists have invented a standard of measure known as the atomic mass unit ("amu"). The mass of the proton is 1.0073 amu, and the mass of the neutron is slightly greater, 1.0087 amu. The mass of the electron is very much less, only 0.00055 amu, or 1/1,836th the mass of a proton. The mass of the alpha particle is 4.0, about four times as heavy as a proton or neutron. However, gold is even heavier. The nucleus of a gold atom consists of 79 protons and 118 neutrons. A single atom of gold weighs 197 amu.

Because gold atoms are fifty times heavier than alpha particles, Rutherford expected that the alpha particles would simply be completely blocked by the gold foil. This was a reasonable prediction, and therefore Rutherford was surprised when it proved to be wrong. What Rutherford actually did witness was that *almost all* of the alpha particles *passed right through the foil*, producing scintillations at virtually the same location on the screen where scintillations would have appeared if the foil had never been inserted in the first place. After further observation Rutherford and his assistants realized that a very small percentage of the particles were in fact deflected, and an even smaller number were redirected at an angle greater than 90°. How could these ricochets be explained?

> Rutherford explained these observations by introducing the assumption that the mass of an atom is concentrated in a very small nucleus. This is [now] one of the cornerstones of atomic physics. In greater detail, his argument was as follows: Since, in the electric discharge, electrons are knocked out of atoms, leaving positive ions, it can be inferred that electrons form the *external* structure of atoms. Rutherford's new assumption was that the center of each atom is formed by a nucleus carrying a positive charge which keeps the electrons assembled. . . . Hence, to a limited extent, Rutherford's atom is similar to the solar system; the electrons travel around the nucleus as the planets travel around the sun. Since the mass of the electrons is very small compared with the mass of the atom, Rutherford assumed that nearly the whole mass of the atom is concentrated in the nucleus.[14]

As an illustration of Rutherford's reasoning, imagine an old sailing ship navigating through the Arctic Ocean in heavy fog. The ship is in constant danger due to possible collisions with icebergs, but there is no way to see the icebergs through the fog. Therefore the clever captain sends an artillery crew to the prow of the ship with instructions to fire a small shell forward at regular intervals. A shell that lands in the water causes only a distant splash, but a shell that strikes an iceberg will detonate, warning the captain to change course.

14. Oldenberg and Holladay, *Atomic and Nuclear Physics*, 85.

Like the icebergs, atoms are invisible—not because of fog but because they are much too small to see even with the most powerful microscope. Rutherford was essentially sending projectiles—alpha particles—into a sheet of gold atoms. He expected the result to resemble the result that would have been produced by the artillery shells if a solid sheet of ice had been concealed in the fog. Instead, it was as if almost every shell caused only a splash of water.

As experimental physicists, Rutherford and his assistants kept careful records and then undertook a mathematical analysis. They considered the precise percentage of alpha particles that were deflected and also the angle of deflection. Because it was clearly the nucleus that caused the deflections, Rutherford was able to calculate the size of the nucleus as a rough proportion to the number of alpha particles that were deflected. The result was astonishing.

> From a mathematical analysis of the scattering, Rutherford was able to deduce that the charge on the nucleus responsible for the observed deflections was equal approximately to one-half the atomic weight of the metal constituting the foil. Further, he was able to estimate that the dimensions of the nuclei were of the order of 10^{-12} to 10^{-13} cm, dimensions which along with the volume of the electrons involved are only 10^{-12} to 10^{-13} of the volumes actually occupied by the atoms. From these figures it follows that only $1/10^{12}$ to $1/10^{15}$ of the volume of an atom is occupied by protons, electrons, and neutrons, with the rest being nothing but empty space.[15]

In plain English, this says that only one-trillionth to one-quadrillionth of the volume occupied by an atom is composed of particles. The other 99.9999999999 percent is sheer emptiness. In this figure the number of repeated nines that are printed is not hyperbole but the result of precise arithmetic. *All atoms, even in heavy elements such as gold, consist mostly of empty space.* Therefore the most misleading aspect of the familiar icon of the atom is its depiction of the atom's components, which resemble a ping-pong ball closely surrounded by three ellipses. A more accurate diagram would look like a small pea placed in the middle of a football stadium with the nearest electron located at some undetermined location in the upper bleachers.

What this means is that even the heaviest, hardest materials with which we are acquainted—steel beams, concrete walls, granite

15. Maron and Prutton, *Physical Chemistry*, 608–9.

boulders—are actually fluffy and delicate. The world we inhabit is really not solid but lacy and finely structured. Objects that seem sturdy and dense only appear to be sturdy and dense because of the way our senses perceive them and because, obviously, our bodies are composed of the same space-filled atomic components that compose the objects in question. The atoms that compose solid objects bond with one another in certain predictable ways because of the regular patterns that are embodied by their electrons. It is interesting that the atom's modes of interaction are determined mainly by the lightest of the three particles in the atom. The seeming solidity of atoms is actually created by their electrons' blazing speed. The regularity of their motion creates the structure—the chemical bonds—of the macroscopic materials that result. *Solid matter is actually composed not of inert substances but of ceaseless, rapidly occurring, patterned processes.*

To grasp this more fully, imagine that you are standing beside a freeway in a large city. You would like to cross to the other side on foot, but five lanes of traffic are flowing in each direction—ten lanes altogether. If the freeway is congested and the cars are hardly moving, you can without much difficulty walk between the cars and make it all the way across. But if the traffic thins out and the cars speed up to even 15 miles per hour, even though the cars are now a bit farther apart, you will hesitate to attempt a crossing. If the cars are traveling at 60 miles per hour you will not make the attempt unless you are sure that the no car is approaching within a considerable distance. But now imagine that the freeway is not crowded at all, but that when a car does appear it is moving at an extraordinary velocity—not 100 miles per hour but faster, even faster than the speed of sound and too fast to estimate the car's distance accurately. The cars seem to appear out of nowhere, and even a glancing collision would instantly be fatal. There is no way you can cross; in these circumstances the freeway has become an impenetrable barrier. In effect, the high speed of the vehicles has made the freeway as solid as any concrete wall.

The barrier created by cars moving at high speed is a helpful analogy for the structural rigidity of solid objects, which is created by the electrons moving about in an atom. In fact, the atomic boundary created by the electrons is significantly more efficacious than the barricade caused by the cars, for the electrons—to the degree to which they can be visualized as moving objects—are moving a great deal faster than any automobile. The energy of their motions is such that even at the center of the earth, where the pressure is about fifty million pounds per square inch, the moving electrons preserve the empty space in the atoms with only a minor deformation of their behavior.

Next consider what happens when you touch an object. Imagine that you are picking up a tea kettle. The handle feels smooth and, if the kettle contains much water, a little heavy. You can guess how much water is in the kettle because of the way the handle presses against the nerve cells in your fingers. But at the atomic level the handle is not smooth and neither is your hand. When you grasp the handle the moving electrons in the handle come into proximity with the electrons in the molecules of the skin of your fingers. The electrons never actually collide with one another; their negative electrostatic charges keep them apart much too forcefully for any physical collision to occur. As the electrons in your skin are pushed away from the handle by electrostatic forces, those electrons bring along the entire molecule of which they are a part, and the molecules in turn carry the cells of the skin far enough to conform (more or less) to the shape of the kettle's handle. All this is actually a rather byzantine process, for there are about one hundred trillion atoms in each skin cell or nerve cell. One hundred trillion is fourteen thousand times as many atoms as there are human beings living on planet Earth—and this is merely the average population of atoms in each cell of your body. Each one of these atoms maintains its basic architecture because of something called "the Pauli exclusion principle," which is discussed below in chapter 7.[16] As your nerve cells are compressed or moved, they send signals to your brain, and you feel the weight of the kettle and the texture of the handle. Your brain interprets the nerve impulses as being caused by something "smooth" and "heavy." It seems to you that these molecular and neurological processes take place with natural rapidity and ease, but of course our perceptions of "ease" and "rapidity" are relative. The speed with which our brains can sense nerve impulses and think thoughts is limited by the same principles of atomic interaction that pertain to the atoms in our fingers and in the handles of tea kettles.

If we ponder the enormous complexity of such a seemingly simple experience, even the ordinary processes of our physical existence ought to awe us and amaze us. Instead, however, persons who are encountering these facts of physics and perception for the first time are sometimes beset by the disturbing thought that all of our sensory experiences are actually nothing but an elaborate and rather arbitrary construction and

16. Ash, "Atoms Touch." I find that Arvin Ash's YouTube videos on physics are especially helpful.

that therefore the whole of our experience is hardly more than a complex fabrication.

We do indeed live in what might be called a "virtual" world. This seems distressing—unless we can find a way to affirm that *experience is just as real as matter*, if not in fact *more* real. I argue in favor of the actual (ontological) reality of experience in chapters 8 and 9. In the meantime, don't worry. The problem is not as serious as it may seem.

IV. THE BOHR MODEL OF THE ATOM

Niels Henrik David Bohr (1885–1962) was a Danish physicist who discovered and promulgated insights of paramount importance pertaining to atomic structure and quantum theory. The influential "Copenhagen Interpretation" of quantum mechanics was pioneered by Bohr and named for the city where he lived and worked. Bohr received the Nobel Prize in Physics in 1922. To some extent he was also known as a philosopher; the principles of quantum theory are so radical that they call for a complex, new perspective related to both metaphysics and epistemology. Whether or not Bohr actually relished the task of philosophical explanation, he embraced the assignment willingly at least occasionally, recognizing it as an inherent aspect of quantum theorizing.[17] I will investigate Bohr's philosophical ideas more thoroughly in chapter 8. At present our focus is on the Bohr model of the atom.

Rutherford's depiction of the atom was an enormous advance as far as it went, but it was by no means a final explanation. Though Rutherford had shown that the electrons create most of the atom's volume by speeding around in the vicinity of the nucleus, he had not explained why they neither fall into the nucleus nor fly off into space. In other words, he had not explained the manifest stability of ordinary atoms. Now that atoms were no longer imagined to be solid, "uncuttable" particles, their stability posed considerable difficulties for then-known physical theories. Classical electrodynamics had demonstrated that when electrons move in a circular pattern they create a magnetic field. (Some of us remember making a magnet in science class by wrapping an insulated wire around a nail and connecting the ends of the wire to a battery.) The maintenance of such a magnetic field is not a freebie; it requires the ongoing input of

17. For a thorough but not excessively technical exposition of Bohr's ideas, see Barad, *Meeting the Universe Halfway*, esp. the introduction and ch. 3.

energy. According to nineteenth-century physics, without the addition of energy the orbiting electron would radiate its energy away, lose speed, and fall into the nucleus.[18] Since negative electrons are very strongly attracted to positive protons, it seemed logical that the electron should be very stable as a permanent addition to the nucleus. Of course such a collapse would obliterate the structure of the atom and destroy matter itself—but clearly that is not what happens. Atoms do not collapse; in fact, they typically endure for eons.

Bohr was aware of these issues, and in 1913 he responded with a new theory. The theory involved three postulates. (1) The orbiting electrons in an atom ordinarily exist in a stable, nondissipative state in which they do not radiate energy. (2) As long as the atom is in this stable state, the motions of its electrons can be analyzed according to the equations of Newtonian mechanics. Strangely, however, any transitions in state, such as those that occur if the electron changes its energy level, cannot be precisely described. (3) Whenever an electron in an atom does transition from a higher energy state to a lower, more stable state, its reduction in energy is emitted as radiant energy in an amount specified by Planck's equation, $E = h\,v$. (This equation was presented at the end of chapter 5, and the principles involved were discussed in the first two sections of the present chapter.) In other words, the amount of energy that is emitted is equal to Planck's constant multiplied by the frequency of the light wave that is radiated away by the transitioning electron. Bohr also provided a fourth postulate—that for a single electron orbiting around the nucleus, the *angular momentum* is a whole-number (an integer) multiple of $h/2\pi$ (Planck's constant divided by two times pi). This postulate explicitly introduces a quantum (integral) requirement to the theory of atomic structure.

The concept of angular momentum is important because the electrons were considered to be moving in a more or less circular orbit. Recall that Newton's first law of motion states that all objects in motion tend to continue at the same velocity and in the same direction. (In practical experience, objects lose velocity because of friction and air resistance.) In other words, moving objects tend to maintain their original momentum. *Angular* momentum refers to an object that is moving in a curved path, usually a circle. For example, angular momentum carries planets in their orbits around the sun for billions of years. It also sustains the rotation

18. Kaplan, *Nuclear Physics*, 85.

of the planets on their axes. Angular momentum causes spinning tops, gyroscopes, and flywheels to keep rotating until friction causes them to slow and come to a stop. Angular momentum was a central factor in Bohr's analysis of the atom's electrons because angular momentum was presumed to be the primary force involved in an electron's continuation in its orbit.

When an object moves in a circular path, there are generally two forces involved. *Centrifugal force* refers to the tendency of an object moving in a circle to fly away from the center of the circle—that is, to follow its natural tendency to move in a straight line as predicted by Newton's first law.[19] *Centripetal force* is what holds the object in its orbit and prevents it from flying away. We ourselves feel these forces when riding in a car that turns a corner sharply. Centrifugal force pulls us outwards away from the direction of the turn, while the seat belt, the car seat, and perhaps the door hold us in place. A well-known example of centrifugal and centripetal forces is found in 1 Samuel 17:30–50, the narrative of David's victory over Goliath. Not being constantly distracted by a cell phone, a laptop, a television, or even a book to read, David had lived an active life outdoors doing such things as defending his father's flock of sheep from predators. Part of this involved becoming an expert with a slingshot—two strips of leather attached to a pouch that held a smooth rock. The pouch containing the rock would be swung as rapidly as possible around the person's head until one strap was released, and the rock would fly out at considerable velocity. The circular, swinging motion of David's arm would impart centrifugal forces to the rock, while the leather straps in effect exercised centripetal force, temporarily restraining the rock from escaping. Swinging the rock in a circle would impart a much higher velocity than could be produced by simply throwing it. David had evidently becoming quite skilled at the use of this sling, as Goliath woefully discovered.

In view of the enduring stability of ordinary atoms, Bohr reasoned that the centrifugal force pulling the electrons outwards must be exactly equal to the centripetal force holding electrons in their orbits—for, if the forces were unequal, atoms would spontaneously tend to disintegrate. In the case of electrons and their nucleus, the attractive force is not gravity, as it is with planets and the sun, but rather the electrostatic attraction between the negative electrons and the positively charged protons

19. Centrifugal force is sometimes referred to as a "pseudo force" because no contribution of additional energy is involved; rather, the object is simply attempting to follow Newton's first law by conserving its own motion in a straight line.

in the nucleus. Physicists have equations for centrifugal force and for electrostatic attraction that allowed Bohr to analyze the electron's motion mathematically. Bohr simply formulated equations in which the two forces were equal. Bohr made the mathematics conform to quantum conditions by including only whole-number values at appropriate places in the equations. He then calculated the energy values that should be produced when an electron makes a transition from one orbital state to another. The values predicted by Bohr's equations conformed precisely to the observed values of the Ballmer series, a pattern of spectral emission lines produced by the transitions of electrons in heated hydrogen atoms.[20] This provided an extraordinary experimental confirmation of the accuracy of Bohr's model. Further tests, including the accurate prediction of other emission phenomena, also verified Bohr's basic concepts, though there were minor discrepancies in some circumstances. Nevertheless it was clear that Bohr was on the right track, and the Bohr model of the atom became a starting point for all further inquiry regarding the principles of quantum theory.

I have described black-body radiation, the photoelectric effect, Rutherford's alpha scattering experiment, and the experimental corroboration of Bohr's model of the atom. There are other experiments that could be mentioned. In particular, readers may wonder why I have not included the much-discussed "two-slit experiment," which is probably the most well-known experimental example of wave–particle dualism. This experiment is described briefly in chapter 7. However, at present we already have enough information to move on to a summary of the basic quantum mechanical principles that were eventually discovered. These principles are the Schrödinger wave equation, Heisenberg's uncertainty principle, quantum entanglement and nonlocality, the Pauli exclusion principle, and the nature of each atom's electron orbitals and energy levels. These principles are discussed in our next chapter. In chapters 7 and 8, I discuss what I believe are some of their most important philosophical implications of quantum mechanics and the contributions of these implications to a new metaphysics or worldview.

20. *Encyclopædia Britannica*, 15th ed. (1974), s.v. "Atomic Structure." See also Maron and Prutton, *Physical Chemistry*, 615–18.

7

Quantum Mechanics: Concepts and Principles

IN THIS CHAPTER I present some of the most distinctive and well-known aspects of quantum mechanics. These aspects are the Schrödinger wave equation, Heisenberg's uncertainty principle, quantum superposition, quantum "jumps" and quantum tunneling, quantum entanglement and nonlocality, and the Pauli exclusion principle. I do not discuss specific instances such as the two-slit experiment and the problem of Schrödinger's cat.[1] I do describe electron orbitals and the distinct energy levels that pertain to the electrons in every atom, since electronic orbitals and energy levels are the basis of chemical bonding and are therefore necessary to explain the existence of matter itself. After describing these quantum principles, in chapter 8 I discuss what I believe are some of their most important philosophical implications.

The Schrödinger equation was formulated in 1925 and published in 1926 by Austrian-Irish physicist Erwin Schrödinger (1887–1961). The equation describes the behavior of electrons and other particles and is also relevant to entire atoms and molecules. Its discovery constituted an essential development in quantum mechanics, and it endures today as an indispensable component of physics and chemistry. The Schrödinger equation is defined as "a linear partial differential equation that governs

1. These two subjects are left out because they are not general principles but particular examples and because our discussion up to this point has not proceeded far enough to consider them satisfactorily. I do discuss the problem of Schrödinger's cat and the "multiverse" in chapter 8.

the wave function of a quantum-mechanical system."[2] Partial differential equations are typically studied by second-year calculus students, but the subtleties of the Schrödinger equation quickly move beyond calculus to a state of arcane complexity. Fortunately, our present purposes allow us to sidestep the mathematical details.

The search in quantum theory for a more accurate mathematical formula arose because Bohr's equations failed to describe several experimental observations. At the time, the most serious failure of the Bohr model was that it could not predict the spectrum of light emitted by atoms that have more than one electron—that is, atoms that are heavier than hydrogen. (The emission of light by heated substances was discussed in our consideration of black-body radiation.) Since all atoms except hydrogen have more than one electron, the search for a better theory was imperative.

Actually, the first notable mathematical improvement to the Bohr model was achieved not by Schrödinger but by German physicist Werner Heisenberg (1901–1976). Heisenberg had completed his doctorate when he was only twenty-one. After that he studied with Bohr for a few months in Copenhagen and then became the assistant of physicist Max Born (1882–1970) in Göttingen. In June 1925, attempting to escape the pollen that was inflicting him with a miserable attack of hay fever, Heisenberg retreated to Heligoland, a tiny, rocky island in the North Sea near Hamburg. By temperament Heisenberg was more nearly a theoretician than an experimentalist, and on Heligoland the insight came to him that the Bohr model's visualization of electrons as tiny planets orbiting a tiny sun was obscuring significant options, so he resolved to pursue mathematical solutions without considering specific physical imagery.[3] Alone on the island, and working obsessively into the night, Heisenberg suddenly experienced a moment when he realized that his mathematics satisfied all of the necessary conditions.

> I worked all night and I made many slips in the calculation. Around two or three o'clock in the morning I saw that the conservation of energy was correct. I was extremely excited, and it

2. Wikipedia, "Schrödinger Equation."

3. Gribbin, "Heisenberg, Hayfever, and Heligoland." This resource provides a very interesting and readable narrative of Heisenberg's discovery. Heisenberg's own account of the early development of quantum theory is found in Heisenberg, *Physics and Philosophy*, ch. 2 (all references to Heisenberg's *Physics and Philosophy* refer to the Prometheus edition, unless otherwise specified).

> was just early in the morning already. I decided that I would go out for a walk and so I did. I rather half-climbed on one of the cliffs of Heligoland just for excitement. And I felt, "Well, now something has happened." So then after a while I went back and I went sound asleep. Then I started writing on a paper.[4]

Heisenberg was at that time attempting to analyze the transition of particles or atoms from one state to another. His method was to compare two tables of values (two mathematical arrays) pertaining to each of the two states. This comparison of arrays is what eventually led to successful results. When Heisenberg returned to Göttingen he eagerly shared his work first with Austrian physicist Wolfgang Pauli (1900–1958) and then with his supervisor, Max Born. Pondering Heisenberg's method, Born realized that the sub-branch of mathematics known as *matrices* would be the best way of analyzing the relationship between two mathematical arrays. The use of matrices was then incorporated into Heisenberg's published work, and thereafter Heisenberg's approach became known as *matrix mechanics*. Heisenberg, Born, and German physicist Pascual Jordan (1902–1980) jointly published a work now known as "the three-man work" (*drei Männer Arbeit*), which was the first consistent articulation of quantum mechanics that did not rely on visualizable depictions such as Bohr's "solar system" atom. Matrix mechanics avoided graphical imagery by treating the physical states of electrons and atoms as matrices that shift quantum-style, all at once, from one array of values to another.

It may not have been fully recognized at the time, but the new mechanics amounted to a sea change in physics, not only for the atomic model but for methods of analysis used throughout the whole discipline. Matrix mechanics jettisoned the physical-mechanical models that had always lain at the heart of physics and relied instead on *mathematical expressions themselves* as a representation of reality. In quantum mechanics, two well-known examples of this reliance on mathematics are found in the notion of "quantum jumps" and in the Heisenberg uncertainty principle. I will describe these two concepts in a moment.

Shortly after the "three-man work" was published, Erwin Schrödinger published his aforementioned equation. Schrödinger's approach immediately garnered more attention than Heisenberg's, perhaps mostly because partial differential equations were, and still are, a principal tool in physics and chemistry, whereas matrices were viewed as a minor (and

4. Heinsenberg, "Session VII," 14.

unfamiliar) branch of theoretical mathematics. After Schrödinger's work appeared, physicists assumed that either his methods or Heisenberg's would turn out to yield more accurate predictions, and a rivalry sprang up between the two camps. The competition soon ended when English physicist Paul Dirac (1902–1984) showed that the two methods were compatible and that they produced identical results.

As noted, Schrödinger's equation is described as a "wave function," and therefore Schrödinger's version of quantum theory is sometimes referred to as "wave mechanics." I will discuss the idea of a wave equation in a moment. First, however, we should note at least in passing that there is a third approach to quantum calculations, known as "Feynman arrows" or "Feynman diagrams." A Feynman diagram is a graphical depiction of the mathematical formulae pertaining to the physical characteristics and processes of subatomic entities. American physicist Richard Feynman (1918–1988) introduced the diagrams in 1948. Feynman diagrams provide simple drawings that simplify what would otherwise require enigmatic mathematical symbols and complicated calculations. The fact that these three very different methods—matrix mechanics, wave mechanics, and Feynman diagrams—all predict the same results suggests strongly that there is a testable, empirical reality at the base of all three approaches. But it is Schrödinger's wave equation that has endured as the most commonly encountered tool in quantum mechanics. The equation leads to accurate predictions about all sorts of phenomena in physics and chemistry.

It would seem obvious that the phrase *wave equation* pertains to a mathematical equivalence that describes a wave. This could be a wave in the ocean, a sound wave, or a radio wave. But, by extension, wave equations need not be applied only to phenomena that we think of specifically as physical waves. They can be used to describe many processes that alternate regularly between two limits or extremes—for example, the waxing and waning angle at which sunlight strikes a certain latitude of the surface of the earth over the course of a year, the phases of the moon, the rising and falling of tides, or perhaps even the fluctuating traffic volume on a certain street over a twenty-four-hour period.

Naturally, not all equations in physics are wave equations. The equations that are most typically used to describe phenomena scientifically do not pertain to waves. As a simple illustration, to calculate how far we can drive during a certain time period, we use the formula

$$distance = speed \times time$$

If you average a speed of 60 miles per hour, in two and a half hours you will have driven 150 miles. Equations such as this one yield precise answers, at least in principle. However, there are other types of mathematics that pertain not to certainties but to probabilities. For example, since there are fifty-two cards and four aces in a standard deck, your chances of randomly drawing one ace out of a whole deck are four out of fifty-two, which amounts to one out of thirteen, or approximately 7.7 percent. We are accustomed to hearing about probabilities in the weather forecast ("a 60 percent chance of rain") or in a baseball game when a hitter with a certain batting average comes to the plate. Obviously, a probability must always be stated as a figure *not less than zero and not more than one*. In practice this is often expressed as a percentage. There is a mathematical sense in which wave equations resemble probability calculations because both are constrained by higher and lower limits. The limit on a physical wave is the mathematical difference between the height of its peak and the depth of its trough. Other points as the wave rises and falls are necessarily lower than the peak and higher than the trough.

In quantum mechanics the calculated result of the wave function developed by Schrödinger is interpreted as the *probability* of finding a particle at a specific point in space. (Technically, it is the square of the wavefunction that expresses this probability.) In other words, the equation is such that its mathematics can produce answers only between zero and one. It may not be recognized at first, but the applicability of Schrödinger's wave equation to all atomic particles and systems of particles suggests that a wavelike vibration or periodicity somehow pertains not only to radiant energy but *systematically, to all of the basic, constitutive elements of nature as a whole, including matter*. In fact, the thought that waves characterize not only energy but matter was first advanced in 1924 by French physicist Louis de Broglie (1892–1987), and the existence of matter waves was corroborated experimentally in 1927 by George Paget Thomson's thin metal diffraction experiment and by the Davisson–Germer series of experiments at Bell Labs between 1923 and 1927.

The electrons in an atom serve as a good example of matter waves. In an atom, though the electrons are usually pictured as particles (that is, as material objects), quantum mechanics considers them to have both particle-like and wavelike properties. The wavelike properties are these: (1) The electrons do not orbit the nucleus in the way first imagined by Bohr, like a planet orbiting the sun, but instead behave like standing waves. In physics, a standing wave, also known as a stationary wave, is a wave that oscillates in time but whose peak amplitude position does not move in a direction through space. Conceptually this resembles a vibrating string; the string as a whole does not move through space, but it vibrates at a given frequency. Higher frequencies are possible as overtones or harmonics, both for a vibrating string and in quantum mechanics. (2) Like waves, the electrons are never located at a definite point in space, but the *probability* of interacting with a certain electron at a single point can be calculated by the Schrödinger equation using the particular quantum numbers of the electron. (I describe quantum numbers below.) (3) The charge on the electron acts like it is smeared out in space in a continuous distribution, proportional at any point to the squared magnitude of the electron's wave function.

In contrast, the particle-like properties of the electron are these: (1) The number of electrons orbiting the nucleus can only be an integer. (2) Electrons jump between orbitals all at once, not gradually as a wave would. The "jumps" are often caused by "collisions" with photons. Photons act on electrons one by one, resembling collisions between two particles. (3) The wave state of an electron has the same specific electrical charge as its particle state, and each wave state has a single discrete spin (spin up or spin down). "Spin" does not mean that particles are actually spinning like little spheres; rather, it refers to certain behaviors that occur when the particles are passed through a magnetic field.[5]

It is more or less well-known that Niels Bohr adopted the word *complementarity* to refer to this strange situation in quantum theory in which two seemingly incompatible physical states—in this case, waves and particles—are both applicable at once. The word *complementarity* is easy to misconstrue and calls for careful interpretation. Complementarity does not refer simply to the obvious sources of differences in perspective that characterize human knowledge. A girl may view her grandmother as a kindly, elderly relative, though the girl's grandfather understandably views the same woman as his wife. In this situation, beyond these specific family relationships there are more neutral terms (*older woman*) that are accurate enough for general use. But *grandmother* and *wife* are not therefore examples of complementarity. Bohr applied *complementarity* to situations in which contradictory terms such as *wave* and *particle* are both necessary because neither term is consistently adequate, and there is no suitable neutral term that can be introduced to describe whatever

5. Wikipedia, "Atomic Orbital."

the electron really is. What it "really" is depends upon the experimental methods that are currently being used. Bohr tended to view questions about the electron's underlying "reality" as superfluous or at least as unanswerable.[6]

Heisenberg's uncertainty principle may to some extent be regarded as a corollary to the wavelike aspect of matter. If we think about a single wave for a moment, it becomes clear that *a wave is innately holistic*, defined by having both a peak and a trough and including all the positions and the continuous change in between. If we ask where a sound wave is located, there can be no precise answer because, by its very nature, a sound wave must be spread through space at least far enough and for enough time to provide at least one cycle of rarefaction and compaction in the air. This minimum space and time cannot be physically divided or compressed without fundamentally altering the pitch of the sound, and if we force the wave into a single point spatially or into a single instant temporally, we eliminate the sound altogether.

This example provides an insight regarding Heisenberg's uncertainty principle, for *we cannot force an electron to occupy a specific location at a specific moment without destroying the wavelike aspects of the electron*—and the wavelike aspects are essential to what the electron really is, what philosophers would refer to as its ontological status. Thus, though in Newtonian physics particles and atoms would be deemed to have a specific location and a specific velocity, in quantum theory the wave equation provides not a certainty of location or velocity but a probability. In mathematical terms, the Heisenberg uncertainty principle is usually stated in the form of a simple equation:

$$\Delta x \times \Delta p \geq h/4\pi$$

where delta-x is the uncertainty of position, delta-p is the uncertainty of momentum, and *h* represents Planck's constant. (Momentum, "p," is equal to velocity multiplied by mass.) In other words, the more precisely we determine the location of a particle, the less precisely we are able to determine its momentum, and vice versa.

The fact that Heisenberg was able to define uncertainty in terms of a mathematical equation, and not merely in terms of ad hoc laboratory phenomena, shows us that in quantum mechanics *the uncertainty principle is an intrinsic aspect of the quantum-mechanical conception of reality.* It is very important to understand that the uncertainty principle is not,

6. For a fuller discussion of complementarity, see Barbour, *Issues in Science and Religion*, 282–83.

as has often mistakenly been believed, merely a practical consequence of the interruptive effects of laboratory measurements or of the inherent imperfections of experimental methods.

> Historically, the uncertainty principle has been confused with a related effect in physics, called the observer effect, which notes that measurements of certain systems cannot be made without affecting the system, that is, without changing something in a system. Heisenberg utilized such an observer effect at the quantum level as a physical "explanation" of quantum uncertainty. It has since become clearer, however, that the uncertainty principle is inherent in the properties of all wave-like systems, and that it arises in quantum mechanics simply due to the matter wave nature of all quantum objects. Thus, the uncertainty principle actually states a fundamental property of quantum systems and is not a statement about the observational success of current technology. Indeed the uncertainty principle has its roots in how we apply calculus to write the basic equations of mechanics.[7]

The disparity between "the observer effect" and mathematically necessary uncertainty has sometimes been described in philosophical terms as the difference between an *epistemological* interpretation (pertaining to the intrinsic limits to our knowledge) and an *ontological* interpretation (pertaining to an actual indeterminacy in nature).[8] The word *indeterminacy* abandons the Kantian notion that it is our knowledge that is imperfectible in favor of the belief that nature herself makes certainty impossible in atomic physics because *in nature there really are no tiny material particles occupying exact places at exact times*. The choice of *indeterminacy* over *uncertainty* thus implies a categorical rejection of philosophical determinism. Niels Bohr himself had at first used the word *uncertainty*, but as evidence of his own awareness of the epistemological-ontological distinction, after about 1935 Bohr tended to drop the use of *uncertainty* and speak instead of *indeterminacy*.[9] Bohr was intensely concerned to correct the error of believing that experimental methods are inconclusive in quantum mechanics simply because the laboratory equipment itself cannot help but alter the behavior of the objects being investigated. Bohr developed the position that the alleged errors caused

7. Wikipedia, "Uncertainty Principle."

8. Werner Heisenberg himself uses the terms *epistemology* and *ontology* pertaining to different ways of understanding the uncertainty principle (Heisenberg, *Physics and Philosophy*, 48).

9. Faye and Folse, introduction to Bohr, *Causality and Complementarity*, 2–3.

by the "observer effect" are based on the flawed assumption that there is actually an objective world external to the observer and that it should be possible to devise scientific tests that can be carried out without actually having any effect on the "objects" being measured.

> The apparent contradiction . . . discloses only an essential inadequacy of the customary viewpoint of natural philosophy for a causal account of physical phenomena of the sort with which we are concerned in quantum mechanics. Indeed the *finite interaction between object and measuring agencies* conditioned by the very existence of the quantum of action entails . . . the necessity of final renunciation of the classical ideal of causality and a radical revision of our attitude towards the problem of physical reality.[10]

Bohr wrote these words in 1935. He went on to develop a holistic approach to quantum experimentation, advocating that the word *phenomenon* (that is, the experimental results) should be used "exclusively to refer to the observations obtained under specified circumstances, including an account of the whole experimental arrangement."[11] That is, in quantum mechanics, the subject of the investigation and the apparatus being used are inseparable, even theoretically.

Quantum superposition is a fundamental phenomenon in quantum mechanics in which a particle or a system of particles exists simultaneously in two (or more) states at once.[12] The prefix *super-* should be taken to mean not "superior in quality" but "beyond" or "additional." The term *superposition* was originally used in physics to refer to the combined effect of two or more waves in a given medium. The interaction or "interference" of two waves was described briefly in our consideration of the phenomenon of interference in chapter 4. When superposition occurs, the amplitude of the resulting wave is the algebraic sum of the amplitude of the contributing waves as they interfere with one another.[13]

However, in quantum physics superposition does not refer to two existing waves that physically interact but to two or more coexisting

10. Bohr, "Quantum-Mechanical Description," 74–75; emphasis original. This article was first published in 1935 as a response to the Einstein–Podolsky–Rosen criticism of quantum mechanics.

11. From *The Philosophical Writings of Niels Bohr*, vol. 2, as quoted in Barad, *Meeting the Universe Halfway*, 119.

12. Bobroff, "Quantum Superposition." The information in this video is helpful and simply presented.

13. Sears and Zemansky, *University Physics*, 495, 895.

possible wave states. The two coexisting states are said to be "superposed," and the result of the addition of the two (or more) will be another valid quantum state. Mathematically, superposition refers to a property of solutions to the Schrödinger equation.[14] The principle of superposition posits that any two or more quantum states can be added together in a way analogous to the way that vector arrows are added mathematically.[15] In math, *vectors* are usually contrasted to *scalars*; scalars are associated with counting or numeration (e.g., thirty apples, or two and a half dollars) while vectors have a directional component and are often represented by arrows (for example, wind blowing from south to north). Vectors can therefore be added to one another using the rules of algebra and analytic geometry. Persons who are not familiar with vectors may find this confusing, and it is important to describe superposition in a less technical way.

Superposition may be viewed as an extension of the uncertainty principle, inasmuch as superposition assumes that particles cannot be said to occupy one unambiguous position or to maintain one unambiguous state of momentum. The technical terms *quantum coherence* and *decoherence* depend on the idea of superpositioning because the wave functions of the two superpositioned states can interfere with each other and create an interference pattern. When there is no coherence, there is no system-wide relationship between peaks and troughs and no regular interference pattern but just random, featureless variations in the resulting wave amplitude. This is decoherence.

Published explanations of superposition, coherence, and decoherence quickly grow specialized, and more than a few persons find the details difficult to grasp. Obviously, saying that things can exist in two states at once seems paradoxical, but the appearance of paradox can be mitigated greatly if materialistic presuppositions are abandoned—that is, if we avoid our natural inclination to assume that particles and systems of particles normally exist as stable material objects. Indeed, the probabilistic nature of the wave function suggests that superposition can be thought of not as two conflicting states of *actuality* but as two coexisting states of *potentiality*. When two states are "superposed," each state represents a *possibility* that may eventually be actualized. The Schrödinger equation calculates the probability of each outcome. Heisenberg wrote,

14. Wikipedia, "Quantum Superposition."

15. Varsity Tutors, "Adding and Subtracting Vectors."

> In throwing dice we do not know the fine details of the motion of our hands which determine the fall of the dice and therefore we say that the probability for throwing a special number is just one in six. The probability wave of Bohr, Kramers, Slater, however, meant more than that; it meant a tendency for something. It was a quantitative version of the old concept of "potentia" in Aristotelian philosophy. It introduced something standing in the middle between the idea of an event and the actual event, a strange kind of physical reality just in the middle between possibility and reality.[16]

One of the charming aspects of Heisenberg's Gifford Lectures[17] is his frequent reference to Greek philosophers—Democritus, Thales, Plato, Aristotle, Heraclitus—but especially Aristotle and Heraclitus. His special interest in the Greeks may have been inspired by his father, August Heisenberg (1869–1930), who taught classical languages and philology in Germany at the high school and university level. At any rate, like Bohr, Heisenberg recognized intuitively that quantum theory manifested a need for philosophical resources other than the conventional physicalism that had been the standard worldview of Newtonian mechanics. I discuss Heisenberg's appreciation of Heraclitus below; at present it is his use of Aristotle's notion of *potentia* that is of interest.

Aristotle was characterized by an unflagging fascination with the processes of nature. Noting that the natural world involves constant transitions from one state to another, Aristotle posited a fundamental dichotomy in nature between *potentiality* and *actuality*. The words *potentiality* and *potency* are translations of the Greek word *dunamis*, which is the source of our words *dynamic*, *dynamo*, and *dyne*. *Dunamis* can also mean potency, capacity, or capability. The Latin translation of *dunamis* is *potentia*, and, giving credit to Aristotle, Heisenberg repeatedly uses the term *potentia* to refer to the possibilities or tendencies that are indicated by quantum states such as superposition.[18]

16. Heisenberg, *Physics and Philosophy*, 40–41.

17. Heisenberg gave his lectures at St. Andrews in 1955–1956; these were published as *Physics and Philosophy*. The Gifford Lectures were first delivered in 1881 after having been established by Adam Lord Gifford (1820–1887) for "'Promoting . . . and Diffusing the study of Natural Theology' in the widest sense of the term, in other words, 'The Knowledge of God'" (The Giffords, "Lord Gifford's Will"). Despite this stated purpose, many of the lecturers, including Heisenberg, have managed to avoid any talk about "natural theology" and focus on other topics.

18. Heisenberg specifically mentions Aristotelian *potentia* at least five times in *Physics and Philosophy* (see 41, 53, 147, 160).

Aristotle's differentiation between potentiality and actuality is paralleled in modern physical science by the common distinction between potential energy and kinetic energy. *Kinetic* energy is commonly illustrated by referencing objects in motion. A moving freight train contains a great deal of kinetic energy. In contrast, *potential* energy refers to energy that has somehow been stored. For example, if you lift a ten-pound dumbbell and put it on a shelf five feet above the floor, you have stored fifty foot-pounds of energy in the weight. (A foot-pound is the energy required to lift one pound a distance of one foot.) Fifty foot-pounds is the equivalent of sixty-eight joules or sixteen calories of energy. The dumbbell may fall from the shelf and strike the floor, in which case its potential energy will be converted to kinetic energy, and this amount of energy will be transmitted into the floor, probably causing a dent in the floor boards or a crack in the ceramic tiles.

Potential energy can also be stored in molecules. Explosives are manufactured by heating chemicals together to form new compounds in which the energy imparted by heat is stored in chemical bonds. The Swedish chemist Alfred Nobel (1833–1896) heated a mixture of nitric acid and glycerol to create nitroglycerin. The energy of the heat added to the compounds is stored in the chemical bonds of the nitroglycerin molecule. Nobel stabilized this notoriously dangerous compound by mixing it with diatomaceous earth and patented the product as "dynamite," becoming wealthy from the proceeds. I mention dynamite not merely to provide anecdotal relief from this technical discussion but as a further illustration of the connection between *dunamis*, dynamism, *potentia*, potency, on the one hand, and power, on the other. In other words, there is a connection between *possibility*, *probability*, and the *capability* or *power* to produce specific outcomes. In Nobel's experience, several tragic accidents pointed to a relationship between the great *power* of nitroglycerin and the *probability* that it would explode. This example, of course, is merely heuristic and not specifically about quantum mechanics. But there is an important point: that the association between *dunamis*, *potentia*, and probability is not simply etymological but empirical. At this point I invite you, the reader, to consider how the connection between power (or capability), potentiality, and probability (or tendency) exists not only in examples from chemistry but in life generally.

When Heisenberg repeatedly referred to Aristotle's notion of *potentia* to describe quantum phenomena, he was not talking simply about stored physical energy. Though Heisenberg's *potentia* does have a connection with a physical setting (various specific quantum situations), not only physical energy is involved; there is a suggestion of *tendency* or *possibility*, in the sense of *a pending outcome that is as yet indeterminate but that tends eventually to come to pass* in one way or another. In other words, superposition may be understood as something like a set of pending options—options that can be analyzed mathematically as probabilities by the Schrödinger equation.

Quantum entanglement and nonlocality refer to instantaneous informational connections between particles even when they are separated by great distances. The 2022 Nobel Prize in Physics was awarded to three researchers for experimental work related to quantum entanglement. Entanglement and nonlocality are usually explained in terms of relations between particles, but they may also be understood as corollaries of superposition or, also, as manifestations of the ontological state of *potentia*.

In quantum theory the notion of *system* is of fundamental importance. The word *system* suggests a group of parts that interact in ways that display a type of consolidation or holistic interaction. For example, the transmission in an automobile contains a system of gears. In a quantum-mechanical context, however, *system* does not refer simply to a physical assemblage, like the movement of high- and low-pressure cells through the atmosphere, or even like the mechanical interactions of the gears in a transmission. In a quantum system a type of wholeness is present that involves the *spontaneous exchange of information*. The parts of a quantum system are said to be *entangled*, which is another way of saying that they mutually affect one another in ways that are holistic. Quantum systems may be said to be *organismic*, but with the emphatic caveat that the word *organismic* does not imply that the system is necessarily alive or biological. The word *organismic* is metaphorical, indicating that the system *resembles* an organism in certain essential respects.

In a quantum system it is not simply that the parts create the characteristics of the whole but that, conversely, the characteristics of the whole also modify the character of the parts. This mutuality of influence is what is suggested by *organismic*. In a grain of table salt, sodium and chlorine atoms alternate with one another in a three-dimensional lattice pattern. Salt is holistic in the sense that it manifests properties that do not pertain to either sodium metal or to chlorine gas. This type of holism can be explained reductionistically because the emergent properties of salt can be explained in terms of the ways the sodium and chlorine atoms bond chemically. But now consider an organ in the body of an animal—say, the lungs. A lung, existing on its own, could never appear in nature spontaneously in the way that atoms do. Lungs appear *only* in the context of a bodily system, and the lungs are reciprocally indispensable to the survival of the body as a whole. In other words, the lungs and the body are mutually necessary to one another. It is this type of mutual necessity or mutual creation that we have in mind when we use the words *system*, *organismic*, and *entanglement* in discussions about quantum mechanics. Quantum entanglement refers to the superposed state of two or more particles whose intertwined state of interaction has turned them into a single quantum entity or quantum system.[19] The state of entangled particles or systems of particles (such as electrons in atomic orbitals) resembles the example of the lung and the body to the extent that changes in one part directly and immediately alter the other parts. This would not be true in the same way for, say, individual atoms in a crystal of salt. It should be noted that entanglement exists because of the immediate, nonlocal communication of information.

Nonlocality refers to a connection or a relation between entangled particles that does not entail physical forces or physical collisions, or any chain-of-events type of causation that is transmitted through space-time. Our usual understandings of any physical relationship depend on the notion of "locality." Local causation is understood to be transmitted by objects or energies that move from place to place through contiguous locations as a "chain of events" in the manner of colliding billiard balls or falling dominoes. Such causation is said to be "local" because it is envisioned as taking place when cause and effect share a common space or locale and because causation is presumably transmitted via a path of adjacent spaces. Even causes that at first appear to be widely separated from their effects, as when light emitted by the sun causes plants to grow on earth, are actually local causes, because the sun's rays pass through an adjoining series of locations in the intervening space at a certain rate—in this case, the speed of light.

In contrast to local causation, quantum theory predicts that immediate (non-mediated) "correlations" may exist between particles—or, more accurately, "particle-like entities"—even at great distances, whenever the particles are co-members of a single quantum system. Such correlated particles are said to be entangled, and the correlation between them does not depend on any form of place-to-place local causation.

19. Ball, "Universe Is Always Looking."

The correlations between entangled particles are not conveyed by any mediating agent and are presumed to be instantaneous. The idea of nonlocality is counterintuitive. Even Albert Einstein clung to the idea that no form of influence can travel faster than the speed of light, and he derided nonlocality by referring to it as "spooky action at a distance."[20] Amazingly, however, during the past two decades quantum nonlocality has been experimentally verified.[21] Entanglement can be observed empirically as a faster-than-light change of state in a particle when its entangled partner particle's state changes. Since, according to quantum physics, particles that are entangled in this fashion have been being created continually ever since the Big Bang, quantum mechanics has become the basis for the idea that the entire universe is to a significant extent entangled in a web of nonlocality.[22]

Quantum "jumps" or quantum "leaps" are a specific example of the quantized behavior of particles, atoms, and molecules. By now we are familiar with the fact that energized particles release their energy in discrete amounts indicated by line spectra and that quantum mechanics therefore uses integers in its basic calculations. These principles lead to the idea that when atomic particles absorb or release energy, they do not move continuously from one location to another but rather seem to disappear from one location and emerge at another.

> One of the basic ideas of quantum theory is that a molecule may not possess any arbitrary quantity of internal energy, but rather can exist only in certain "permitted" energy states. If a molecule is to absorb energy and be raised to a higher energy level, it must absorb a quantity appropriate for the transition. It cannot absorb an arbitrary quantity of energy determined by the experimenter and linger in an energy state intermediate between its permitted levels. This quantization of molecular energy, coupled with the concept that photons possess definite quantities of energy, sets the stage for selectivity in the absorption of radiant energy by molecules. When molecules are irradiated with many wavelengths, they will abstract from the incident beam

20. For more information, see Wikipedia, "Einstein–Podolsky–Rosen Paradox," and Wikipedia, "Hidden-Variable Theory."

21. In fact the 2022 Nobel Prize in Physics was awarded jointly to Alain Aspect, John F. Clauser, and Anton Zeilinger "for experiments with entangled photons, establishing the violation of Bell inequalities and pioneering quantum information science" (The Nobel Prize, "Nobel Prize in Physics 2022").

22. In these two paragraphs I am quoting my own article: see Conner, "Quantum Non-Locality," 259–60. See also Nadeau and Kafatos, *Non-Local Universe*.

> [only] those wavelengths corresponding to photons of energy appropriate for permitted molecular energy transitions, and other wavelengths will simply be transmitted [passed through without any interaction].[23]

The crucial statement in this passage is that particles and systems of particles do not "linger in an intermediate energy state." Consider what this means for a single electron in an atom. When radiant energy passes through an atom, *if a wavelength matches the energy for a transition allowed by quantum conditions*, the electron will absorb that energy and *appear immediately* in its newly energized state. This is not what happens with macroscopic objects. For example, a sports car may accelerate from 0 to 60 miles per hour in three seconds—but obviously the car requires a measurable amount of time to gain speed and an ongoing input of energy supplied by the car's engine. The equations of classical mechanics do not allow for acceleration that is instantaneous. But at the level of particles or individual quantum systems, energy is absorbed or emitted only in discrete packets. Therefore there can be no gradual absorption or emission of energy; systems simply change from one state to another. Though the popular expression "quantum jump" is used to refer to any relatively abrupt change in circumstances, technically the phrase, even when used as a metaphor, should be applied only to either-or situations in which there is no in-between possibility.

Quantum tunneling is a phenomenon conceptually similar to quantum jumps or leaps. In quantum tunneling, a particle appears in a new location even though, according to the calculations of classical physics, it lacked the energy to get there. The particle could be a moving electron or perhaps a nuclear particle that appears outside the nucleus.[24] Quantum tunneling occurs in some types of radioactive decay, in the sun,[25] and in human inventions such as flash memory chips.[26] Tunneling is often illustrated by referring to the analogy of a ball rolling up a hill.[27] Classical mechanics requires that a rolling ball that does not have enough energy to reach the top of the hill will simply stop and roll back down towards its starting location. Similarly, a ball that lacks the energy to penetrate a wall

23. Day and Underwood, *Quantitative Analysis*, 293–94.
24. See Ball, *Beyond Weird*, 51–52.
25. Siegel, "Sun Wouldn't Shine."
26. Wikipedia, "Applications of Quantum Mechanics."
27. Wikipedia, "Quantum Tunnelling."

will either ricochet or be absorbed into the wall. However, in quantum mechanics, there is a small probability that some particles can "tunnel" outside of the limits of their current energy state, simply appearing on the other side of the metaphorical hill or on the outside of the metaphorical wall. The "hill" and the "wall" are actually "energy cages" created by physical forces that, according to classical physics, ought to contain the particle within a certain region.

Though the principle of quantum jumps does not allow a particle to "linger in an intermediate state," among scientists the *concept* of quantum jumps itself has ironically lingered in an intermediate state as an idea, for there is still some doubt among physicists regarding its interpretation. That a particle can simply disappear from one position (or energy state) and instantly reappear in another has remained debatable. Though most of his colleagues accepted the notion of quantum jumps, Erwin Schrödinger himself denied that the jumps occur instantaneously, even though it was Schrödinger's own mathematical work that predicted the jumps. Experiments are still being carried out to test this question.[28] In any case, the truth of quantum theory does not rise or fall with the outcome of these experiments.

It should be noted that skepticism concerning the immediacy of the transitions seems to be based primarily on the ingrained image of particles as tiny physical objects that, if they really were enduring material entities, could not simply disappear and reappear somewhere else. It should also be noted that, as I have already said more than once, if we discard the notion of stable particles of matter and imagine instead that enduring objects are composed of serially ordered processes, the conceptual difficulties ascribed to quantum mechanics, and particularly to quantum jumps, are eliminated. The particle is no longer viewed as a material body that is stuck in an energy cage. Each particle is a series of events, and in quantum tunneling the next event simply takes place in a new location.

The Pauli exclusion principle is the final foundational rule in quantum mechanics and chemistry that we will consider.[29] The principle is

28. Ball, "Quantum Leaps." The designers of this experiment have a clever way of attempting to avoid the influence of measurement. However, they clearly presuppose that an electron is a physical particle that must continually exist somewhere in space during transitions from one energy state to another. It is axiomatic in quantum mechanics that experiments tend to find what the experiment is designed to detect.

29. Richard Feynman describes the exclusion principle in his inimitable way in Feynman, *QED*, 113–14.

usually applied particularly to the possible states of electrons in atoms and molecules, but it pertains to other atomic systems as well. The principle itself is named for Austrian physicist Wolfgang Pauli, who first proposed it in 1925. The principle states simply that "no two electrons in an atom may have all four quantum numbers the same."[30]

> The principle states that in a given atom no two electrons can be in identical states (or have identical quantum numbers specifying energy, angular momentum, orientation, and spin). To this remarkable and far-reaching principle can be attributed the periodic chemical properties of the elements. When another electron is added to a given atom, it must assume a state different from all electrons already present, even if this necessitates a very different energy (as occurs in starting a new "row" of the familiar periodic table).[31]

I submit that the Pauli exclusion principle simply cannot be explained by appealing to mechanistic or reductionistic factors. Pauli's rule cannot be anticipated from the properties of an individual electron; it rests on the implied assumption that systems of electrons—such as the organization of electrons in an atom—must be treated as an ontological whole. In the words of German-American physicist Henry Margenau, "Mechanistic reasoning, already far behind, has gone out of sight as a result of this latest advance [the exclusion principle]."[32] Reductionistic philosophy contends that the characteristics of every whole can be explained by referring to the characteristics of the parts, but in atomic structure it is clear that the system as a whole somehow guides each electron's functioning in the larger structure, not just in a general way but in a way that assigns a different state to each electron. If we imagine a nucleus—any nucleus, even uranium, with ninety-two protons—and then imagine a cloud of electrons hurled at this nucleus, the result is that ninety-two of the electrons will spontaneously organize themselves into a system of orbitals surrounding that nucleus in such a way that every single electron has its own unique energy configuration. This organization seems remarkable enough even if we simply ponder it after the orbitals have already formed, but the deeper question is *how the electrons know, in advance, to organize themselves in this way*. It is as if

30. Maron and Prutton, *Physical Chemistry*, 621–22.

31. Barbour, *Issues in Science and Religion*, 295.

32. Margenau, *Nature of Physical Reality*, 443, as quoted in Barbour, *Issues in Science and Religion*, 296.

the electrons can somehow anticipate the system that they are about to produce. The answer I am proposing involves the role of mathematical (nonphysical) information as a causal factor. We will consider this issue further in chapters 8 and 9.

The Pauli exclusion principle is easy to verbalize, but background information is needed in order to understand it. This background information must include a brief explanation of quantum numbers and chemical bonding. I apologize to any readers who may find this information tedious or elusive. But take heart: it is not necessary to grasp and retain all the details, and readers who are averse to technical-sounding science may skip to the final three paragraphs of this chapter. The discussion that now follows is intended to reveal the way in which the details behind the Pauli exclusion principle illustrate the deep and sometimes subtle connection that exists between mathematics and the formation of atoms. Atoms are what make it possible for the earth and life to exist. Moreover, the Pauli exclusion principle serves as a paradigm in our investigation of the role of mathematics in the behavior of nature. Therefore the intimate relation between mathematics and atomic structure is a crucial subject in this book.

The principle that "no two electrons in an atom may have all four quantum numbers the same" may sound like an arbitrary rule in an unnecessarily complicated card game, but in fact Pauli formulated the exclusion principle in response to empirical observations—namely, the observed repetition of certain chemical properties in the periodic table of the elements, and the appearance of a "fine-line structure" in the spectrum emitted by certain elements under certain conditions. To repeat an earlier explanation, when light emitted by a heated element in a gaseous state is passed through a prism or a diffraction grating, the observed result is that only certain wavelengths ("colors") are produced. This appears as a line or a cluster of lines on a screen or a detection device. Since the position of a line on a spectrograph is a result of the energy of the light which produced it, the fine-line structure suggested to physicists that electrons in the atom may acquire subtly different energy levels. Further analysis led, at length, to the discovery that each electron could be described by associating it with four values. As is standard in quantum mechanics, the values needed are integers. These integers came to be called *quantum numbers*. The only non-integer is the spin quantum number, which is a simple fraction: either ½ or –½. "Detailed spectroscopic study has in fact shown the four quantum numbers are required to characterize each electron in a given atom. These are: (a) the principal quantum number, *n*; (b) the azimuthal quantum number, *l*; (c) the magnetic quantum number, *m*; and (d) the spin quantum number, *s*."[33]

Quantum mechanics often seems extraordinarily complicated, but occasionally its stipulation that only integers may be used to describe basic transitions means that certain explanations turn out to be very simple. This is actually the case regarding the values that are allowable for quantum numbers. Here are the names of the quantum numbers and the rules that pertain to each one.

1. The *principle quantum number*, *n*, can have any positive whole-number value larger than zero—that is, 1, 2, 3, 4, 5, and so on.
2. The *azimuthal quantum number*, *l*, can have any whole-number value beginning with zero, as long as *l* is less than *n*. That is, the maximum value of *l* is *n*—1. The azimuthal quantum number is also known as the *angular momentum* quantum number.
3. The *magnetic quantum number*, *m*, can have any value that *l* has, but also including negative numbers—that is, +*l* or –*l*. For example, if *l* = 3, *m* can equal +3 or −3.
4. The *spin quantum number*, *m*, must have a value of either +½ or −½.

For our present purposes there are four key facts relevant to understanding the physical significance of the quantum numbers. First, the numbers are used to provide a unique identification tag for each electron in every atom. Remember, in a quantum system such as an atom, no two electrons may have identical quantum numbers; each electron has its own set of unique numbers within the system of which it is a member. Another way of explaining this ID tag is to note that the first and second quantum numbers correspond to the atom's position in the periodic table of the elements. This is explained at greater length below.

Second, as the first and second quantum numbers increase, there is an increase in the energy state of the electron being described and a corresponding change in the region in which the electron is likely to be located. In physics, an increase in energy level generally corresponds to a decrease in stability.

Third, the quantum numbers are actually applied by inserting them into the Schrödinger equation in order to calculate the probability

33. Maron and Prutton, *Physical Chemistry*, 620. A helpful discussion of the meaning of the quantum numbers follows on 621.

of locating the electron in a certain region relative to the nucleus. The mathematical techniques of three-dimensional geometry are used to describe these regions.[34] The electron's region in turn has implications related to the energy emitted or absorbed by certain transitions in the electron's state. The Schrödinger equation is involved in the calculation of this energy.

Fourth, the quantum numbers play an explanatory role pertaining to the shells, subshells, and orbitals that are customarily utilized in chemistry as a way of identifying specific electrons in an atom or describing the electrons' arrangement.[35] As an analogy, if we compare an atom with an apartment building, shells may be likened to the building's street address, subshells can be likened to a floor number, and orbitals may be likened to room numbers. This analogy illustrates the organizational hierarchy between shells and orbitals, but the comparison with room numbers fails if it seems to suggest that electrons remain quietly in tiny, precisely circumscribed spaces. In an atom, electrons change routinely from one orbital to another, and even when they remain in their most stable orbital—their "ground state"—their position is not limited by any physical boundaries resembling walls, a floor, or a ceiling. Rather, the most precise aspect of their existence is their energy level, which is required by quantum principles to be a specific amount determined by calculations involving integers. The word *shells* is therefore metaphorical and should not be taken to imply that the electrons reside in concentric layers like the layers of an onion. In fact, the Schrödinger equation reveals that the various orbitals have widely varying shapes. Some are spherical, but others are shaped like crescents, rain drops, or a torus (a donut).[36] Furthermore, the boundaries of these shapes are not clear-cut; graphical depictions of the shapes simply show where the electron is *most likely* to be located.

The fact that the orbitals of electrons in an atom are based on probability waves explains why the word *orbitals* is not merely a more technical-sounding synonym for *orbits*. Orbits, such as the orbits of planets around the sun, are precisely defined and allow astronomers to predict

34. In three-dimensional geometry, the variables *x*, *y*, and *z* are typically used to represent the horizontal, vertical, and depth (front to back) dimensions of space; or "polar coordinates" may be used, which specify distances and angles related to a central point or origin in space.

35. Maron and Prutton, *Physical Chemistry*, 624–32. See also Kittsley, *Physical Chemistry*, 148–52, and Haas, "Quantum Numbers."

36. For an image, see PoorLeno, "Hydrogen Density Plots."

the location of planets, the occurrence of eclipses, and so on, hundreds of years in advance. In contrast, the positions of electrons in orbitals are predictable only within a range of probability.

The quantum numbers are directly involved in the determination of which orbital is occupied by each electron in an atom. The orbital structures determine many of the actual physical and chemical properties of each element. Quantum mechanics therefore pertains directly to the macroscopic properties of matter.

We come now to the close of this chapter on quantum theory. In my imagination I can hear a collective sigh of relief from readers who, despite having only a modest interest in science, have resolutely stuck with the discussion up to this point. The rules for the calculation of quantum numbers are not exactly what most people think of as thrilling reading. Even scientists themselves are not always enthralled by the details of experiments and the plodding elaboration of theories. And yet I submit that, regarding the discoveries of twentieth-century science, some awareness of the details of the experiments and of the intermittent discovery of improvements to the theories is necessary in order to gain an adequate appreciation both for the dogged devotion and the intellectual stature of the scientists, and for the final, astonishing importance of what they discovered.

James Clerk Maxwell had published his treatises on electricity and magnetism in 1871, and by 1880 many scientists believed that physical science was almost complete. It was expected that only a few details were left to iron out. Then, within a span of just a few decades, discoveries followed, one after another, that were destined not merely to add a few details to science but to change our entire understanding of physical reality. In 1889 scientists still believed that light was transmitted by a luminiferous ether and that space and time were absolute receptacles that contained the entire universe. Atoms, though poorly understood, were believed to be indivisible material particles. By 1929—a period of only forty years—all of these ideas had been discarded as utter anachronisms.

Einstein, Rutherford, Bohr, Schrödinger, Pauli, Heisenberg, Dirac, and most of the other leading physicists of the first half of the twentieth century understood that their discoveries called for more than new scientific methods and new experiments. What was also needed was a completely new outlook on the basic principles of physical reality—in other words, a new philosophy of nature. This new philosophy is our topic in chapters 8 and 9.

8

New Physics and a New View of Nature

QUANTUM MECHANICS WAS ORIGINALLY viewed as a mystifying subject—and it still is. From the very beginning it was considered to be at odds not only with the established methods and theories of science but also with commonsense understandings of physical cause and effect. We have noted already that Max Planck himself, who initiated the whole discussion, hesitated to publish his findings not because he doubted his own experiments but because he could scarcely believe the conclusions to which those experiments led. A few years later, as Niels Bohr was about to publish his first papers, Ernest Rutherford sent a word of caution: "There appears to me one grave problem in your hypotheses which I have no doubt you fully realize, namely, how does an electron decide what frequency it is going to vibrate at when it passes from one stationary state to another? It seems to me that you would have to assume that the electron knows beforehand where it is going to stop."[1]

Werner Heisenberg's initial breakthrough regarding quantum theory came upon him as a kind of middle-of-the-night conversion experience while he was alone on a tiny island in the North Sea. Erwin Schrödinger refused to accept the reality of the quantum jumps that his own equations predict. Einstein never accepted quantum theory's principles of entanglement and nonlocality, even though his own paper on the photoelectric effect had paved the way for the discoveries of those principles. As questions proliferated, Bohr distanced himself from any

1. Rutherford, quoted in Moore, *Niels Bohr*, 40; also cited in Nadeau and Kafatos, *Non-Local Universe*, 33.

metaphysical interpretations, saying, "There is no quantum world. There is only an abstract quantum mechanical description. It is wrong to think that the task of physics is to find out how Nature *is*."[2] Richard Feynman repeatedly joked about the inherent incomprehensibility of quantum mechanics. "An atom," he said, "behaves like nothing you have ever seen before. There is one simplification, at least. Electrons behave in this respect in exactly the same way as photons; they are both screwy, but in exactly the same way."[3]

It is a surprising fact that most contemporary physicists are no more confident about the interpretation of quantum theory than their predecessors were in the 1930s. In 2011 a group of physicists attending a conference in Austria on "Quantum Physics and the Nature of Reality" were polled about the meaning of the theory's main ideas. The pollsters reported that "nearly 90 years after the theory's development, there is still no consensus in the scientific community regarding the interpretation of the theory's foundational building blocks."[4]

The problem is not merely a lack of consensus but a deep bewilderment about the true meaning of quantum mechanics itself. Science writer Philip Ball suggests that quantum physics is not simply weird but "beyond weird."[5] The 2022 Nobel Prize in Physics was awarded jointly to Frenchman Alain Aspect, American John F. Clauser, and Austrian Anton Zeilinger, who were all cited for experiments verifying the reality of the "totally crazy" phenomenon of quantum entanglement.[6] Remarkably, all three physicists view the subject of their work as thoroughly perplexing. Aspect described entanglement as "weird," adding, "I am accepting in my mental images something which is totally crazy."[7] Clauser said, "Why this happens I haven't the foggiest. I have no understanding of how it works but entanglement appears to be very real. . . . Most people would assume that nature is made out of stuff distributed throughout space and time. And that appears not to be the case."[8] If physicists who have devoted years of research to entanglement and nonlocality still find the subject counterintuitive, who then *can* claim to understand it? Most

2. Petersen, "Philosophy of Niels Bohr," 12; emphasis added.
3. Feynman, *Character of Physical Law*, 122.
4. Moskowitz, "Physicists Disagree."
5. Ball, *Beyond Weird*.
6. Borenstein et al., "Physicists Share Nobel Prize."
7. Borenstein et al., "Physicists Share Nobel Prize."
8. Borenstein et al., "Physicists Share Nobel Prize."

persons—not only laypersons but even scientists—might well agree with physicist George Greenstein, who laments that the meaning of quantum mechanics is still as impenetrable as ever and suggests that what is needed is a new "experimental metaphysics."[9]

It will sound bold if not indeed brash, but an outline of such an "experimental metaphysics"—or, at least, an *empirical* metaphysics—is precisely what the present chapter intends to provide. I argue that the needed experimental metaphysics depends on two key principles. The first is that reality is founded not upon "stuff distributed throughout space" as Clauser puts it, but upon discrete processes or events that come into being, culminate in a point of definiteness, and then contribute to their successors. All of the mysterious phenomena of quantum physics become more understandable if we are willing simply to jettison, once and for all, the preconceived notion of solid material particles that endure through time and accept instead the premise that enduring objects are composed of serially ordered routes of events.[10] The second principle is that it is the nature of these events to transmit not only physical influence but information—and, specifically, information that is *abstract* or *conceptual*. I argue that both of these two principles are supported not only by philosophical considerations but by the discoveries and concepts of special relativity and, especially, of quantum mechanics. In all of these assertions I gain inspiration from the philosophical system of Alfred North Whitehead, though this book is not simply an introduction to Whitehead's philosophy, and the "experimental metaphysics" that is presented here really is based on special relativity and quantum mechanics.

Obviously it is not possible here to formulate a complete metaphysical system in which all significant questions are anticipated and addressed. One difficulty is that the development of a comprehensive philosophy of nature would require an entire volume devoted to that topic alone. This book addresses various topics. Another problem is that the attempt to extract detailed philosophical conclusions from the discoveries and principles of quantum mechanics cannot be regarded as a foolproof enterprise; the interpretation of quantum theory is at many points inherently ambiguous. Therefore, eschewing any attempt to provide precise definitions or detailed explanations, what I now offer is

9. Greenstein, *Quantum Strangeness*, 5–6, 102.

10. As I acknowledged in chapter 1, this idea was developed most fully by the mathematician-philosopher Alfred North Whitehead. I describe Whitehead's ideas more fully in chapter 11.

simply a summary—but a summary that I believe is detailed enough to make a significant contribution not only to quantum theory but to our overall understanding of nature. Thus this chapter's conclusions contribute to this book's primary goal—the articulation of a more integrative worldview in which the sciences and the humanities are not merely compatible but interrelated.

In chapter 5 I argued that there are four characteristics of nature that are suggested by the special theory of relativity. It will be helpful to review those characteristics now, before we expand them with ideas derived from quantum mechanics. From special relativity we concluded, first, that space and time are not uniform or constant throughout the universe but may more accurately be understood as modes of relation. Second, the phrase *frame of reference*, which is indispensable in any interpretation of special relativity, strongly suggests that something like perspective or viewpoint is inherent within nature. These first two principles, taken together, suggest that the physical universe is profoundly relational, for attributes such as velocity and mass that were long thought to be objective properties are now known to vary depending on the frame of reference of the observer. It is clear that observation (implied by "frame of *reference*") entails relation. Third, the relationality of nature suggests in turn that nature must be fundamentally dynamic and processive rather than static. And fourth, the fact that time does not pass for photons or electromagnetic waves suggests that there is such a thing as nontemporal process—that is, that there is a dynamic type of reality that somehow exists apart from the passage of time. It is important to keep these ideas in mind as we supplement them with philosophical ideas that, I am suggesting, we gain from quantum theory. These ideas can be summarized as follows.

The first principle derived from quantum mechanics is that *nature is radically processive*. This assertion simply echoes the third principle that we gained from special relativity. The new word *radically*, however, represents an expansion of the notion of process presented in connection with special relativity and signifies that *process is at the very heart of nature*—in other words, that process, and not static endurance, is the foundational reality upon which nature is constructed. *Radically* thus indicates that quantum mechanics provides empirical evidence supporting the claim that was made in chapter 2—that the traditional materialism of science and of daily experience must be abandoned as a basic category in metaphysics. The concept of matter will always be a practical idea, but if we want a deeper perspective on nature we must remember that the apparent solidity of matter is actually based on constant processes that involve constant interaction and extraordinarily high rates of change. In chapter 6 I presented an illustrative analogy pertaining to a person who is unable to cross a freeway because of the presence of high-speed traffic. Quantum physics reveals that the structure of matter is maintained by high-speed electrons whose motion cannot be predicted precisely, even in principle. Furthermore, the electrons themselves are not the tiny solid spheres or "corpuscles" that traditional materialism postulates, for the electrons' wave nature is a foundational aspect of their behavior. Quantum physics, including actual experiments, discloses that this wavelike quality pertains not only to electrons but to all forms of matter. Even the most stable objects we can imagine are, at the atomic level, composed of a ceaseless activity and dynamism. Quantum mechanics makes it necessary to say that, in the philosophy of nature, *vibration* and *fields* should replace *matter* as the central categories. The traditional notion of matter as a self-identical substance that endures through time is not merely unneeded but inaccurate. In the words of process philosopher Alfred North Whitehead, "The process itself is the actuality, and requires no antecedent static cabinet."[11] Nature at the atomic level is composed not of substances but of processes or events.

It is important to understand that the type of process philosophy called for by quantum mechanics and special relativity does not simply envision a universe filled with particles that are ceaselessly in motion ("in process"). Rather, it really is "the *process itself*" that "is the actuality." This statement is not intuitively easy to accept. There is an obvious objection of "You can't have just *vibration* alone; there has to be *something there* that is vibrating." In process philosophy the response is that this objection simply begs the question. The vibrations or waves actually *are* the fundamental reality. Though there are particles associated with waves[12] especially when waves are disturbed in certain ways, it would be a mistake to imagine that light waves are really little particles called photons that vibrate back and forth like waves as they travel along through space at the speed of light. The wave itself is purely processive. Heisenberg describes the ancient atomism of Democritus and then explains how modern atomic theory is different:

11. Whitehead, *Adventures of Ideas*, 276.

12. The particle (the "gauge boson") associated with electromagnetic fields is the photon; the particle associated with gravity is the graviton.

> The indivisible elementary particle of modern physics . . . is not a material particle in space and time but, in a way, only a symbol on whose introduction the laws of nature assume an especially simple form. Modern atomic theory is thus essentially different from that of antiquity in that it no longer allows any reinterpretation or elaboration to make it fit into a naïve materialistic concept of the universe. For atoms are no longer material bodies in the proper sense of the word.[13]

Writing in 1938, Whitehead essentially agrees with Heisenberg:

> The notion of empty space, the mere vehicle of spatial interconnections, has been eliminated from recent science. The whole spatial universe is a field of force, or in other words, a field of incessant activity. The mathematical formulae of physics express the mathematical relations realized in this activity.
>
> The unexpected result has been the elimination of bits of matter, as the self-identical supports for physical properties.[14]

If, in our outlook on nature, we focus not on the seeming stability of material particles but on the inherently processive character of waves, a different picture emerges. When waves travel through space, they are imagined as a disturbance in a field—and the field is not a substantive one, as the negative results of the Michelson–Morley experiment demonstrate. Even if the photons associated with light waves could be thought of as the supposedly necessary "thing that vibrates," it is important to remember that a photon has no mass and no dimensions[15] and is therefore not a very good stand-in for matter. David Tong, a theoretical physicist at the University of Cambridge, elaborates on this point:

> And so there's kind of a question that emerges, which is, how should you think of this relationship between the fields on the one hand and the photons on the other. And I think [there are] two logical possibilities for the way this could work. It could be that you should think of electric and magnetic fields as comprised of lots and lots of photons, rather like a fluid is comprised of lots and lots of atoms, and you think the atoms are the fundamental object. Or alternatively, it could be the other way around, it could be that the fields are the fundamental thing.

13. Heisenberg, *Philosophical Problems*, 55.

14. Whitehead, *Modes of Thought*, 136.

15. The photon may under certain circumstances be considered to have a cross-section area, but this is related to its tendency to interact with certain particles and not with any actual volume.

> And the photons come from little ripples of the fields. So they were the two logical possibilities.
>
> And the big development in, well, it sort of starts in 1927. But it takes a good 20 or 30 years until this is fully appreciated. The big appreciation, then, is that it's the fields that are really fundamental, that the electric and magnetic field is at the basis of everything. And little ripples of the electric and magnetic field get turned into little bundles of energy that we then call photons due to the effects of quantum mechanics.
>
> And the wonderful big step, one of the great unifying steps in the history of physics, is to understand that that same story holds for all other particles.[16]

Fields, obviously, are not a kind of material substratum or "fabric."[17] Fields are inherently dynamic, processive, immaterial entities. A field can be thought of as something like a map of potentialities—a correlation between (1) specific locations in space-time and (2) values that would pertain in that location if a physical actuality occurred at that location. Philosophically, fields can be thought of as *a physically extended tendency towards certain types of influence* (for example, electromagnetic or gravitational influence). I discuss the nature of fields more fully later in this chapter, but at present my point is that the particle-creating status of fields entails an implicit repudiation of the conventional notion that nature is ultimately founded upon the machinelike motions of tiny material particles.

This rejection of materialism is supported by many themes in theoretical physics, including current research on the nature of the proton. High school physics students learn to think of protons as tiny, solid spheres, all with the same mass and one unit of positive electric charge. More advanced students find out that the proton is not a simple particle but an assemblage of three elementary particles called quarks. However, the quarks are not simply held together in some sort of permanent structure.

16. Strogatz, "Quantum Field Theory." This is a transcript of a live podcast hosted by Steven Strogatz for *Quanta Magazine* on August 10, 2022. For a more extended discussion of the originating role of fields, see d'Espagnat, *Veiled Reality*. This resource is cited in Clayton, *Mind and Emergence*, 27.

17. Popular accounts of the general theory of relativity persist in using the word *fabric* to refer to the space-time continuum. This word is strictly a metaphor and a rather misleading one. There is no material substrate.

> But decades of research have revealed a deeper truth, one that's too bizarre to fully capture with words or images. . . . The proton is a quantum mechanical object that exists as a haze of probabilities until an experiment forces it to take a concrete form. And its forms differ drastically depending on how researchers set up their experiment. Connecting the particle's many faces has been the work of generations. "We're kind of just starting to understand this system in a complete way," said Richard Milner, a nuclear physicist at MIT.[18]

As an example we can consider the traditional depiction of electrons that surround the atomic nucleus. Thanks to quantum theory, we now know that the position of each electron is impossible to predict precisely, so that electrons' possible orbitals are portrayed as cloudy regions of varying intensity around the nucleus.[19] The varying density of the cloudiness in the depiction suggests that each electron is more or less likely to be found in certain areas. But it is wrong to assume that the electron already exists somewhere and that our uncertainty about its position is simply a matter of human ignorance. Rather, electronic events come into being one by one, sometimes because of the influence of their predecessor event, sometimes because they are forced to occur by the influence of their environment, or, perhaps, usually because of both factors. And this process certainly does not occur only because of experimenters working in laboratories. It is universal. Whenever and wherever it exists, the electron is never a solid particle at all. It is a serially ordered history of events. In process philosophy, whether it occurs in a laboratory or in some other encounter with the environment, electronic events and all other events involving "particles" actually come into being in this way. This understanding of particles as serially ordered events fits nicely with the theory that "virtual particles" are constantly appearing and disappearing in empty space.[20] This idea may seem bizarre, but one of the goals of this book is to show that process metaphysics obviates the weirdness that still clings to quantum mechanics as long as a materialistic outlook is maintained.

18. Wood, "Inside the Proton." This online article contains several very helpful animated graphics by Merrill Sherman.

19. See Wikipedia, "Atomic Orbital."

20. For an accessible explanation of virtual particles, see Fermilab, "What Are Virtual Particles?" Though the narrator wholly affirms the existence of virtual particles, he acknowledges that the idea of particles appearing and disappearing seems "mind-bending" and "crazy sounding."

The native dynamism of particles such as the proton seems bizarre mostly because most people, including more than a few physicists, intuitively embrace the mental imagery of solid, particulate components—tiny material spheres—as the basis for the natural world. The intuitive appeal of this corpuscularism is not simply a result of the influence of Newton or Descartes or Democritus; rather, as I argued in chapter 2, we learn virtually from the time of our infancy that the objects around us take up space, that they are solid, stable, and have weight and inertia—that they do not act unless acted upon. But in fact the proton, initially viewed as a paragon of enduring solidity, is not a stable elementary substance at all, or even a collection of stable substances, but rather a series of repeating phases of activity existing in "a haze of probabilities." If this probabilistic, hazy, animated depiction of the proton continues to seem bizarre, weird, or inexplicable, it is because a wholesale shift in thinking is required in order to accept the idea that the world is composed at bottom not of material objects but of processes. Because this shift is so counterintuitive, most people who are process philosophers acknowledge that at some point they had to experience a moment of what might be called intellectual conversion.

A second principle from quantum mechanics is that *each process must achieve a conclusive state of definiteness in order to influence other processes.* It seems obvious that a thing that is still internally in process is indeterminate and that it therefore cannot have a determinate influence on other entities. This is most consistently true regarding events at the atomic level. In the macroscopic world, processes often exert an influence even before they are completed. An avalanche certainly knocks down many trees and alters large sections of the surrounding terrain before it has come to a halt. Human mental and social processes may produce many side effects long before all of the components of the activity are completed. But at the atomic level *the inherent uncertainty of the wave function must arrive at a state of definiteness in order to act upon its environment.* Scientists sometimes refer to this conversion as "the collapse of the wave function."[21] In this regard, at the atomic level nature is analogous to a player rolling dice in a board game. In order to see how

21. The phrase *collapse of the wave function* was first proposed by mathematician-physicist John von Neumann in 1932. The words themselves are metaphorical and have an imprecise meaning, but the phrase is useful as a way to refer to the quantum-level transition from possibility to definiteness. See Ball, "Experiments Spell Doom."

many spaces the player's token will move, the tumbling dice must come to a stop and be observed.

As I have already suggested, another way of understanding the emergence of definiteness in the midst of quantum uncertainty is to interpret particles, with their definite properties, as the outcomes of fluctuations in fields.

> In 1927, Paul Dirac used quantum fields to successfully explain how the decay of an atom to a lower quantum state led to the spontaneous emission of a photon, the quantum of the electromagnetic field. This was soon followed by the realization (following the work of Pascual Jordan, Eugene Wigner, Werner Heisenberg, and Wolfgang Pauli) that all particles, including electrons and protons, could be understood as the quanta of some quantum field, elevating fields to the status of the most fundamental objects in nature.[22]

As an aid to understanding the relation between fields and particles, consider the following thought experiment. The star nearest to our own sun is Proxima Centauri, which is about 4.25 light-years away. For billions of years light from Proxima Centauri has been passing by the earth, with a very small proportion of the star's total output actually striking the earth's surface. But on April 24, 1990, the Hubble Space Telescope was launched, so that a tiny amount of the light that would formerly have passed by the earth now encounters the orbiting telescope. In fact some of Proxima Centauri's light has actually been captured by the Hubble to create images that have been relayed back to earth for human inspection.[23] When light waves are interrupted by a physical object—in this case, an orbiting telescope—we then speak of the light as individual photons, in that each photon acts as discrete, well-defined packet of energy that acts upon the object it has encountered. These considerations can be applied also to the reception of electromagnetic waves by an antenna that receives signals, to microwaves that heat food, and so on. In the case of the Hubble telescope, the telescope in effect forces the light waves, which are intrinsically processive, to stop and be measured. The same thing can be said about any other entity that light encounters, such as asteroids, planets, other stars, interstellar clouds of dust, or the retina of a human eye.

22. Wikipedia, "Field (Physics)."

23. ESA/Hubble, "Proxima Centauri."

In chapter 5 I used the example of two persons who are driving in a car and trying to choose between two restaurants. They are finally forced to choose a destination when they come to a T-intersection in the road. In effect, a physical condition—the road—forces them to become definite—that is, to do what we would describe at the human level as "making a decision." In a way this resembles what happens to electromagnetic waves when they encounter an object. Light, of course, does not make conscious choices. But it is forced to relinquish its vibratory nature and become a definite, specific packet of energy when it acts causally.[24] And causation is precisely what is detected whenever an experimental observation is made. As George Greenstein puts it, in quantum theory there really is a sense in which "a measurement does not *reveal* a property of the microworld; rather, the measurement *creates* that property."[25] In the words of physicist John Wheeler, "In actuality it is wrong to talk of the 'route' of the photon. For a proper way of speaking . . . it makes no sense to talk of the phenomenon until it has been brought to a close by an irreversible act of amplification. . . . 'No elementary phenomenon is a phenomenon until it is a registered (observed) phenomenon.'"[26]

If the photon doesn't actually travel along a route from one place to another, then what does happen? In order to clarify this situation let us recall that electromagnetic waves that "travel through space" are in themselves nontemporal and, in the direction of travel, nonspatial, for the distance in which (from our frame of reference) they seem to move is in their own frame of reference reduced to zero, the result of relativistic foreshortening.[27] The inapplicability of time and space to this wave is not merely apparent; it is a factual aspect of the ontological status of electromagnetic waves. Another way of looking at the nontemporality

24. My description of the production of definite events by disturbances in electromagnetic fields is in basic agreement with quantum field theory, insofar as I understand it. See Wikipedia, "Quantum Field Theory," in which we read, "QFT treats particles as excited states (also called quanta) of their underlying quantum fields, which are more fundamental than the particles."

25. Greenstein, *Quantum Strangeness*, 96; emphasis original.

26. Wheeler, as quoted by Philip Ball from an unstated source in Ball, *Beyond Weird*, 90. See also Davies, *Mind of God*, 224.

27. These comments pertain to electromagnetic waves traveling through a vacuum. The speed of light is slightly reduced when light travels through a transparent material such as glass or water because the atoms in the material "scatter" (have momentary interactions with) the light. This scattering is what creates a material's refractive index. Refraction is what makes it possible to create lenses in objects such as telescopes and eyeglasses. See also Feynman, *QED*, 109.

of light waves is to refer to the Leibnizian notion, which I presented in chapter 5, that space and time are not absolute receptacles but modes of relation. In this view *space is not omnipresent and uniform but occurs as a way of distinguishing the positions of at least two objects relative to one another.* As Heisenberg puts it, "in empty space, direction and position cannot be defined."[28] Similarly, as a mode of relation time occurs as a way of establishing the sequential order—the before-or-after status—of events, *relative to one another.* This mode-of-relation concept of space and time avoids the error of the traditional belief that time is an absolute receptacle that passes at a uniform rate throughout the universe. The relational concept of space and time implies that as long as the electromagnetic field does not include a physical object it remains undisturbed or unrelated and, therefore, is in itself nontemporal and nonspatial. This clarifies Wheeler's seemingly enigmatic comment that a process can only be referred to as a "phenomenon" when it is "brought to a close"—that is, when an encounter with its environment prompts it to become something definite.

So far I have proposed that two implications of quantum mechanics are that nature is composed of processes rather than substances and that each process must reach some point of final definiteness in order to serve as a cause. These two principles go a long way towards explaining one of the most debated enigmas of quantum mechanics—the enigma of the influence of the experimental apparatus on the results of the experiment. The example of the Hubble telescope causing light to "stop and be measured" illustrates the same sort of observer effect that quantum experimenters were forced to recognize as the impact of experimental instruments on laboratory measurements. Bohr and several other early quantum scientists realized that the seemingly incongruous aspect of the observer effect depends on the common but mistaken assumption that the object or objects being tested are composed of stable material particles that have static properties that remain unperturbed while being measured. If, instead thinking of the phenomena being investigated as static objects "out there," those phenomena are thought of as dynamic processes (e.g., waves), it becomes understandable that *the experiment itself must alter the situation by forcing the process to come to a stop* and reach some definite, measurable state. American physicist-theologian Ian Barbour (1923–2013) explains,

28. Heisenberg, *Philosophical Problems*, 58.

> The "influence of the observer" . . . does not consist in disturbing a previously precise though unknown value, but in forcing one of the many existing potentialities to be actualized. The observer's activity becomes part of the history of the atomic event, but it is an objective history, and even the spontaneously disintegrating atom, left to itself, has its history. Heisenberg writes similarly that "the transition from the 'possible' to the 'actual' takes place during the act of observation."[29]

Sometimes it is actually suggested that it is uniquely the human act of measurement that causes the wave function to collapse, but in this book that idea is rejected. Wave functions collapse whenever the energy transmitted by the wave reacts in a specific way upon its environment. Thus, wave functions have been collapsing ever since the dawn of time.[30] As we consider the ubiquity of processes and the requirement that each process must reach a state of definiteness in order to serve as a cause, a picture of nature emerges in which reality is composed of a vast, dynamic matrix of specific events. Thus nature is characterized by atomicity—but it is an atomism of events or occasions rather than of solid particles or matter. I will now offer a more detailed description of these events or occasions.

We human beings cannot directly perceive the individual atomic processes of which nature is composed. In daily experience, what we do perceive is a world populated by enduring objects and regular patterns. As we interact with our environment, these stable objects and regular patterns produce repeatable physical results, sometimes referred to as the "laws of nature." In traditional Western worldviews based on stable matter or substances, it is necessary to explain how and why change happens. In contrast, in a metaphysics based on process and novelty it is necessary to explain how stability emerges—that is, how stable objects and recurrent patterns develop from a welter of individual occasions. In the

29. Barbour, *Issues in Science and Religion*, 304. Barbour earned a PhD in physics in 1950 at the University of Chicago, where he was a teaching assistant for Enrico Fermi (1901–1954), and a bachelor of divinity degree from Yale University in 1956. Heisenberg's statement is from his *Physics and Philosophy*, 54 (Harper ed.).

30. The idea that human activity is the unique cause of the collapse of the wave function probably arose because the phenomenon was first observed as something caused by physicists in the context of laboratory experiments. This context seems to have led to the odd idea that human awareness itself is what causes the "collapse." See Morris, "Time-Jumping." Though this article is informative, there are several points at which my interpretation of quantum theory diverges from Roger Penrose's. My explanation of human consciousness is presented in chapter 12.

process-oriented philosophy of nature that I am proposing, material stability and dependable patterns arise because of the third, fourth, and fifth principles that, I am arguing, are discoverable in quantum mechanics. These principles are that *the processes found in nature occur not randomly or incoherently but in relation to their relevant predecessors*; that *the causal influence of events on their successors is not merely physical but informational*; and that *the availability of an informational component is what allows nature to produce outcomes that are characterized by wholeness (holism), genuine novelty (outcome-oriented originality), and increasing complexity (levels of organization)*. I will examine these principles in the order in which I have just listed them. They require not only explanation but also several disclaimers and words of caution.

The third principle discernible in quantum mechanics is that *the processes found in nature occur not randomly or incoherently but in relation to past events and, especially, to their most relevant predecessors.* Events occur as novel integrations of the conditions, tendencies, and influences bequeathed to them by their precursors. Whitehead wrote, "Also the processes of the past, in their perishing, are themselves energizing as the complex origin of each novel occasion. The past is the reality at the base of each new actuality."[31]

In Western philosophy the attempt to explain the relation between past and present—in other words, the nature of cause and effect—has had a rather peculiar history. British philosopher John Locke (1632–1704) argued that all human knowledge is based ultimately on experience. In epistemology Locke is therefore regarded as an empiricist rather than a rationalist. Locke's well-known statement that the mind is a "blank slate" (*tabula rasa*) was expressed in opposition to the rationalistic Platonic doctrine of anamnesis or remembrance, which held that the mind contains innate knowledge even prior to any experience. Scottish philosopher David Hume (1711–1776) accepted Locke's empiricism but with the seemingly odd additional doctrine that, actually, *we have no direct experience of causation*. Rather, Hume said, it is simply by repeated observation that we form the habit of believing that certain events are caused by certain other events. If a bottle falls on a hard floor, the bottle breaks. If a living plant receives water, it grows, but if its soil dries out, the plant withers. Our assumption that the first event causes the second is merely that: an assumption. Hume himself provided the example of a vibrating

31. Whitehead, *Adventures of Ideas*, 276.

string. We see the string vibrating and hear a sound, but our belief that there is a causal relation between the two is based only on a mental association. Hume maintained that we have no direct experience of the causal influence itself (we do not see, hear, or feel it); rather, it is merely by inference or habit of thought that we come to assign a causal role to the processes involved.[32] Studying these ideas of Hume, Kant (1724–1804) declared that he had been "awakened" from his "dogmatic slumbers" and reacted by writing the *Critique of Pure Reason*, arguing that we can have no direct knowledge of "the thing in itself"—that is, of the true nature of reality. Even today, Kant's conclusions about this subject still linger as a nagging encumbrance blocking the path to constructive metaphysics. On the other hand, in my experience many laypersons are tempted to dismiss arguments such as Hume's and Kant's as evidence that academic philosophy has little direct bearing on the practical issues of daily life. It is common sense that bottles are real and that if you accidentally drop one and break it, then *you*, collaborating with gravity, are the cause of its destruction.

By mentioning this example of common sense I do not mean to commend the layperson's naïve realism at the expense of scholarly philosophy. I have a deeper point in mind—a point about the shared assumptions upon which the arguments of Locke, Hume, and Kant are based. All three assume that there is some stable, substantive reality—whatever it might be—out there, as it were, external to the observer, and that things in the external world can be objectively observed without significantly affecting their essential being. In other words, Locke, Hume, and Kant presuppose that in perception there is an unavoidable difference between subject (the observer) and the object being observed. But, as Hume realized, a logical consequence of this separation between the knower and the known is that the observer actually can have no direct experience of causation, that is, of the processes themselves that produce the physical changes that are, supposedly, observed objectively.

Scientists, too, even if they are not much concerned with Hume and Kant, have historically assumed that experiments can be carried out in ways that procure objective results and that appropriate procedures can therefore provide definitive conclusions about the world as it really is. In the macroscopic world, in which the objects being studied are composed of extraordinarily large numbers of atomic events, experiments based

32. Hendel, *Hume Selections*, 157–58.

on the assumptions of Newtonian mechanics are very useful. But at the atomic level, the observer effect—the interaction between the phenomenon being studied and the laboratory apparatus—is inescapable. Even more profoundly, quantum indeterminacy itself rules out the existence of any previously established determinate reality, sitting there placidly, waiting to be measured. Scientists were generally quite confident about the objectivity of the scientific method—until quantum mechanics came along and undermined the dichotomy between subject and object, which had served as the central presupposition undergirding the supposedly objective reliability of all scientific experiments.

But does quantum uncertainty truly undermine experimentation at the atomic level? Not necessarily. Quantum theory's puzzle concerning the classical ideal of experimental objectivity becomes profoundly less puzzling if, in our understanding of the experimental setting, we replace stable substances with a *field of potential influence* in which past events contribute in nondeterministic ways to the present events which observe.

As an illustration, let us consider process philosophy's description of the ontological status of an electron. Quantum mechanics has shown that the electron can no longer be pictured as it once was—as an enduring little material sphere with a constant (nonrelativistic) mass and a certain location at some point in space. In an ontology based on processing events, this materialist picture of the electron must be drastically revised. If process rather than substance is the underlying reality, the electron must be reconceptualized as *a serially ordered route of finite occasions* or, in colloquial language, as a "chain of events." Each new electronic event is derived primarily from its predecessor event or its prior "link in the chain," but other events, such as the behavior of nearby charged particles or the absorption of a photon, also contribute to the formation of the new event. It is important to stipulate that the influences of a new event's forebears are not deterministic, for, as we know, the electron's properties (such as position and momentum) must be expressed not as certainties but as probabilities. I will present a philosophical explanation of quantum uncertainty in our consideration of the informational aspect of the electronic event's formation. First, however, let us briefly consider the conceptual advantages that are conferred by this process-based depiction of the electron.

The main advantage is that the in-your-face incongruities of quantum phenomena largely disappear. For example, quantum jumps and quantum tunneling are hard to imagine if the electron is conceptualized—as it was prior to quantum mechanics—as a continuously existing material entity. How can a solid particle abruptly disappear from one location and appear in another? This sounds more like magic than physics. But in a process-based ontology the new location and the new energy level are thought of simply as new conditions or new states pertaining to a newly occurring event. There simply is no preexisting object or "antecedent static cabinet" whose path must be traced or whose motion from place to place must be accelerated and decelerated or whose established properties must somehow instantly be transmuted. Though derived in indispensable ways from its past, the new electronic event is, ontologically, a fresh start in a fresh location.

Another conceptual advantage in an event-based ontology is that the enigmatic wave–particle dualism becomes more nearly a *contrast* than an incongruity, so that Bohr's notion of complementarity loses much of its aura as a quirky but unavoidable consequence of the fact that the wave model and the particle model are too dissimilar to integrate into one model. As Heisenberg stated, the conceptual tension present in the notion of complementarity is finally justified only under the aegis of the success of the mathematics. "Generally the dualism between two different descriptions of the same reality is no longer a difficulty since we know from the mathematical formulation of the theory that contradictions cannot arise."[33] Despite this hopeful claim that "contradictions cannot arise," many physicists have continued to accept the complementarity principle with a vague sense of uneasiness. The eminent Irish physicist John S. Bell (1928–1990) remarked, "For my part, I never got the hang of complementarity, and remain unhappy about contradictions."[34]

In a process-based philosophy of nature, this sort of "just-because-it-works" reliance on mathematics is not necessary. If an electron is viewed as an event rather than as an enduring object, it makes sense to think that the event can simultaneously have both wavelike and particle-like aspects depending on its context and on the methods being used to measure it. Ian Barbour notes that Bohr himself espoused a neo-Kantian view of human knowledge in which the "given structures of the mind" create epistemological limitations that make it impossible for human beings to conceptualize the atomic level of reality in an objective way.

33. Heisenberg, *Physics and Philosophy*, 50.

34. Bell, *Speakable and Unspeakable*, 194 ("Six Possible Worlds of Quantum Mechanics").

> Bohr shares Kant's scepticism about the possibility of knowing the world in itself. He holds that classical concepts are "forms of perception" imposed by man. If we try, as it were, to force nature into certain conceptual moulds, we preclude the full use of other moulds. Thus we must choose between complete causal *or* spatio-temporal descriptions, between adequate wave *or* particle models, between accurate knowledge of position *or* momentum. . . . This reciprocal limitation occurs because the atomic world cannot be described in terms of classical concepts, which, according to Bohr, are the only ones available to us.[35]

I am asserting that classical concepts are in fact *not* the only concepts available to us. Quantum mechanics invites us to adopt a view of nature in which reality is dynamic, relational, and processive, a view in which fields of potentiality give rise to novel events whose outcomes contribute to future events. As a general rule, the phase of origination or formation manifests wavelike properties, and the definite outcome, which influences subsequent events, manifests particle-like properties. But the rhythm of this becoming-and-influencing-and-becoming is ceaseless and seamless so that the rhythm of becoming in its wholeness is a single model that does not require us to maintain the constant intellectual tension of wave–particle complementarity.

In a process-based ontology, the question does arise as to how the chain of influence from past to present can be sustained. In substance-based ontologies, a main advantage is that physical endurance is self-explanatory; it is simply the nature of matter to remain self-identical as time passes. In fact, enduring self-identity can to some degree be regarded as the main explanatory benefit of the doctrine of substances. In contrast, in a process-based ontology the persistence of objects and stable patterns may seem to be endangered. What is it that ensures consistency and continuity in process philosophy's chain-of-events picture of the world?

A first response is that relatedness is an observed aspect of nature, and nature itself is in many ways dependable and consistent. From both special relativity and quantum theory we have gained the insight that relatedness is a fundamental aspect the cosmos. But, again, what is it that causes this relatedness to occur in dependable patterns? It is a fair question to ask *why* relatedness should be universal. And, again, in answering we can appeal to the basic difference between philosophies of substance and philosophies of process. From antiquity we have the doctrine that

35. Barbour, *Myths, Models, and Paradigms*, 76.

one substance cannot be present in another substance. This dictum is more or less self-evident in the definition of substance itself as a stable reality, an inner essence, that maintains each object's identity over time. Before the twentieth century the concept of the atom as something that cannot be cut or reduced to components may be regarded as the scientific equivalent of the philosophical notion of substance. Even when an atom does bond with other atoms, something of the atom's original identity is preserved, and this identity reappears if the compound is chemically broken down. As a parallel, in substance-oriented philosophies two substances may be blended, but they cannot actually enter into one another—or else the basic identity of both would be altered, in effect creating a new substance.

In process philosophy this whole outlook is overturned. It is the very nature of novel events to *include* their antecedents; this is the meaning of radical relatedness and of being derived from the past. It is to be expected that preceding events not only *can* be but *must* be present in the events that follow. This is the meaning of radical relatedness. In ballistics, this ongoing influence is what is known as *trajectory*. In physics it is exemplified by the *conservation of energy*. In thermodynamics it is illustrated by *entropy*, the spontaneous flow of heat into the environment. In genetics it is observed in *inherited traits*. For our purposes here, perhaps the best example of all is found in atoms themselves, whose orbitals vibrate in an endless alternation between uncertainty and certainty, and yet through the constancy of this very vibration almost all the individual atoms of planet Earth have maintained their continuous identity. The oldest known rock on Earth is part of the Acasta Gneiss found in northwestern Canada, which has been dated to an age of four billion years. Obviously the atoms in this gneiss—primarily silicon, oxygen, aluminum, potassium, and sodium—have stably been in place ever since the rock was formed.[36] In all of these examples the present does not repeat the past precisely, but it nevertheless resembles the past closely enough that functional identity is maintained.

Ultimately, in the philosophy that I am presenting, there is a sense in which the principle of radical relatedness must simply be accepted as a given—a self-evident truth, a primal condition. Readers are invited to appreciate the fact that all worldviews ultimately rest on certain insights or

36. The individual atoms in this gneiss and of all terrestrial atoms lighter than cobalt were created by nuclear fusion in the cores of stars that subsequently exploded long before our solar system was formed.

convictions that must be accepted as initial premises. Whitehead wrote that any fundamental ideas that attempt to explain the nature of things "are not, and should not be, the result of an argument. For all argument must rest upon premises more fundamental than the conclusions. Discussion of fundamental notions is merely for the purpose of disclosing their coherence, their compatibility, and the specializations which can be derived from their conjunction."[37]

For conventional scientific materialism, it is such an initial premise that asserts that matter is the foundation of physical reality. In materialism it is therefore meaningless to ask what is inside an elementary particle; it simply is what it is. But I am arguing that the philosophy of nature best suited to quantum mechanics maintains as an initial premise that the universe is composed of an ever-emerging system of novel events that are innately related to the past and innately influential in the future. In quantum physics this continuity of relatedness is in basic agreement with the quantum discovery that was stated just a few pages ago, that the field is more primordial than the particle. In other words, the basic nature of a field is to be a scenario of influence.

The fourth principle to be gained from quantum mechanics is that *the causal influence of events on subsequent events is not merely physical in character but informational or conceptual.* In my experience, this is the principle most likely to inspire doubt or misunderstanding, and therefore I will examine it more closely.

I stated in chapter 3 that "nature traffics in information," but what is now meant by "information"? At the conclusion of chapter 3 I presented a heuristic typology of information consisting of three kinds: information that is purely physical, information that is conceptual, and information that is purely abstract. I defined physical information as information that is embodied in physical objects or events. For example, an iron cylinder embodies information about its own weight—information that can be converted into intelligible symbols by a scale but remains embodied in the cylinder even when the cylinder is not being weighed.

I defined conceptual information as information that involves generalizations or abstractions but that can usually be derived from or applied to specific circumstances. Conceptual information is information that may be associated with specific conditions but is sufficiently abstract to be applied in more than one setting. In the sentence "The dog is black,"

37. Whitehead, *Adventures of Ideas*, 294.

the words *dog* and *black* serve as abstractions or symbols that are often used in a specific situation but can be used in widely varying circumstances. Conceptual information includes high-level abstractions. The phrase *divine providence* is not derived from any single, concrete example, but the words are meaningful enough to be pondered, discussed, affirmed or denied, and even acted on. Conceptual information is generally the chief tool of human thought. However, though the phrase *conceptual information* may sound cerebral, conceptual information is not restricted to highly evolved mentality. Biologist Ursula Goodenough observes that "all creatures evaluate," noting as an example that even an amoeba is able to discriminate between things it should approach because they can be eaten and things from which it should flee because they pose a threat.[38] The amoeba's primitive ability to sort objects in its environment makes use of what I am calling conceptual information. The information is said to be "conceptual" because it is able to identify the common traits that are shared between specific objects and situations well enough to envisage categories that then serve to guide behavior. Quantum mechanics helps us to discern that this use of information to sort possibilities into types is operative even at the atomic level. I will elaborate on this point below.

The third type of information is information whose definition is purely ideational. This is information that is not derived from any physical circumstance and need not be applied concretely in order to be meaningful. The most obvious example of this type of information is mathematics. The operations of addition, subtraction, multiplication, and division, and properties such as being commutative, associative, or distributive are not derived from any specific situation and have validity regardless of circumstantial considerations. This would also be true of the axioms and theories of algebra, geometry, trigonometry, calculus, matrix theory, and other branches of math. There is a universal and non-particularized quality to information of this type.[39] It seems reasonable to think that if nature does make use of information qua information—and not merely as information implicitly accompanying collisions or forces—then the information so used might mostly reasonably be information of this mathematical or "pure" type.

38. Goodenough, *Sacred Depths*, 106.

39. These comments about the nature of mathematics would not be accepted uncritically by all authorities. The philosophy of mathematics is a rich and complex subject that is clearly beyond the scope of this book. For introductory summaries, see Horsten, "Philosophy of Mathematics," and Wikipedia, "Philosophy of Mathematics."

Over the centuries many persons have wondered why nature embodies certain recurrent patterns of order or natural laws. Mathematics has often been cited for its remarkable role in describing this orderliness and, perhaps, even as the source of it. Pythagoras (ca. 570–490 BCE) and Euclid (fl. 300 BCE) were noteworthy in this regard, and in chapter 3 I briefly discussed the theories of Plato and Aristotle pertaining to mathematics. Also in chapter 3 I quoted Richard Feynman's comments on Newton's law of gravitation.

> Nevertheless it is kind of mathematical, and we wonder how this can be a fundamental law. What does the planet do? Does it look at the sun, see how far away it is, and decide to calculate on its internal adding machine the inverse of the square of the distance, which tells it how much to move? . . . If this were the only law of this character it would be interesting and rather annoying. But what turns out to be true is that the more we investigate, the more laws we find, and the deeper we penetrate nature, the more this disease persists. Every one of our laws is a purely mathematical statement in rather complex and abstruse mathematics.[40]

In 1960, Nobel laureate Eugene Wigner (1902–1995), one of the pioneers of quantum theory, published an article titled "The Unreasonable Effectiveness of Mathematics in the Natural Sciences."[41] Though Kepler, Galileo, Newton, and numerous others demonstrated centuries ago that there is a connection between nature and mathematics, they made no attempt to explain *why* that connection exists. Wigner's article draws attention to this lack of explanation. His use of the word *unreasonable* is meant to suggest not "immoderate" or "perverse" but that the innate connection between mathematics and nature is not logically necessary and therefore seems arbitrary and enigmatic. He continued, "The miracle of the appropriateness of the language of mathematics for the formulation of the laws of physics is a wonderful gift which we neither understand nor deserve. We should be grateful for it and hope that it will remain valid in future research and that it will extend, for better or for worse,

40. Feynman, *Character of Physical Law*, 31, 33.

41. Wigner, "Unreasonable Effectiveness of Mathematics." Wigner Jenő Pál was born in Budapest and studied in Germany. In 1930 he was recruited to teach at Princeton University and, as a Jew, remained in the United States due to Nazi anti-Semitism in Europe. Wigner became a US citizen in 1937 and changed his name. He won the Nobel Prize in Physics in 1963.

to our pleasure, even though perhaps also to our bafflement, to wide branches of learning."[42]

I am now proposing that quantum mechanics does offer an inherent but not widely recognized explanation for this connection between nature and mathematics. The explanation is that *mathematics is an integral contributor to the ontological status—the very being—of reality at the atomic level.* One of the primary advantages of a processive view of nature is that *it is a great deal easier to imagine that conceptual or mathematical information is a constitutive element of a process than it is to imagine than it is a constitutive element of a material particle.* How could information that is purely mathematical be inserted into a particle?

As an analogy, when an event is being planned, it is a part of the planning process to apply determinative information to the event. It will be held on a certain day at a certain time and at a certain place, certain food items will be available, and so on. My point is that we cannot plan or create a material particle this way. In physicalism, a particle will *already* be characterized by mass, velocity, and spin, but it is not possible somehow to insert information qua information into the particle as a way of influencing what it will become. But in quantum theory, the phenomenon of superposition does allow differing potentials to be co-present in mathematical or conceptual form. This is easy to imagine if we view the particle as a series of events, analogous to the event-planning activity mentioned above. It is not easy to imagine if we are committed to viewing the particle as an instance of self-identical material substance. Thus it is the lingering effect of materialistic assumptions that causes superposition to seem paradoxical.

The notion that nonphysical information is a constituting aspect of nature is a departure from classical physics and, in fact, from commonsense perceptions of physical reality. Newtonian physics holds that math is involved in nature simply because bodies have measurable qualities such as mass and velocity and that, for reasons that remain unexplained, these measurable qualities follow regular patterns ("laws"). Quantum mechanics suggests something much deeper—that events at the atomic level somehow incorporate information that is purely mathematical, without

42. Wikipedia, "Unreasonable Effectiveness of Mathematics." Philosopher Hilary Putnam (1926–2016) identified Wigner's outlook on mathematics as a type of realism that maintains that mathematical facts and relations exist independently of human thought. In this book I assume that most people would view this type of realism as common sense—so that, for example, "2 + 2 = 4" is true whether or not any human beings are aware of that fact.

depending on any sort of physical intermediary. Heisenberg points to this mathematical aspect with words that are stark but evocative:

> The atom of modern physics can only be symbolized by a partial differential equation in an abstract multidimensional space. Only the experiment of an observer forces the atom to indicate a position, a colour, and a quantity of heat. *All* the qualities of the atom of modern physics are derived, it has no *immediate and direct* physical properties at all, i.e., every type of visual conception we might wish to design is, *eo ipso*, faulty. . . . As Bohr has stressed, it is no longer correct to say that the qualities of bodies have been reduced to the geometry of atoms. [The qualities of bodies] can only be reduced to the mathematics of the atom.[43]

As we know, quantum indeterminacy arises because Schrödinger's partial differential equation describes a probability rather than a certainty. But the mathematics in this case is not merely descriptive—describing an entity that already exists—but creative, contributing potentialities and limitations for an event that has not yet become definite. Thus the mathematics is significant not just epistemologically but *ontologically*. Heisenberg wrote, "[In] the experiments about atomic events we have to do with things and facts, with phenomena that are just as real as any phenomena in daily life. But the atoms or the elementary particles are not as real; they form a world of potentialities or possibilities rather than one of things or facts."[44] Heisenberg is not using *potentialities* and *possibilities* the way we normally do, to refer to the limits of our knowledge. If we say that there is a 60 percent possibility of rain tonight, we are not referring to the actual physical processes that cause rainfall but to our best guess—considering current weather conditions—as to whether it may rain. In contrast, Heisenberg's *potentiality*—a more suggestive way of talking about indeterminacy—is an intrinsic aspect of the ontological constitution of the atomic phenomena themselves.

43. Heisenberg, *Philosophical Problems*, 38–39. By "color" Heisenberg does not mean a visible tint but rather the electromagnetic wavelength emitted by the atom under certain conditions. By "geometry of atoms" Heisenberg is referring to the overall shape of the atom, as if it had a stable three-dimensional structure. Heisenberg noted that the ancient philosopher Democritus had concluded that the only necessary property of atoms was that each type of atom takes up a certain amount of space and must have a distinctive shape. In contemporary science these ideas have been discarded.

44. Heisenberg, *Physics and Philosophy*, 160 (Harper ed.). This citation is of the pagination in the original hardback edition.

Early on, this affirmation of potentiality as a fundamental aspect of nature at first led certain physicists to make some rather unguarded overstatements.

> [Pascual] Jordan declared, with emphasis, that observations not only *disturb* what has to be measured, they *produce* it. In a measurement of position, for example, as performed with the gamma ray microscope, "the electron is forced to a decision. We compel it *to assume a definite position*; previously it was, in general, neither here nor there; it had not yet made its decision for a definite position. . . . If by another experiment the *velocity* of the electron is being measured, this means: the electron is compelled to decide itself for some exactly defined value of the velocity. . . . We ourselves produce the results of measurement."[45]

In the view that I am now proposing, contrary to Jordan, the word *disturb* is in fact preferable to the word *produce*. When a particle or an electromagnetic wave has an encounter, the encounter itself does not produce the results *de novo*; it merely prompts a definite outcome to emerge from a pending pattern or tendency that has been derived from the events of the relevant past. That outcome is not finally in the hands of the experimenter but is constrained by its own internal process within limits that are usually very narrow. In fact those limits are specified by the mathematical expression of the uncertainty principle, which I presented in chapter 7. Furthermore, Jordan's use of "decision" and "compelled to decide" are too anthropomorphic. The word *decision* generally signifies a conscious process in which various alternatives are evaluated. Electrons are not conscious, and they do not think. I will argue later that there really is a sense in which atomic events are self-caused, but it would be premature to elaborate on that idea now. Finally, Jordan's statements, like many statements made about the observer effect, focus unduly on the effects of human experimenters and the laboratory apparatus. Our example of light from Proxima Centauri encountering the earth illustrates that events at the atomic level are constantly being "forced to a decision" all around us, not only in experiments being conducted by human beings. Actually this "forced-to-a-decision" aspect of physics has been happening throughout the cosmos ever since the beginning of natural history.

45. Bell, *Speakable and Unspeakable*, 142 ("Bertlmann's Socks and the Nature of Reality"), quoting Jordan, whose comment originally appeared in Zilsel, "P. Jordans Versuch."

Because he appears to substitute mathematics for physical properties, Heisenberg has been described as an "anti-realist."[46] In science, *realism* is usually understood to mean that entities exist independently of the human mind.[47] Regarding ordinary objects, this of course would also be the commonsense view. Apparently, among many physicists, "anti-realism" is not a favorable description. However, anti-realism should remind us of an important point. The habitual use of the word *realism* to convey the uncritical supposition that purely physical factors are the true measure of reality exemplifies the logical fallacy of question begging, especially in the light of quantum theory. Quantum physics has shown that it is not at all self-evident that reality is based solely on physical forces and materials. Accusing Heisenberg of anti-realism is therefore based on the assumption—a mistaken one, I am arguing—that the "real" includes only the physical. Heisenberg's critics fail to notice that Heisenberg actually was not claiming that ordinary physical objects can be reduced to mathematics. He was making a statement specifically about the nature of atomic processes—that certain equations are our only feasible way of representing these processes. I will discuss the concept of scientific realism more fully in chapters 9 and 13.

In any case, if we accept the idea that reality is composed not of enduring material but of events, it becomes a good deal easier to see how mathematical information can be included as an integral aspect of natural processes. We do have to accept the idea that each event has some innate ability to receive and utilize information qua information, not just information transmitted by physical forces and collisions.[48] In order to integrate purely mathematical information, the information-utilizing aspect of an event must be more nearly mental than physical, or more conceptual than concrete. We might refer to this ability to receive non-embodied data as the *information processing aspect* of events at the atomic level. However,

46. Rovelli, "Consciousness Is Irrelevant." This is an interview of Carlo Rovelli, Italian theoretical physicist, by Alexis Papazoglou, editor for *IAI News*, the online magazine of The Institute of Art and Ideas. In the interview Heisenberg's alleged "anti-realism" is mentioned repeatedly.

47. Or variations on this theme. See d'Espagnat, *Conceptual Foundations*, 232–39 (ch. 19). I discuss realism at greater length in chapter 13.

48. This is essentially the understanding of information in logical positivism, a school of philosophy that claims that statements must always be either analytic (derived from logic or accepted definition) or synthetic (derived from observation—that is, measurable). I discuss the role of logical positivism in quantum theory in chapter 9.

information processing is an established phrase in computer science, and that type of processing is not what we have in mind.

Whitehead adopted the terms *mental pole* (or *conceptual pole*) and *physical pole*[49] to differentiate between the informational and the mechanistic factors in each event. Specifically, he chose the terms *pole* and *dipolar* (rather than, say, *dualistic*) to emphasize that neither pole can exist independently of the other. In individual events the conceptual elements and the physical elements are analogous to the poles of a magnet in which the existence of one always entails the existence of the other. Whitehead's contrast between the mental and the physical is thus not a dualism like Descartes's, for in Whitehead's philosophy the core function of actual occasions—"the final real things of which the world is made up"[50]—is to *integrate* mental and physical influences, and *without this integration of mental and physical, there is no actuality*. Descartes's philosophy proposes the opposite situation, for thinking substance and material substance are wholly distinct; they are understood to be so dissimilar that there can be no basis for their interaction. In the discussion that follows, I use the terms *mental pole* and *conceptual pole* to refer to the aspect of a becoming event that is capable of receiving and applying the type of information that requires no physical embodiment, such as information that is mathematical.

In a philosophy of nature, and especially in physics, the principal drawback to the adoption of terms such as *mental* and *conceptual* is that they immediately suggest conscious thought processes. I cannot emphasize too strongly that consciousness and thinking are not what is being suggested by the phrase *conceptual pole*. The problem we face is that it is difficult to come up with a better terminology. It is unavoidable that the words and concepts that we employ when we refer to the use of information are drawn almost entirely from human experience. As noted in chapter 2, even as children we acquire a generic type of materialism through our earliest experiences of the objects in the world around us. We subsequently internalize patterns of speech and thought that make it seem intuitively inappropriate to apply words such as *mental* or *conceptual* to nonhuman nature, except possibly pertaining to the behavior of more highly developed animals. After we reach adulthood, this intuitive materialism is reinforced by the attitudes and intellectual standards of

49. Whitehead, *Process and Reality*, 45, 108.

50. Whitehead, *Process and Reality*, 18.

an industrialized, technology-centered world. Among many professional scholars, a routine refusal to consider the role of a rudimentary conceptual or mathematical capacity throughout nature has become a virtual prerequisite for intellectual respectability.

These challenges to the existence of a conceptual pole have been exacerbated by the fact that proposed philosophical alternatives to materialism have often been presented in ways that are oversimplified, romanticized, or psychologized. Instead of winning adherents, the frequently made claim that quantum theory somehow justifies the belief that the universe is conscious, or that consciousness is "explained" by twentieth-century physics,[51] has aroused skepticism among many thoughtful persons. It is assumed that the only alternative to materialism is panpsychism, often inflated into the notion of some sort of cosmic consciousness or interpreted as "the mind of God."[52] In this book, panpsychism, cosmic consciousness, and the belief that God is a personal being with a mind and a will are all roundly rejected (see chapter 15) and the assumption that we must choose between pure physicalism and panpsychism is held to constitute a very misleading false dichotomy. As John S. Bell has put it, "So I think it is not right to tell the public that a central role for conscious mind is integrated into modern atomic physics. Or that 'information' is the real stuff of physical theory."[53]

If we liberate the notion of a conceptual pole from the encumbrances of anthropomorphism and psychologism, the hypothesis that each event has at least some vestige of a mental or conceptual capacity can join the principle of process and the principle of relatedness as a rudimentary building block in the event-based ontology now being presented and in an overall philosophy of nature. The inclusion of the conceptual or mental pole allows us to say, without physicalist restrictions, that nature does indeed traffic in information. From the standpoint of quantum mechanics, the functioning of the mental-conceptual pole in atomic processes confers several distinct advantages.

51. See Oberlander, "New Theory in Physics"; Antic, *Physics of Consciousness*; Patterson, *Quantum Physics*; Azarian, "Romance of Reality"; Nadeau and Kafatos, *Non-Local Universe*, 198; Segall, *Physics of the World Soul*, chs. 5 and 10.

52. See, for example, Davies, *Mind of God*.

53. Bell, *Speakable and Unspeakable*, 170 ("Speakable and Unspeakable in Quantum Mechanics").

ADVANTAGES OF THE DOCTRINE OF A CONCEPTUAL POLE IN ATOMIC EVENTS

(1) The functioning of the conceptual pole provides a new approach to the age-old question as to why nature is mathematical. In quantum mechanics we can ask, Do the quantum numbers just *describe* the patterns that apply to atomic electrons? Or do quantum numbers or the patterns that they represent actually have a *causal influence* on the formation of the orbitals? These questions point to the central problem that has troubled physicists and chemists virtually ever since quantum mechanics was first developed. The obvious temptation is to assert that quantum numbers and the calculations involving them are merely descriptive and not, in and of themselves, causally influential. The main advantage of this assertion—if it actually is an advantage—is that it serves to protect the intuitive physicalism that is still maintained by most scientists as a self-evident orthodoxy. Other than this, however, the relegation of math to a merely descriptive role creates serious problems. The central problem is that if we rule out any influence of the math itself, we are thrown back into the untenable position of having to explain the quantized nature of atomic reality by appealing to something else—presumably, physical particles or some other purely physical factor.

The presence of a conceptual pole helps us to understand that mathematical information describes phenomena because mathematical data contribute to the actual outcomes of natural processes. Technically the math itself does not act as an independent cause. Rather, the mathematical data are received and incorporated by relevant events, and the events contribute to subsequent events after they have achieved culminating definiteness. Because of the presence of the "conceptual pole," mathematics or patterns are involved not merely descriptively but causally. If events have an innate ontological capacity to assimilate mathematical information, then Feynman's rhetorical question about how the planet can respond mathematically to the gravitational attraction of the sun becomes much more answerable, and Wigner's "unreasonable" effectiveness of mathematics becomes not simply more reasonable but to-be-expected.

(2) The functioning of the conceptual pole suggests a basis for the natural appearance of integers in the quantum description of nature. Events that are caused entirely by the combined effect of physical forces and material collisions quickly lose their quantized dependence on integers in favor of net impacts and massive averages. This is why,

in the macroscopic world, experimental measurements do not generally involve calculations that rely on integral values. Fractions appear and become more and more lengthy. But at the quantum level a dependence on whole numbers is pervasive. The attempt to explain the quantized nature of atomic phenomena typically cites the tendency of standing waves to produce harmonics—that is, whole-number multiples of the originating frequency. The standard example is a vibrating string.[54] But this example, of course, is only an analogy. A vibrating string is a physical object composed of a large number of atoms. Its frequency is a function mainly of three factors: the string's length, its mass, and the tension at which it is stretched between the string's two ends. These same factors determine the string's harmonics or overtones. In contrast, electromagnetic waves and matter waves are not composite, solid objects like strings, and they have no mass or stretching tension to determine their frequency. Their frequency is determined by their energy—or, it can just as correctly be said that their energy is determined by their frequency. What this means is that in quantum mechanics *the physical characteristics of a vibrating string are ousted in favor of a mathematical relationship*. In the words of John S. Bell, "It is the mathematics of this wave motion, which *somehow controls the electron*, that is developed in a precise way in quantum mechanics. . . . In the case of the waves of wave mechanics we have no idea what is waving . . . and do not ask the question. What we do have is a mathematical recipe for the propagation of the waves."[55]

In fact, in quantum theory it is sometimes more useful to view a photon or an electron as a packet of information than as a physical corpuscle. I submit that the example of a vibrating string actually raises questions more difficult than the ones it answers. The applicability of whole numbers in atomic events makes much more sense if we abandon our habitual reliance on physical examples and accept the proposition that the forming event has an innate ability to process mathematical information. Then—though no detailed argument can be presented here—it seems reasonable that the simplicity of whole numbers would result in a natural ease of assimilation.

An even broader question asks why it is that the frequencies and energy levels produced by certain physical processes—say, the emission

54. Sears and Zemansky, *University Physics*, 967.

55. Bell, *Speakable and Unspeakable*, 187 ("Six Possible Worlds of Quantum Mechanics"); emphasis added.

of an alpha particle by uranium-238[56] or the red glow of stimulated neon atoms—emit consistent energies everywhere in the universe. In chapter 3 I mentioned the difficulty of explaining why the specific types of particles (electrons, protons, and so on) are all identical to one another. Quantum theory prompts us to ask a parallel question about energy levels and wavelengths. Translated into the example of the vibrating string, the question is how it can be that the vibrating string occurs in nature only as part of a tuned instrument. As a figure of speech, it is as if all the atoms and particles in the cosmos were parts of pianos that are always tuned to A at 440 hertz. This is very difficult to explain unless we accept the idea that throughout nature's web of events, mathematical values are transmitted along serially ordered routes from one event to the next.

(3) The presence of the conceptual pole provides an ontological basis for quantum uncertainty. Newtonian mechanics demonstrates conclusively that if natural causation is based only on processes that are purely physical, then, in principle, precise measurements and correct calculations are always able to predict definite results. This was the worldview of Laplace, who regarded the mathematics as descriptive or predictive—but not causative, because *conceptual* information (including mathematical information) is simply conceptual or abstract; numbers do not serve as causes. We remember that in Aristotelian terms, there is a difference between formal causes, which may be mathematical in content, and efficient causes, which are mechanistic. In quantum physics, various mathematical values are held in contrast in a state of superposition. This is what allows mathematical information to provide potentiality or possibility instead of certainty.

These observations help us to understand why the Schrödinger equation yields a probability wave instead of a mathematical certainty. The calculated values are *possibilities* for actualization (or *potentia*, in Heisenberg's terminology), *not actual physical states*. The ontological state of potentiality is what allows for uncertainty. This helps us to understand problems such as Schrödinger's proverbial cat, for the contrasting states (a living cat and a dead cat) pertain not to physical actualities but

56. Interestingly, the radioactive decay of uranium-238 illustrates both the certainty of the amount of energy that will be released and the great uncertainty of the time when the alpha particle will be emitted. The energy produced by U-238 alpha emission is 4.26975 million electron volts. The half-life of U-238 is 4.468 billion years, which means that from a human perspective individual atoms of U-238 decay only very rarely.

to *potentialities*.[57] But this explanation depends on the ontological structure of a conceptual pole to hold two or more states in tension. The states would be incompatible *if they were physically actual*. This same principle applies to the so-called "many worlds" interpretation of quantum mechanics, which claims "that the universal wavefunction is objectively real, and that there is no wave function collapse. This implies that all possible outcomes of quantum measurements are physically realized in some world or universe."[58] This patently absurd scenario is the result of a failure to allow for any genuinely ontological state of potentiality—the copresence of conflicting states, which is possible in a worldview based on processes in which there is a mental pole but not possible in a worldview based on physical substance or mechanistic materialism.

(4) Similarly, the functioning of the conceptual pole helps us understand how imaginary numbers are operative in quantum mechanics. George Greenstein quips, "If you want to do quantum theory, you need to think about things like the square root of minus one."[59] Complex num-

57. The Wikipedia article on "Schrödinger's Cat" summarizes eight approaches to this problem. In my view, the problem arises because of a failure to distinguish between the ontological status of the completed actual occasions that comprise macroscopic objects (such as the poison gas and the cat) and the ontological status of superposition, which is a state of nonactualized potential, much like Whitehead's genetic description of the phases of concrescence. The radioactive substance in the experiment does not activate the Geiger counter until the wave function in one of its nuclei "collapses" and thus creates an emission that interacts with its environment. As I indicated in the second principle listed above, each event must achieve a definite state (that is, not a state of superposition) before it can influence its environment. In chapter 5 I illustrated this situation with an analogy about two persons who are driving in a car and choosing between two restaurants. So, the answer to the Schrödinger's cat problem is based on (1) an ontology of processive events rather than an ontology of matter or substances and (2) an ontology of indeterminate states and indeterminate information rather than on the traditional belief that nature consists entirely of material particles and therefore can never sustain a tension between conditions that are physically contradictory. All of this is explained more fully in chapter 11.

58. Wikipedia, "Many-Worlds Interpretation." The article goes on to say that the many-worlds interpretation implies that there are most likely an uncountable number of universes. "It is one of a number of multiverse hypotheses in physics and philosophy. MWI views time as a many-branched tree, wherein every possible quantum outcome is realized. This is intended to resolve the measurement problem and thus some paradoxes of quantum theory, such as Wigner's friend, the [EPR paradox], and Schrödinger's cat, since every possible outcome of a quantum event exists in its own world." In the philosophy being developed in this book, there is no need for the many-worlds theory because the conflicting possibilities are just that—possibilities that are envisaged conceptually within nature itself, where not all are actualized physically.

59. Greenstein, *Quantum Strangeness*, 74.

bers, which are combinations of real numbers and imaginary numbers, are essential to the concept of superposition, which was presented in chapter 7.

Nonmathematical readers are reminded that in math there is a distinction between real and imaginary numbers. Imaginary numbers are represented as multiples of the square root of negative one. Obviously, one multiplied by one equals one, but because a negative number multiplied by a negative number yields a positive number, negative one multiplied by itself also equals one. Thus there is no "real" number that, when multiplied by itself, equals a negative number. However, in math and science it is necessary to represent the square roots of negative values, so mathematicians have invented the symbol *i* to denote the square root of minus one. Using this symbol, the square root of negative nine is written as "3*i*." Mathematicians have no problem with *i*, but *i* makes some physicists vaguely uncomfortable since, being "imaginary," *i* cannot be applied to any actual object or process. This inapplicability to physical processes is not a problem in non-quantum branches of physics, for the math eventually works out so that *i* is used only in calculations and is never directly applied to physical objects. But, once more, quantum mechanics is different. Complex numbers are used to specify the configuration of two or more superposed states.

> The square root of negative one doesn't correspond to any physical quantity, but that doesn't mean it has no place in the physical sciences. . . . In electromagnetism and most other fields of physics, imaginary numbers are merely a mathematical convenience. All the relevant phenomena can still be described using nothing but real numbers. Quantum mechanics is an exception: The *observable quantities and probabilities* are by necessity all real, but the *underlying quantum states and governing equations involve imaginary numbers*. . . . A quantum state, for example, doesn't contain enough information to prescribe the outcome of every possible measurement on the state; rather, for most measurements, it offers only a probability distribution among the possible outcomes.[60]

Heisenberg elaborates:

> The atom of modern physics shows a distant formal similarity to the [*i*] in mathematics. Though elementary mathematics maintains that among the ordinary numbers no such square

60. Miller, "Quantum Mechanics"; emphasis added.

> root exists, yet the most important mathematical propositions only achieve their simplest form on the introduction of this square root as a new symbol. Its justification thus rests on the propositions themselves. In a similar way the experiences of present-day physics show us that atoms do not exist as simple material objects. However, only the introduction of the concept "atom" makes possible a simple formulation of the laws governing all physical and chemical processes.[61]

In other words, the image of an atom as an enduring material object makes it possible for us to derive and apply the rules and equations that are basic to physics and chemistry—and yet this image is essentially only an abstraction or a symbol. In the context of our present discussion the significance of all this is that if a forming event does have an innate ability—a conceptual pole—to incorporate information that is purely abstract and not necessarily descriptive of a physical entity, then such information might easily be understood to include imaginary numbers. The conceptual pole, being "mental," can, in a manner of speaking, easily include imaginary numbers.

(5) The functioning of the conceptual pole makes quantum entanglement more intelligible. In chapter 7 I noted that Einstein objected to the "spooky action at a distance" that, he thought, was demanded by the instantaneous correlations that quantum mechanics attributes to entangled particles. When two entities are entangled (this can be specific particles, photons, or even entire atoms), certain information about changes in one is shared instantaneously with the other, even across large distances. Einstein and others protested that no signal between two objects can travel faster than the speed of light. However, since the time of Einstein entanglement has been experimentally verified. The question that still remains is *how* correlations can be instantaneous. The concept of entanglement requires us to discard our persistent habit of thinking that all relations must be transmitted locally—that is, via a contiguous route through physical space. I suggest that since the information being shared is purely abstract, conceptual, or mathematical, it requires no physical manifestation and is therefore not constrained by the speed-of-light limitation. This is yet another quantum phenomenon that points to the functioning of a mental or conceptual pole at the atomic level.

(6) The existence of the conceptual pole helps us to make sense of the Pauli exclusion principle. This principle, which was discussed in

61. Heisenberg, *Philosophical Problems*, 56.

chapter 7, states that no two electrons in an atom or molecule can have all four electronic quantum numbers the same. The Pauli principle is in effect even in the largest atoms, which have over a hundred electrons.[62] It provides an explanation of the arrangement of the periodic table of elements and is a crucial rule in theories of chemical bonding. It is important to understand that there is no way to reword the Pauli exclusion principle by referring solely to physical conditions. The application of the principle is irretrievably mathematical. This suggests a revision of the question that Rutherford posed to Bohr—to wit, "But how does the electron *know* whether its quantum numbers are unique?" The only answer that does not devolve into technical jargon and double talk is that electronic events have a conceptual pole in which the effective content of the Pauli principle is processed at some very primitive level.

(7) The presence of the conceptual pole clarifies the ontological status of fields in physics. Feynman explains, "Physicists use the word 'field' to describe a quantity that depends on position in space and time. Temperatures in the air provide a good example: they vary according to where and when you make your measurements."[63] It is significant that in describing the field, Feynman does not mention the air itself, but rather the temperatures—numerical measurements. It is also significant that, as was the case with the vibrating string, Feynman's example of atmospheric temperatures relies on a physical-mechanical model—the collective motion of very large numbers of atoms that constitute the atmosphere. This is quite different from fields of electromagnetism or gravity, in which no physical medium is involved, and yet the field is nevertheless said to have specific values at specific locations even when no measurements have been made. This results in a puzzle. How is it possible for fields—which are basically like a table of values—to impart a physical force?

Imagine two positively charged bodies. They can be tiny particles, ball bearings, or metal paper clips; it doesn't matter. Because the two objects have identical charges, they will experience an electrical force of repulsion. As one physics textbook puts it, "Like the force of gravitational attraction, this force is of the action-at-a-distance type, making itself felt without the presence of any material connection between [the two

62. The heaviest atom that has been produced in macroscopic quantities is Einsteinium, with ninety-nine protons and ninety-nine electrons. The existence of even larger atoms is accepted, however (see Wikipedia, "Transuranium Element").

63. Feynman, *QED*, 123.

objects]. No one knows 'why' this is possible—it is an experimental fact that charged bodies behave in this way."[64]

The field itself is definitely not a physical substrate. In fact fields are more like an abstract map or a matrix in which each location is associated with a specific value pertaining to the strength of the force being considered (such as gravity or electrostatic force). The values are calculated by equations based on location (usually the distance from the object exerting the force). In some circumstances, time is involved. In *quantum* field theory, factors other than simple location are introduced, and the pertinent concepts become increasingly elusive.

Once again, the hypothesis that all events have a conceptual pole helps us to answer the question as to why fields are efficacious. The mathematical data comprising the field are internalized by the conceptual pole of each forming event in the field so that each event integrates the field's information into the outcome. (Remember, particles are understood to be serially ordered routes of events.) In the field, information is conveyed that may be described as purely mathematical. This information is somehow present even in "empty space."[65] Because the forming event has a conceptual pole, it is able to incorporate this data into its own culminating definiteness. It is the culminating definiteness of the event that constitutes the physical effects that we associate with the field, typically in the form of a particle—which may be long-lived or extremely evanescent.

Physicists now believe that every fundamental particle occurs in the context of its own particular type of field. An example that has been in the news relatively recently is the Higgs field.[66] This field was first postulated in 1964 by Peter Higgs and others as a field that imparts mass to gauge bosons. The Higgs field is a field of energy that is believed to permeate the universe. The field's physical manifestation is a fundamental particle known, not surprisingly, as the Higgs boson, which is used by

64. Sears and Zemansky, *University Physics*, 540.

65. Technically it is not the field itself that imparts the information but the relevant predecessor events, including the events that compose the object that is the source of the field being considered. The details of the process are not important here. The point is to recognize the critical role played by the conceptual pole. It is significant that quantum theorists and Whitehead both suggest that events take place even in empty space. See Hamer, "Empty Space," and Whitehead, *Process and Reality*, 177.

66. The existence of the Higgs boson—the particle produced by the Higgs field—was verified in 2012 at the Large Hadron Collider near Geneva. Peter Higgs and François Englert were awarded the Nobel Prize in Physics in 2013 for their theory and its predictions.

the field to interact continuously with other particles such as the electron. Particles that interact with the field increase their mass and, in a way analogous to an object passing through heavy syrup, the particles slow down as they pass through the field. As a result of having its mass increased, the particle is prevented from traveling at the speed of light.[67] We should notice that Peter Higgs and others first described the existence of a field and then reasoned from the nature of that field to the proposal that a corresponding particle must exist. The originating role of fields in physics is a good deal easier to understand if we incorporate the functionality of the conceptual pole in atomic events.

So far we have considered four philosophical principles that, I have argued, can be derived from quantum mechanics. These are (1) that nature is radically processive; (2) that each process culminates in a conclusive state of definiteness before it can influence other processes; (3) that the processes found in nature occur not randomly or incoherently but, by their very nature, in derivation from their relevant predecessors; and (4) that the causal influence of events on their successors is not merely physical but informational. In support of the fourth principle I have postulated that each atomic event includes an information-processing conceptual pole that can receive and integrate information qua information, including mathematical information. Because this fourth principle is likely to seem less plausible than the others, I have supported it with seven examples drawn directly from quantum mechanics.

I indicated earlier in this chapter that there are five philosophical principles to be derived from quantum theory. We are now ready to express the fifth. It is that the availability of an informational component enables nature to produce outcomes that are characterized by wholeness (holism), genuine novelty (outcome-oriented originality), and increasing complexity (levels of organization). The most creative function of the conceptual pole is that it allows numbers and, in principle, other abstract data to be used not only to *describe* outcomes and *replicate* patterns but to *provide novel potentialities*. Because of the conceptual pole, each event includes some rudimentary but significant ability to compare or contrast potential outcomes. We see this phenomenon of comparing and contrasting exemplified in superposition. The conceptual pole processes the informational component behind superposition, the coexistence of two or more quantum states at the same time. We can also say that the

67. Wikipedia, "Higgs Field."

conceptual pole provides the ontological justification for Heisenberg's repeated references to Aristotle's use of the word *potentia*.[68]

In our own minds, the ability to entertain alternatives is the basis for making choices. Depth psychology reminds us that we effectually make some choices not deliberately but unconsciously, yet still involving mental activity and a functional comparison of alternatives. It is obvious that in order for pure information to contribute to a pending outcome, it must be present as a possibility or potentiality and not as an already-existing, physical state of affairs. It is a central premise of this book that the evaluative, imaginative function is not restricted to human minds. We witness it throughout nature. I have mentioned the functioning of choice-making even in an amoeba. Now we are considering its role in events at the atomic level—though at that level, it cannot accurately be called a "choice" or a "decision." It is more nearly the co-presence of information about possibilities pertaining to quantum states. The bottom line is that quantum physics suggests that something like the entertainment of possibilities exists throughout nature, starting at the very ground floor.

Physicalists and scientific reductionists may scoff at the idea that nature is fundamentally holistic and goal-oriented. However, if we stick steadfastly with the materialist-reductionist position, then unavoidably we are also stuck with the familiar lamentation that quantum mechanics is "weird," "totally crazy,"[69] and irremediably counterintuitive. In contrast, the position taken in this book is that it is the physicalist-reductionist worldview that, ironically, turns out to be weird and implausible. In fact, as you read this page you are an example of the purposiveness and meaningfulness that reductionistic materialism is obliged to explain away. This, I say, is a patently untenable position. As physicist Henry Stapp has written,

> In short, the quantum conceptualization is not *intrinsically* counterintuitive, problematic, or weird. It becomes these things only when viewed from a classical [reductionistic] perspective that is *counterintuitive* because it denies the causal efficacy of our intentional efforts, is *problematic* because it provides no logical foundation upon which a rational understanding of the occurrence of subjective experience could be built, and is *weird* because it leaves out the mental aspects of nature and chops the

68. See chapter 7, especially notes [X-REF] and [X-REF].

69. These are the words of the 2022 Nobel laureates themselves, quoted earlier in this chapter in Borenstein et al., "Physicists Share Nobel Prize."

> body of nature into microscopic, ontologically separate parts that can communicate and interact only with immediate neighbours, thereby robbing both conglomerates and the whole of any possibility of fundamental wholeness or meaningfulness.[70]

In conclusion, the differences between classical mechanics and quantum mechanics can be summarized as follows: (a) where classical mechanics emphasized stable material substances, quantum mechanics emphasizes processes; (b) where classical mechanics emphasized analysis by reduction or division into smaller, simpler parts, quantum mechanics acknowledges the indivisible wholeness of waves, presupposing both the wave's intrinsically processive, dynamic nature and its intrinsic physical togetherness or wholeness (peak, trough, and all values in between); (c) thus where classical mechanics tends to view forms of togetherness in terms of aggregates, quantum mechanics perceives the formation of systems or wholes; (d) where classical mechanics pursued mechanistic certainty à la Laplace, quantum mechanics expects uncertainty or indeterminacy; and (e) where classical mechanics assumed that processes can be visualized in terms of physical collisions and forces, quantum mechanics holds that picturable images must ultimately be surrendered in favor of mathematical descriptions.

It is now possible to suggest an outline of the phenomenology of the atomic-level events that serve as the foundation of all natural processes. Points 5 and 6 are explored more fully in chapters 9 and following.

1. Fields serve as the creative matrix in which events are born. Fields consist of a physical continuum in which forces between objects take on differing values. Fields may therefore be thought of as the physical phenomenon behind radical relationality.
2. All fields have both physical properties and mathematical properties. These two types of properties taken together constitute two types of information that are integrated in the creation of actual events.
3. The mathematical properties define the vibrations or oscillations that characterize physical actuality.

70. Stapp, "Minds and Values," 149; emphasis original. Stapp is a mathematical physicist who was employed at the Lawrence Berkeley Laboratory in California and worked briefly with Wolfgang Pauli, Werner Heisenberg, and J. A. Wheeler (see Wikipedia, "Henry Stapp").

4. The vibrations add to one another according to various quantum principles in ways that contribute to entanglement, system formation, and wholeness.
5. The tendency towards wholeness fosters more complex modes of existence.
6. Wholeness and complexity at the molecular level allow living organisms and the evolution of species to occur. Over time, purposes and goals become more prominent and more complex.

Because it integrates conceptual and physical information, the dipolar, physical-mental ontology of events resolves the conflict between medieval realism and nominalism, and it overcomes the breach in Cartesian philosophy between thinking substance and physical substance. It does this by discarding the philosophy of substance and embracing instead a philosophy of processes in which mental and physical functionalities constitute dipolar events. This explains the integration of mind and body, of purpose and mechanism. Empirically, this integration of the mental and the physical is embodied in an increasingly organismic structure in the entities involved.

The notion that potential outcomes are being evaluated or contrasted depends on the functioning influence of goals or purposes. But how can an event at the atomic level possibly have goals? That question is better suited not for physics but for philosophy. We turn in chapter 9 to a philosophical consideration of purposes or goals.

9

Physics and Purposes

We are considering a concept of nature in which the traditional notion of material substance is rejected in favor of an outlook based on relational process. Each process is deemed to be distinct or atomistic and is required to culminate in a conclusive state of definiteness before it can influence other processes. Processes occur not haphazardly but in a matrix of derivation from their relevant predecessors, and each process influences its successors in ways that are not only physical but informational. Each event thus consists of a dipolar integration of physical and conceptual elements, though in many physical processes the role of the conceptual pole may be negligible.

We come now to a subject that can no longer be avoided—the subject of the presence of goals or purposes in each event. This is the divisive issue of teleology, an idea that seems to inspire wholehearted endorsement in some persons, but disdainful repudiation in others. In my experience, even those who are willing to entertain the idea of a conceptual pole are leery of the suggestion that teleology exists throughout nature.

There are good reasons for this skepticism. The most obvious is that, with the exception of certain animals, there is no immediate evidence of purposiveness in the nonhuman world around us. Clouds, rocks, and waterfalls show no signs of being motivated by goals. A second difficulty is that we intuitively tend to associate purposes and goals with the rational decision-making of human beings. If we think about it, goal-oriented behavior does seem to be displayed by many organisms, such as plants whose roots seek out water and insects who adroitly dodge the flyswatter, but it is hard to imagine that life-forms such as plants and insects make conscious choices. A third hurdle is that the idea of teleology has a notorious history of being conscripted for use by persons who have a vested interest in some ideology or dogma. The appearance of purpose or "design" in nature has been cited as evidence that God exists, or that the universe is conscious, or that nature itself embodies some benevolent plan or ultimate goal. There are reasonable ways to avoid such unwanted ideological baggage, and they are discussed later in this chapter. First, however, readers may very well be wondering why it is necessary to broach the topic of teleology at all. Why should we even bring up the subject of goals or purposes in connection with physics?

The answer, in a nutshell, is that our notion of a mental or conceptual pole necessarily leads to the notion of goals or purposes. If it is true that "nature traffics in information," there must be some reason *why* the information that is used *is* used as it in fact is. In chapter 3 I noted that the word *information* is derived from the root verb *to inform*, and that informing or being informed typically presupposes an intentional or teleological setting. If events at the atomic level are to make use of conceptual information, the conceptual or mathematical information itself cannot determine how it will be used. The only possibility that remains is that *the event itself makes this determination*. The functioning of abstract or mathematical information in nature necessarily involves some proclivity or inclination towards evaluated ends or outcomes.

Another way of approaching the issue of teleology is to consider the character of physical law. Why are the "laws of nature" efficacious? As James Gleick puts it, "A law is not a cause; yet it is more than merely a description."[1] *Why* are "laws" effective in nature? The answer now being proposed is that each event has a conceptual pole that processes a type of information that is not transmitted sheerly by physical forces or collisions. Of course this conceptual pole does not process the laws of nature by using the mathematical symbols that human beings use; rather, the mathematical relationships themselves are transmitted from event to event in informational form throughout the causal matrix of the universe. The presence of a conceptual pole in each event thus explains why so many aspects of nature can explained mathematically. But, as Gleick points out, "the law itself is not a cause." The only available explanation is that *each event must function as its own cause*. That is, the event is ultimately self-caused.

1. Gleick, introduction to Feynman, *Character of Physical Law*, ix.

Self-causation necessarily includes an orientation or a directionality towards certain ends or outcomes. The math or math-like information itself cannot make a choice. Only the event itself can do that.

The view taken in this book is that the so-called laws of nature are not really the strict physical predictors that many people deem them to be, but are rather extremely high probabilities based on the laws' description of the aggregate behaviors of extremely large numbers of single processes at the atomic level. At a more fundamental level, however, there is a profound sense in which nature moves from statistical law involving large numbers of entities to lawlike regularity or repetition at the atomic level; this is because nature is composed of discrete atomic processes with a specific ontological structure (a unity derived from the interaction of a physical pole and a conceptual pole) in which it is inherent to take mathematical relationships into account. Through the conceptual pole, mathematical patterns of relationality or wholeness are an intrinsic aspect of the atomic ontological structures. Because of the presence of the conceptual pole, mathematics or patterns are involved not merely descriptively but causally. In superposition, mathematical relationships are used to represent differing possible outcomes. Again, it is not that the event takes out its calculator and manipulates symbols. Mathematical information can serve as an actual cause through the agential role of the conceptual pole, which evaluates the mathematical or conceptual potentialities present in superposed states. This evaluation seems not to involve complex mathematical processes; we have already noted that quantum states depend initially on integers. There is thus a rational aspect to nature, if "rational" is understood to mean not "conscious, logical thought" but "ratio" in the mathematical sense, involving the comparison of values or outcomes. It is interesting at this point to notice that the word *value* means both *a mathematical quantity* and *something worthy of being sought or achieved.*

We can explore the functioning of mathematical or conceptual information by referring to Aristotle's fourfold analysis of causation involving efficient cause, material cause, formal cause, and final cause. We may easily think of mathematics or patterns as a formal cause, but, as I have said, formal causes alone are not plausible as active agents. For example, the number three may describe either an existing quantity of items or it may pertain to an unactualized possibility (such as a state transition represented by a different quantum number), but "threeness" itself is not an active agent. We need an explanation in which mathematical values and relationships can serve a role that is not merely descriptive or predictive but *influential* or *efficacious*. Therefore if the number three is to enter into the formation of an event, there must be an aspect of the event itself that can evaluate various possible results and recognize "three" as one possibility. The effectual pursuit of a particular result is what is usually identified with Aristotle's notion of a final cause. Since *potentia* or pure possibility cannot serve on its own as a causal agent, the forming event itself must furnish the final cause. Thus a doctrine of *causa sui*, self-causation, is necessary to make sense of superposition, of the Pauli exclusion principle, of the predictive role of complex numbers (which cannot themselves be descriptive of an existing physical condition), and so on.

Aristotle himself uses the Greek word *telos* primarily to refer to consciously entertained purposes—"that for the sake of which," as a person walks for the sake of her health—but *telos* can also refer more broadly to any culmination or result even in the absence of purpose or intentionality.[2] For example, in the New Testament *telos* often signifies any termination or outcome, as in Matthew: "But Peter was following Him at a distance as far as the courtyard of the high priest, and entered in, and sat down with the officers to see the *outcome* [*telos*]" (Matt 25:58 NASB). This expansion of the meaning of *telos* is suggestive. In this book *final cause* and *teleology* do not refer primarily to conscious choices and deliberate decisions but to physical tendencies derived from abstract information, somewhat like the idea of attractors in chaos theory,[3] or related to the mathematical probabilities provided by the Schrödinger equation. Electrons and atoms certainly do not envision results consciously but, as we have seen, neither is their behavior reducible to the precise, mechanistic predictability

2. Souter, *Pocket Lexicon*, s.v. "telos."

3. Chaos theory, which was first applied in the field of meteorology, involves the use of mathematics to describe the tendency of certain seemingly random dynamical systems to evolve towards definite outcomes. The math appears to explain the quasi-predictability of complicated phenomena that initially appear to be unpredictable. Chaos theory applies to systems characterized by large numbers of interacting factors but whose eventual state is presumed to be produced deterministically. We have noted that strict determinism cannot be applied to quantum states pertaining to individual particles and single quantum systems. Therefore chaos theory's notion of "attractors" is actually not a good example of final causation. Though attractors are sometimes spoken of as if they serve a causal function—the word *attractor* itself appears to imply teleology—chaos theorists do not typically attempt to describe the ontological structure of the individual components of the systems they study, nor do they attempt to explain how an attractor, which is essentially mathematical in character, can serve as a cause.

presupposed by classical mechanics. In a state of superposition, some outcomes are more probable than others. I am suggesting that the Greek word *telos* can refer not only to purposive choices but to any specific outcome that is the final result of nondeterministic processes at the atomic level.

At some point we must try to imagine what it would be like for an atomic process to "envision" or "evaluate" an outcome. As we attempt this we are forced to recognize that the only words that our language offers to describe this type of process are words derived from human experience—from our own conscious processes of decision-making. Thus our language itself cannot help but foist an inappropriate aura of personification or anthropomorphism onto the rudimentary processes of nature. This creates a severe encumbrance pertaining to any attempt to describe the functioning of purposes, goals, or final causes at the atomic level. Therefore I must ask readers to remember that all language is, in the end, metaphorical or symbolic[4] and that this metaphorical status is immediately evident whenever we speak of the "choices," "preferences," or "envisagement" that are displayed by quantum phenomena in "the collapse of the wave function" (a phrase which itself is highly metaphorical). As I have noted previously, scientists themselves routinely employ metaphorical language. In chemistry it is an accepted practice to use words such as *electrophilic* and *nucleophilic* (the *-philic* suffix means "loving" or "favorable regard" in Greek), it is accepted that the notion of a particle's "spin" actually has no reference to a physical object that is spinning,[5] and it is accepted that the standard description of quarks consists of various "colors," and "flavors" that include *up*, *down*, *charm*, *strange*, *top*, and *bottom*. My point is that when we use the language of human experience to describe final causation at the atomic level, that language, like many of the terms of chemistry and physics, is metaphorical, and highly so. On the other hand, however, *teleology at the simplest levels of nature is*

4. Often we speak of language that is "literal." In the sentence "The cat climbed into the oak tree," *cat*, *climb*, and *oak tree* are associated directly with objects or activities that are directly observable. Therefore we deem the language literal. But literal language is not a concept that is easy to defend. Every person who hears or reads the sentence about the cat will necessarily bring their own subjective imagination to the words, and each of the words *actually applies generically to a whole class of things*. This is but to say that the words are useful precisely because they are symbols that can be applied in more than one setting. But this implies that words, by their nature, do not correspond "literally" to single objects.

5. See Becker, "Quantum Particles."

not merely a figure of speech. There really is, at the ontological level, a small but functional aspect of outcome selection even in atomic events. I ask readers who feel skeptical about this statement to read at least to the end of this section before surrendering to disbelief. We have no way of describing atomic teleology or goal-seeking literally—but then, neither can we describe an electron or a photon literally. This failure of literal language does not mean that our descriptions are merely imaginary, conjectural, or artificial. After all, some metaphors are more apt than others.

If you immediately feel incredulous when you encounter the idea that purposiveness exists throughout nature, ask yourself what motivates this incredulity. I have acknowledged that even as children we observe that physical objects have no inner feelings or goals, but on the other hand children find it easy to believe that even plants and tiny animals do have a sense of purpose. Children do *not* find it obvious, as we adults often do, that the behavior of plants and animals must be reduced to the mechanistic interactions of tiny physical components. The entrenched status of the physicalist outlook is a result of an inculcated scientific materialism, a habit of thought that has evolved into a required password for admission into the society of intellectual respectability. But, I have argued, the claim that nature is based on mechanistic materialism is supported only by the superficial evidence of the outward appearance of macroscopic objects. At the atomic level, this evidence disappears.

We must recognize that, if we do try to illustrate the purposiveness of subatomic events by referring to human purposiveness, we should cite the simplest human experiences imaginable. If we try to describe primordial teleology, it would be better to avoid reference to human experience altogether, but we are so accustomed to assuming that nature is purposeless that examples of purposes in the simplest levels of nature are hard to come by. Nevertheless, if both human beings and atomic events have an ontological aspect that is purposive, it follows that there really is a continuity or commonality that reaches all the way from the atomic level to the level of human experience. Naturally the differences between the two are vast. When we think of human purposes we usually think of everyday experiences such as finishing our work, deciding what we want for dinner, or planning an activity. Examples like these clearly cannot be applied to quantum processes. When we do find it necessary to illustrate quantum purposiveness by appealing to human experience, we must choose the simplest, most primitive examples imaginable, and the examples must be highly physical, involving the least reliance on any conscious deliberation or cognitive originality. Western epistemologies have focused on detached, visual examples that are typically broken down into smaller components for purposes of analysis. Our procedure here must be the opposite, focusing on examples that are holistic and physically engaging, vague but insistent, even visceral.

With all of these caveats in mind, I suggest that the goal-seeking aspect of atomic phenomena can be characterized in four ways. First, there is an incipient aspect of each atomic-level process that is intrinsically *future oriented* or, one might even say, *anticipatory*. It is the very nature of goal-seeking to presuppose some transition from present to future. Second, each atomic-level process tends to seek *intensity* through the *contrast produced by vibration*. A wavelike alternation between contrasting boundaries tends intrinsically to be engaging. As Whitehead puts it, "In this vector transmission of primitive feeling the primitive provision of width for contrast is secured by pulses of emotion, which in the coordinate division of occasions appear as wave-lengths and vibrations."[6] (By "coordinate division," Whitehead means the description of the relations between events; this parallels Descartes' analysis of space as a system of "coordinate axes.") Third, each atomic-level process is *radically relational*, and the novel relations that are procured tend towards ends that embody *wholeness* and *increasing complexity*. As the number of relations increases, systems become more complex. *The bottom line is that vibration produces intensity based on contrast, and relatedness produces complexity based on novel patterns of combination*. Finally, it is the nature of each actual occasion to self-terminate, or to be terminated if it is disturbed by its environment. This is what is signified by the collapse of the wave function. I listed this condition in chapter 8 as a second principle suggested by quantum mechanics; I now list it again, as an aspect of the teleological character of each event. Termination, or coming to a state of definiteness, allows the actual occasion to have an influence upon its environment. Whitehead wrote that "value is the outcome of limitation."[7] This statement has an obvious colloquial meaning in economics; if diamonds were as common as sand or gold were as commons as iron, neither would be very valuable. Their value depends on their limited quantity. But Whitehead's meaning is not merely economic; it is metaphysical. Metaphysically, each process's termination establishes definiteness or physical instantiation, which is

6. Whitehead, *Process and Reality*, 163. Whitehead's use of the words *feeling* and *emotion* should not be interpreted as referring to conscious human experience.

7. Whitehead, *Science and the Modern World*, 94.

necessary in order for aims to be realized and values to become actual. These four points—(1) orientation towards the future, (2) intensity of feeling found in vibration and contrast, (3) relatedness and a tendency towards organismic complexity, and (4) culmination in a state of definiteness—constitute the basis for purposiveness as a metaphysical category. This is another way of saying that valuation and meaning tend to emerge spontaneously throughout nature.

(1) To say that teleological or goal-seeking behavior is future oriented is virtually a tautology, but it is still important to unpack the meaning of this statement. It is the function of the teleological process to envision, compare, or evaluate the desirability of pending alternatives and then settle on one outcome. There is an obvious sense in which goal-seeking occurs in all living things. I have already cited Ursula Goodenough's example of the amoeba: "All creatures evaluate. The amoeba perceives and then moves toward a food source; it perceives and then moves away from a toxin."[8] The amoeba can hardly be said to be conscious, and certainly it does not think thoughts as we do. I cite the nonconscious nature of the amoeba's behavior as an illustration of something that, I am saying, happens not only in living organisms but at the very ground floor of nature. At the atomic level the purposive aspect of the process is even less complex, less subjective, and more nearly physical than conceptual or mathematical, but it is nevertheless outcome oriented. Even in electronic and photonic events there is a functionality that assesses alternatives; we see this in the termination of superposition, in which certain possibilities or potentialities are effectually eliminated, allowing the phenomenon to be measured. In fact this is the source of the "measurement problem" in quantum physics; like the choice forced upon a driver who encounters a T-intersection, the experimenter's act of measurement forces a process that was hitherto indeterminate to become determinate. My point at present is that every atomic event in nature has, as it were, some innate directedness towards the future and some tendency to seek certain results.

(2) I have said that each atomic-level process tends to seek intensity through the contrast produced by vibration. We note once again that even matter has a wavelike character. Vibration is innately associated with contrast through the juxtaposition of peak and trough, back-and-forth, out and in, waxing and waning. It is evident that animals enjoy

8. Goodenough, *Sacred Depths*, 106.

forms of vibration, including music.[9] But the esthetics of contrast are apparent throughout nature, ranging from the visual appeal of flowers whose contrasting colors are attractive to insects to the simple contrast between positive and negative charges in a salt molecule. Obviously this esthetic contrast does not require a conscious appreciation of fine art. I am affirming that there is a type of esthetic intensity embedded in nature itself, originating in the alternation of vibration. I discuss this esthetic quality below in association with the notion of *panexperientialism* (which is not, it should be noted, the same thing as panpsychism).

In human experience the intuitive appeal of vibration or alternation is best illustrated by activities involving repetitious physical pulsation or rhythm. Persons who are seated may occasionally be observed to move one of their legs rapidly up and down by pushing upward from the ankle and then releasing. Sometimes we drum our fingers on a table or desk. When the persons engaged in such activities are asked about it, they usually look a little embarrassed and say that they didn't realize they were doing it. If an explanation is requested, the person may blame "nervous energy" or "an old habit." Nervousness and habits may indeed be involved, but these replies do not actually explain the soothing or satisfying aspect of the activity, which has its primal roots in the engaging, almost hypnotic quality of vibration or regular patterns of alternation.

Drumming, rhythmic chanting, dancing, and singing are ubiquitous in human societies. Music, a more highly developed expression of vibration, is a daily aspect of human experience, illustrated by tunes hummed unconsciously, by advertising jingles that we can't manage to forget, and of course also by the complex compositions performed at concerts. Intuitively, we use music to mark life's significant transitions. "When they had sung the hymn, they went out to the Mount of Olives" (Matt 26:30 NRSV). Seldom if ever do we ask ourselves *why* these patterns of rhythm and tonality are innately enjoyable or worthwhile—but they are. Whether we are scientists, philosophers, psychologists, or ordinary music lovers, we would do well to wonder why various chords or combinations of notes evoke differing emotions: joy, melancholy, wistfulness, tension, serenity.[10] The frequencies that prompt these emotions

9. We might expect that birds and other animals that can make rhythmic sounds would enjoy vibration, but even rats, who are relatively quiet animals, are included (Devlin, "Slaves to the Rhythm").

10. Actually it would probably be a mistake to entrust the question to professional scholars, who would likely reduce the whole inquiry into an erudite reexamination of

have a systematic mathematical relationship: the chromatic half steps are related to one another by the numerical ratio of one to the twelfth root of two (or 1 to approximately 1.0595).[11] The frequencies of all the notes in a classical symphony can be calculated by using this formula. Pondering this, we may discover a new appreciation for Pythagorean mysticism.[12] There is a teleological aspect to all esthetic experiences, for such experiences are characterized by a direct, intuitive sense of appreciation—an act of assessment or preference. In fact, virtually every experience involves some evaluative response; even the feeling of tedium or boredom amounts to an assessment. In opposition to Descartes, I am proposing that our sense of the esthetic originates not in a spiritual-intellectual capacity that is functional only in "thinking substance" (the mind) but in the teleological-appetitive aspect of each dipolar event's mental-physical integration of vibratory and rhythmic patterns. This physical-mental integration has gradually become more and more sophisticated in various life-forms through the ages of natural history. Using music as an example is especially helpful because it reveals that mathematics, often regarded as unrelated to beauty or feeling, has a direct applicability, à la Pythagoras, to the creation of harmony, rhythm, and esthetic expression. We observe this involvement of mathematics at its simplest level in quantum mechanics and in the mathematical foundations of repeated patterns ("laws") in physics.

We often fail to recognize that sounds are not the only experiences that involve vibration. In fact, *all* experiences do. It seems to us that eyesight conveys not vibration but static images. However, what the rods and cones of our retinas encounter is actually a stream, focused by the eye's lens, of varying electromagnetic waves. The millions of energy events[13]

---

the old idea that tonal intervals based on simpler frequency ratios sound more pleasant. This is a bit like describing the grandeur of the Alps in terms of plate tectonics and the resistance of granite to erosion.

11. This is so because the frequencies of octaves occur as multiples of two (A = 220, 440, 880, etc.), there are twelve chromatic half steps in each octave, and so the ratio between the two frequencies of consecutive half steps is a constant.

12. Pythagoras (ca. 570—ca. 495 BCE) is said to have believed in the "harmony of the spheres," the idea that the movements of the stars and planets follow mathematical equations that also pertain to the frequencies of musical notes. Pythagoras held that the heavenly bodies produce an inaudible music. Pythagoras is also said to have taught that the seven Muses were actually the seven planets singing together (Wikipedia, "Pythagoras").

13. A single human retina contains about 6 million cones and 120 million rods. Rods detect light and dark ("black and white"), and cones discriminate color

in our retinas are converted into signals that are transmitted by the optic nerve to the visual cortex, which converts the pulsing signals into static visual images. This welter of pulsing signals pertains not only to vision and sound. The truth is that *all* of our sense experiences—touch, smell, and taste, as well as vision and hearing—are like this, originating in vibration and variation. In addition to the standard five senses, a particularly good illustration is provided by the experience of sexual arousal, which generally finds expression in rhythmic or vibratory movement.

It should be noted that this and the other physical urges or desires that contribute to our survival do not originate in conscious decisions or rational analysis; they are physical appetites that are somehow in effect from the time of our birth and, no doubt, even in utero. In other words, the desires or urges that sustain life are rooted in our physical genetics. Though they have logical explanations ("I eat so I won't starve"), the basic urges of life—eating, breathing, sex, quenching thirst, protecting ourselves from heat and cold—are bodily in origin and are traceable backward in history through earlier and simpler species on the tree of evolution. The sensory experiences associated with these urges stem from the vibratory quality of nerve impulses, deriving ultimately from the gradual elaboration of the primordial wavelike character of energy events. Our minds may decipher the vibratory signals received from our senses in terms of static objects, but the static aspect results from the way our mental processes have evolved. In this way we objectify, analyze, and respond to our environment. But this way of objectifying our surroundings is an evolved adaptation. It is not the only way of experiencing the world.

Jill Bolte Taylor (b. 1959) is a neuroanatomist who suffered a massive cerebral hemorrhage in December 1996. Taylor recovered and wrote a book about the experience. Remarkably, when Taylor realized that she was having a stroke, her inner reaction was not one of panic but of euphoria.

> I look down at my arm and I realize that I can no longer define the boundaries of my body. I can't define where I begin and where I end. Because the atoms and the molecules of my arm blended with the atoms and molecules of the wall. And all I could detect was this energy. Energy. And I'm asking myself, "What is wrong with me, what is going on?" And in that moment, my brain chatter, my left hemisphere brain chatter went

(Wikipedia, "Photoreceptor Cell").

> totally silent. Just like someone took a remote control and pushed the mute button and—total silence.
>
> And at first I was shocked to find myself inside of a silent mind. But then I was immediately captivated by the magnificence of energy around me. And because I could no longer identify the boundaries of my body, I felt enormous and expansive. I felt at one with all the energy that was, and it was beautiful there.[14]

We routinely assume that our way of perceiving a world of distinct objects "out there" is accurate and that this view of the world corresponds with reality. But the discoveries of quantum mechanics indicate that the world we thus objectify is incomplete. It could be argued that Taylor's experience of an expansive, vibrant, energy-permeated, interconnected world with no well-defined boundaries is actually in closer alignment with the reality of atomic phenomena. Taylor's perception of beauty and her feeling of being captivated illustrate that there is something innately alluring or enjoyable about experiences of vibrancy or rhythm and about the experience of feeling connected within a larger unity. The position taken in this book is that this universal enjoyment of vibration and wholeness is not an emergent phenomenon reserved for the biologically-neurologically advanced but something that we inherit from the remote-and-yet-ever-present primordial ontology exemplified in quantum phenomena.

(3) My third proposition about final causation is that each atomic-level process is radically relational and that the novel relations that are actualized tend towards ends that embody wholeness and increasing complexity. Human beings and many animals have an inborn urge to form social relationships, but social behaviors are not restricted to larger animals that are more highly evolved. Bees, which create highly organized hives, share specific information about the location of pollen sources by doing two types of dances—the round dance and the waggle dance.[15] Even trees, which have no central nervous system, communicate and provide support for one another,[16] sometimes collaborating symbiotically with the mycelium of mushrooms.[17] When tomato plants

14. Pangambam, "My Stroke of Insight." Taylor's TED Talk video is online at Taylor, "My Stroke of Insight."

15. Wikipedia, "Waggle Dance."

16. Grant, "Do Trees Talk."

17. Holewinski, "Underground Networking."

are attacked by caterpillars, the plants emit a pheromone that attracts a species of wasp that preys on the caterpillars.[18] Slime molds, which are the simplest of organisms, demonstrate an uncanny ability to detect and respond to environmental changes with a type of purposiveness that has been described as a form of intelligence.[19] These phenomena and others too numerous to list illustrate the natural tendency to form larger wholes by involving the transmission of information.

This tendency toward increasing organization exists even in nonliving systems. This is evident in the most fundamental quantum events. The transition from a superposed state to an objectively measurable state typically involves some influence from the environment on the event, and a concomitant influence of the event upon the environment. It is generally believed that if the superposed state of a quantum system can be protected from environmental interruptions, its superposition and entanglement can continue indefinitely.[20] When a quantum object is disturbed by an environmental influence, the influence constitutes a novel mode of relatedness that often, though certainly not always, uses information to create increasing complexity or wholeness. We see this in the applicability of the Pauli exclusion principle, whose assignment of unique quantum numbers to orbital electrons reminds us that the atom itself manifests a type of organization or wholeness displaying novel complexity. The quantum numbers are a way of describing the fact that the electrons are organized into orbitals in which each electron has its own unique status. As we have noted, the whole structure of matter depends on this principle. This is but to say that relationships that are guided by information (illustrated by what we human beings understand as the Pauli principle) tend to form increasingly complex systems—systems that can be described as "organismic." Atomic orbitals are not simply repetitious in the way that the arrangement of atoms in a salt crystal is repetitious; in an atom, each electron exhibits a type of uniqueness, and the orbitals as a whole display a systematic arrangement. This tendency towards wholeness, which Teilhard de Chardin called "complexification,"[21] is carried

18. University of California, Davis, "Caterpillars Foiled."

19. Jabr, "Slime Molds Redefine Intelligence."

20. Zyga, "Physicists Find Quantum Coherence."

21. The term *complexification* can be used mathematically, but I associate the concept with the French Jesuit paleontologist Teilhard de Chardin, who coined the term in reference to the evolution of life in a world that originated in matter (Teilhard de Chardin, *Phenomenon of Man*, 48). *Man* in the title is the choice of the translator. Teilhard's

a step further at the atomic level by the formation of bonds with other atoms to form compounds. The relationships between these compounds lead eventually to nature's slow but persistent journey from minerals, water, and sunlight towards living organisms.

The atom itself is an outstanding example of holism. Only certain electrons in any atom are available for the formation of chemical bonds. These electrons are known as *valence electrons*. For reasons that lie beyond the scope of this discussion, metals with unpaired valence electrons are magnetic. Metals with paired valence electrons are not. The first thing that we may notice about this is that the quantum properties of the atom's invisible system of electrons determine its observable properties as a metal—not only whether it is able to be magnetized but its ability to conduct heat, to form chemical bonds, and so on. But there are deeper, more philosophical implications regarding the systemic relation between paired electrons and magnetism. We know already that it is a mistake to envision the electrons as tiny planets moving about in peaceful, predictable orbits. It would be closer to the truth to depict them as a group of kindergartners running chaotically about a small room in a state of unpredictable exhilaration. The question that must now be asked is, In this vigorous welter of activity, how does one valence electron detect the identity of its partner and that partner's magnetic orientation?

Strict materialists—who attempt to reduce all causation to efficient causes—may say that the specific partner electron is identified by its specific energy level, but this only takes us back to the initial question of why energy levels are quantized and organized in the first place. Another proposal might be that the paired electrons resonate somehow and thus identify the partner electron that is resonating in a similar way. But for an electron to resonate in a way that is detectable by its partner requires the transmission of energy. The energy of electrons in an atom is regarded as nondissipative; this means that the electrons are not constantly sending physical signals to one another. Of course their electrostatic charges exert a force on their neighbors, but this force is the same for each electron. It cannot serve as a basis for telling the electrons apart. The electrons do contribute to the magnetic fields that they create, but the question then again arises as to how the two valence electrons single one another out—or, if the last valence electron is unpaired, how it "knows" that it is unpaired. Materialists might say that the simple difference between an

---

title was *Le Phénomène Humain*. See also Morowitz, "Teilhard, Complexity."

odd and even aggregate number of electrons accounts for a metal's magnetic or nonmagnetic character, but this explanation won't do. It is only the outer valence electrons that are involved in the process. How does each electron distinguish between the valence and nonvalence status of all of its comrades? The running kindergartners certainly would not be able to do this, unless perhaps we put shirts of a distinctive color on each individual. But colored shirts are not an option for electrons.

The truth is that the difference between valence and nonvalence electrons is an attribute created by the organized or organismic ontological status of the whole assemblage of electrons. Moreover, information about this whole cannot be communicated on a piecemeal basis. It must be communicated as a conceptual or mathematical information derived from the wholeness of the entire system. The pattern of organization itself is what exerts the influence. Atomic physics thus requires the utilization of information about this pattern as a whole. The creation of this information necessarily involves an abstractive process. The process includes information about the specific relations between various parts. This same abstractive process pertains generally to the Pauli exclusion principle, a principle that applies not simply to each particle but to the various relations among the particles. The organized status that ensues is not a property of an aggregate, like the total weight of marbles in a bag. It is a systemic attribute, created by the organized pattern of the system as a whole.

Clearly this pattern cannot function as it does unless it is communicated in a way that is not merely physical but is at least partly conceptual. This requires that each entity must possess something like what Whitehead called a "conceptual pole" or a "mental pole." This holistic influence depends on the ability of each actual occasion to create and utilize information that is partially abstract or nonconcrete.

(4) The achievement of definiteness terminates every actual occasion, or else it never achieves a state in which it can influence subsequent occasions. In other words, it never becomes real. I have already discussed this in chapter 8 as the second principle derived from quantum mechanics. In physics a wave function "collapses" when it acts upon its environment. This is another way of saying that in a teleological event no goal can be achieved unless a point of definiteness is reached. Whitehead referred to John Locke's phrase *perpetual perishing* in order to make this point. Whitehead's technical term for the final phase of each occasion is "satisfaction." This does not mean that each occasion has a satisfactory or

successful outcome, but simply that the conditions necessary for physical definiteness have been reached.[22]

The claim that some sort of primitive purposiveness exists even at the level of particles is easily misunderstood, and it is therefore necessary once more to state clearly what this claim about purposiveness does not imply. I am not implying that atomic events visualize potential outcomes cognitively, the way human beings do. My guess is that thoughts and emotions are not possible apart from an enduring biological structure including, especially, the activity of a complex central nervous system. Also, I am not advancing a revamped form of vitalism, the idea that the origin of life and the activities associated with it are dependent on a force or principle distinct from the forces and principles we associate with physics and chemistry. More specifically, I am not asserting that purposiveness or teleology somehow obviates or transcends the second law of thermodynamics. Since purposes are usually viewed as a source of increasing order, it may seem that I am asserting that the effects associated with entropy are evaded. Such an evasion is, as far as I know, impossible. Biological complexification does indeed entail increased order, but this increasing order is achieved by extracting energy from the organism's environment. What I *am* asserting is that this extraction of energy is accomplished through the conceptual pole's ability to make use of information as a way of absorbing or collecting and concentrating energy. This process enables nature to create novel complexity. This growing complexity depends on the implementation of goals at a primordial level. Nature does employ information to create growing complexity based on an innate lure towards novel relatedness.

This inclination of nature, under the right conditions, to foster organismic wholeness is an indisputable fact, too long dismissed or explained away under the aegis of the tiresome formula "the survival of the fittest." Darwin's ideas are used—or abused—as a basis for the assertion that there are no actual purposes or goals in nature; nature is viewed simply as one, big, aimless welter of activity in which any biological novelties that are dysfunctional go away because they tend to be eliminated from the gene pool. The organismic orderliness observed in the bodies of the survivors may appear to be the result of purposiveness or goal-seeking, but actually this order is said to be merely epiphenomenal, the purposeless result of a physical process of elimination. The seeming order

22. Whitehead, *Process and Reality*, 60. On this point see Halewood, "Death, Entropy, Creativity."

produced by evolution is only mechanical, just as the grains of sand on a beach are sorted by the motion of waves. We have heard this argument so often that it may seem unshakable, but, perhaps surprisingly, there is actually no genuine evidence for the physicalist-reductionist position. Materialistic reductionism is little more than an embellished version of naïve realism,[23] inspired by an uncritical endorsement of the solid appearance and predictable motion of physical objects, and sustained by the indisputable successes of Newtonian mechanics. But these Newtonian successes, when adopted as the basis for a philosophy of nature, paint a one-sided picture of the world in which life, feeling, and novelty are marginalized, ignored, or disallowed altogether. In this perspective, nature, as Whitehead said, is not alive but lifeless.[24] What the physicalist-reductionist approach to evolution obsessively ignores is the obvious fact that "fitness for survival"—the key to Darwin's ideas—is itself a teleological phenomenon, dependent even in the simplest organisms on the functioning goal of self-preservation.

In his classic monograph on Aristotle's philosophy, W. D. Ross states that "one of the most conspicuous features of Aristotle's view of the universe is his thorough-going teleology." But then Ross writes: "But it is not so clear what interpretation is to be put on this view. Does he mean (1) that the structure and history of the universe [are] the fulfilment of a divine plan? Or (2) that it is due to the conscious working towards ends of individual beings? Or (3) that there is in nature an unconscious striving towards ends?"[25] For scholarly reasons, Ross rules out the first and second possibilities. Then he writes,

> On the whole, it would seem that view (3) is that which prevails in Aristotle's mind. . . . The notion of unconscious teleology is, it is true, unsatisfactory. If we are to view action not merely as producing a result but as being aimed at producing it, we must view the agent either as imagining the result and aiming at reaching it, or as the tool of some other intelligence which through it is realising its conscious purposes. Unconscious teleology implies a purpose which is not the purpose of any mind, and hence not a purpose at all. But Aristotle's language suggests that he (like many modern thinkers) did not feel this difficulty, and that, for

23. In philosophy, *naïve realism* is the simplest form of the belief that sense perception conveys a direct awareness of things as they really are: that we do perceive the "external world" as it actually exists (Flew, *Dictionary of Philosophy*, s.v. "Naïve Realism").

24. Whitehead, *Modes of Thought*, chs. 7 and 8.

25. Ross, *Aristotle*, 181.

> the most part, he was content to work with the notion of an unconscious purpose in nature itself.[26]

I cite Ross here because he gives us a succinct typology of explanatory options regarding the presence of purposes in nature. Ross declares that a purpose that is unconscious is not a purpose at all, whereas, of course, in this book *that is exactly the kind of purpose that is being affirmed*, apparently in agreement with Aristotle. When we *think* about *thinking*, it is natural to think that thinking is the sine qua non of purposiveness. But it is not.

In this chapter I have cited some of our simplest experiences of vibration, contrast, and wholeness in order to illustrate the more primitive roots of human experiences in physical nature. I have mentioned the quantum-mechanical concept of superposition as a phenomenon in which conceptual information and physical information are integrated to form a determinate outcome, and I have argued that some notion of primitive goal-seeking is necessary in order to make sense of each event's use of conceptual information. I suspect that many readers may find these ideas unfamiliar or doubtful, if not simply implausible. But these ideas in fact gain considerable support and amplification from several eminent philosophers. As a way of clarifying and reinforcing the proposals now being advanced, in chapters 10 and 11 I summarize the ideas of four philosophers, emphasizing the ways in which their ideas anticipate or undergird the philosophical principles that, I have argued, are implied by the discoveries of special relativity and quantum mechanics. These four philosophers are Heraclitus of Ephesus, Henri Bergson, William James, and Alfred North Whitehead. However, before embarking on chapter 10, I conclude the present chapter by summarizing the ideas of two philosophically minded physicists: Bernard d'Espagnat (1921–2015), and Wolfgang Pauli (1900–1958). I note also that Pauli's ideas were formed in collaboration with psychiatrist-psychologist Carl G. Jung (1875–1961).

APPENDIX: THE THOUGHT OF D'ESPAGNAT, PAULI, AND JUNG

From the beginnings of relativity and quantum theory it was recognized that the new discoveries extended beyond physics towards a radically revised outlook on nature as a whole. Quantum theory, especially, seemed

26. Ross, *Aristotle*, 182.

to cast doubt on certain long-standing assumptions about the foundations of reality. We have already considered some of the views of certain physicists regarding issues of a quasi-philosophical nature, but those physicists' views did not include any extended conversation with academic philosophy as such. One scientist who did pursue philosophy seriously and whose work should therefore be mentioned is French physicist Bernard d'Espagnat.

A second-generation quantum theorist, d'Espagnat received his PhD in 1950 for studies guided by Louis de Broglie, who had first proposed the existence of matter waves. D'Espagnat's analysis of the philosophical foundations of quantum physics focused on the disparity between the *realist* and *instrumentalist* interpretations of quantum mechanics—that is, on the question of whether basic quantum concepts refer to an underlying physical reality or are merely rules for predicting the outcomes of experiments.[27] D'Espagnat defined *realism* as the belief that the properties of material objects and physical systems exist whether or not they are observed by human beings—in other words, that the physical world really exists independently of any human perception of it. The opposing position, "instrumentalism," was associated by d'Espagnat with the twentieth-century school of philosophy known as "logical positivism,"[28] which I mentioned in chapter 2. Logical positivism asserted that in order to be meaningful, statements must be either logically necessary or empirically verifiable. D'Espagnat noted that in quantum mechanics the realist assertion that *physical reality would exist even if no observer existed* would itself be adjudged meaningless by logical positivists, for their definition of truth is based on the principle of verification *by observation*, and if no observers exist, there can be no act of verification.[29] Logical positivism is thus a type of instrumentalism. The instrumentalist position can also be discerned in the thought of Niels Bohr, who, as I noted in chapter 8, distanced himself from any metaphysical interpretations, saying, "There is no quantum world. There is only an abstract quantum

27. D'Espagnat's philosophical views are presented in d'Espagnat, *Conceptual Foundations*, esp. chs. 9 and 19–25. See also d'Espagnat, *Veiled Reality*, and d'Espagnat, "Quantum Theory and Reality."

28. Logical positivism, also called logical empiricism, arose in Vienna in the 1920s. It was distinguished by the claims that the scientific method is the definitive basis of knowledge and that traditional metaphysical dogmas should be dismissed as meaningless. Logical positivists included Moritz Schlick, Rudolf Carnap, Philipp Frank, Kurt Gödel, Otto Neurath, Carl Hempel, Hans Reichenbach, and A. J. Ayer.

29. D'Espagnat, *Conceptual Foundations*, 232.

mechanical description. It is wrong to think that the task of physics is to find out how Nature is."[30] Interestingly, the instrumentalist position also closely resembles the immaterialism of philosopher George Berkeley (1685–1753), who argued that what we are actually aware of in experience is not physical objects but our *ideas* of those objects.[31]

D'Espagnat undertook a thorough analysis of the realist-instrumentalist debate, concluding that though the physical world does exist, quantum physics does not allow that world to be understood as a settled, objective array of neutral facts that can be impartially verified by an observer. Rather, the world itself somehow includes the activity of observers and, moreover, there is no single, comprehensive view of things since a variety of perspectives and methods is unavoidable, and reality itself is inherently veiled ("*le Réel voilé*")—not only practically but as an implication of quantum theory. However, despite the significance of these insights, d'Espagnat kept his focus on epistemology and never developed a new theory of reality in which the necessary role of subjective factors such as observation, perspective, appearance, and interpretation is reconciled with the final physical facticity of the world itself.

The view taken in this book is that no significant progress can be made in the philosophical interpretation of twentieth-century physics unless the physicalist-materialist paradigm is resolutely jettisoned, once and for all, in favor of the view that nature is composed of a network of events and dynamic relationships, and that each event consists inherently of a synthesis of physical and informational (conceptual) factors. The culmination of the synthesis of the physical and the conceptual in finalized outcomes establishes the physical and informational conditions that provide the basis for the events that are yet to come in the future. Quantum superposition, in which possibilities seem from an observer's perspective to be suspended in a state of paradoxical incompatibility, illustrates the source of the conflict between the instrumentalist position, which notices the perspectival quality of the results obtained by differing methods of measurement, and the realist position, which notices the concreteness of the final outcome. In this book, it is held that in the superposed state, the synthesis of physical aspects and informational (mathematical or conceptual) aspects is still in process, ontologically. As we have noted already, this process is an internal one that cannot be

30. Petersen, "Philosophy of Niels Bohr," 12.

31. Downing, "George Berkeley," sec. 2 ("Berkeley's Critique of Materialism in the *Principles* and *Dialogues*"). See also Berkeley, "Three Dialogues," 217.

observed or measured without forcing the process to terminate—and the ontological status of the terminated state is different from the ontological status of the event while it is still in process. The failure to understand the significance of the unobservable quality of the internal process—because, ontologically, it is indeterminate—is the source of the so-called measurement problem in quantum mechanics. As long as we cling to the notion of static matter as the de facto paradigm in our understanding of nature, the essential dynamism and the inclusion of information that are manifest in this internal process will remain inexplicable, the differences between realism and instrumentalism will remain irreconcilable, and reality will continue to seem enigmatic and "veiled," as d'Espagnat said.[32]

One physicist who made a substantive contribution to the need for a new philosophy of nature was Wolfgang Pauli, the discoverer of the exclusion principle and one of the leading innovators of twentieth-century physics. In 1928, Pauli became professor of theoretical physics at the Federal Institute of Technology in Zurich. Tragically, in 1930 Pauli's mother committed suicide, and in that same year Pauli and his first wife were divorced. Seeking help, in January of 1932 Pauli consulted psychiatrist-psychotherapist Carl Jung, who also lived near Zurich. A noted authority regarding the interpretation of dreams, Jung's work with Pauli led them into an extended discussion of the images and themes in Pauli's dreams that Jung identified as "archetypal." *Archetype* is a technical term in Jungian psychology referring to pervasive, foundational symbols that stem from the collective unconscious; Jung posited the collective unconscious as a psychic network of feelings and images underlying human experience. The role of the archetypes in human experience resembles the role of instincts in animals, providing an unconscious source for the mental representation of objects or situations that are associated by the psyche with certain feelings and motives. Archetypes are described as a kind of extensive, innate knowledge derived from the current evolution of the human brain.[33] Jung regarded the existence of the archetypes as an empirical phenomenon, for he had observed their functioning in the dreams and behaviors of clients with widely diverse backgrounds and in various parts of the world.

32. Contrary to d'Espagnat, in science the interpretive options are not exhausted by the dichotomy between instrumentalism and realism. Physicist Ian Barbour lists four possibilities: naïve realism, positivism, instrumentalism, and critical realism (Barbour, *Myths, Models, and Paradigms*, 34–38). I discuss these models more fully in chapter 13.

33. Wikipedia, "Jungian Archetypes."

Though thoroughly scientific in outlook, Pauli was open-minded about the idea that mathematical, conceptual, and ideational factors are not merely descriptive of nature but are actively operative in determining the outcome of natural processes. Pauli's openness helps to explain his discovery of the exclusion principle in physics, which describes the way in which quantum numbers guide the behavior of electrons in atoms. Working with Jung as he attempted to understand his own dreams, Pauli soon adopted many of Jung's ideas. From 1937 to 1953 Pauli and Jung engaged in a warm correspondence that reflects a sophisticated understanding of both psychological and scientific issues. The letters have been published as *Atom and Archetype* and occupy over two hundred printed pages.[34]

One idea that the two discussed at length is Jung's concept of *synchronicity* (*Synchronizität*). Both in his own life and in the lives of his clients, Jung had observed many examples of seemingly coincidental concurrences between physical events and specific thoughts or mental images. Around 1930, Jung coined the term *synchronicity* "to describe circumstances that appear meaningfully related yet lack a causal [i.e., physical] connection."[35] Synchronicity thus refers to personal experiences of seemingly unaccountable synchronization between physical events and psychological content. Jung described an example.

> A young woman I was treating had, at a critical moment, a dream in which she was given a golden scarab [a piece of jewelry in the shape of a beetle]. While she was telling me this dream I sat with my back to the closed windows. Suddenly I heard a noise behind me, like a gentle tapping. I turned round and saw a flying insect knocking against the window pane from the outside. I opened the window and caught the creature in the air as it flew in. It was the nearest analogy to a golden scarab that one finds in our latitudes, a carabaeid beetle, the common rose-chafer (*Cetonia aurata*), which contrary to its usual habits had evidently felt an urge to get into a dark room at this particular moment.[36]

34. Jung and Pauli, *Atom and Archetype*.

35. See Wikipedia, "Synchronicity."

36. Jung, *Synchronicity*, 22, and see also 109. The experience with the scarab beetle is also reported in Arts of Thought, "Carl Jung."

Jung noted that up until then the woman had been blocked psychologically by an overly rational view of life and that the experience with the beetle opened her to the deeper meaning of her dreams.[37]

What does this have to do with physics?[38] We have seen that quantum entanglement involves the instantaneous communication of information without involving physical ("local") transmission through space. These correlations seem quite likely to have been interpreted by Pauli as a phenomenon in physics that has something in common with the concept of synchronicity. In fact, it is reasonable to think that the exclusion principle that is now named for Pauli suggested to him that abstract or conceptual information can be and is transmitted throughout nature by some means.

Jung developed the idea of synchronicity more fully in collaboration with Pauli through their letters, culminating in their joint work *The Interpretation of Nature and the Psyche* (*Naturerklärung und Psyche*), which contains one paper by each of the two. The book sets forth what is now known as "the Pauli–Jung conjecture," which posits the existence of a foundational "psychophysically neutral reality" that manifests both mental and physical aspects. "Pauli and Jung held that this reality was governed by common principles ('archetypes') that appear as psychological phenomena or as physical events. They also held that synchronicities might reveal some of this underlying reality's workings."[39] The Pauli-Jung Conjecture may be viewed as a type of dual-aspect ontology. In the philosophy of mind, dual-aspect theories propose that the physical and the mental are two manifestations of a single, more fundamental reality. Dual-aspect theory is also referred to as "dual-aspect monism," with the word *monism* indicating that the physical and the mental, though functionally different, are ontologically inextricable; each is mutually implicated by the other.[40] Because the two do not interact like physical parts, and because synchronicities always seem related in an intelligible or meaningful way to the circumstances in which they occur, Jung and

37. I extend thanks to my wife, the Rev. Claudia C. Schmitt, Jungian analyst, for her invaluable guidance regarding the thought of Carl Jung.

38. For an extended examination of Jung and Pauli's collaboration on the subject of synchronicity, see Halpern, *Synchronicity*. This book is interesting and informative, but unfortunately Halpern tends dismiss many of Jung's ideas from a perspective of traditional scientism ("Jung never proved his hypothesis"; "Neuroscience has shown no indication of a collective unconscious" [Halpern, *Synchronicity*, 185]).

39. Wikipedia, "Wolfgang Pauli."

40. Wikipedia, "Double-Aspect Theory."

Pauli proposed that the explanatory approach to these correlations is not quantitative probability or statistics, as in quantum physics, but qualitative meaning.[41]

It is interesting that at the end of his life, Pauli himself experienced an incident of synchronicity. In 1958 he was diagnosed with pancreatic cancer and admitted to the Rotkreuz Hospital in Zurich. When his final assistant, Charles Enz, visited him, Pauli asked him, "Did you see the room number?" It was number 137. Throughout his life Pauli had been fascinated with the question of why the fine-structure constant, a dimensionless fundamental number in physics, has a value nearly equal to 1/137. Pauli died in room 137 on December 15, 1958.[42]

Jung and Pauli are, of course, not the only two people who have reported experiences of synchronicity. In fact, I myself have had several. One seems especially remarkable. For three years I served as the senior minister of the University Park United Methodist Church in Denver. One Sunday during the summer of 1992, I had prepared a sermon in which I was attempting to critique the attitude, frequently observed in our society, that life involves inescapable conflicts and that we must be competitive and uncompromising in order to achieve any consistent success. As a way of illustrating the initial idea of struggle and conflict, I had recruited two musicians from the choir to perform Simon and Garfunkel's song "The Boxer," which brings the requisite metaphor into view. However, at the wrong time earlier in the service, the sounds of "The Boxer" could be heard in the sanctuary. From the pulpit I turned to the associate pastor and asked her to go to the choir room and tell the two performers to rehearse more quietly. She shrugged and pointed to the front row, where both men were speechlessly sitting and looking more than a little astonished. The music we were hearing was coming through the public address system, which possibly was picking up a radio station—though this had never happened before. The chances of a PA system randomly receiving a broadcast of a specific song on a specific occasion at which that song was to be sung seem impossibly remote. It would therefore be natural to conclude that a prankster had tapped into the speaker system, except that the amplifier and controls were located in plain view behind the pulpit, and the audibility of other intermittent radio noise ruled out

41. Wikipedia, "Double-Aspect Theory."

42. ChemEurope, "Wolfgang Pauli."

any obvious chicanery such as the use of a prerecorded tape. All of this was witnessed by approximately two hundred people.

The relevant issue is, of course, not whether events like this take place but, even when they do, what significance they have. It is one thing to notice that there is much that we cannot explain—something only the most positivistic skeptic would deny. It is entirely another to claim that abnormal and quirky inexplicabilities demonstrate the existence of a personal God or, indeed, of occult powers and intelligences of any kind whatsoever. Jung and Pauli identified a pattern in nature whereby meaningful coincidences occur without apparent physical explanation. In Whitehead's terms, there are mental or conceptual prehensions that somehow contribute to unexpected physical results—such as a public address amplifier playing a certain song. Such events do seem to indicate that conceptual information is used in a teleological way in nature, and not necessarily in ways that are consciously controlled by human beings. It seems likely that a great many things like this may happen without being noticed.

At any rate, in my view, though the Pauli–Jung conjecture is very suggestive and constructive as far as it goes, it has some shortcomings. Like d'Espagnat's conclusions, the Pauli–Jung conjecture remains primarily epistemological without making any extensive advances into metaphysics. Obviously both Pauli and Jung were already deeply committed to their own fields and could hardly be expected to become philosophers. Interpreters of the Pauli–Jung conjecture might also point out that one should not attempt to describe a reality—in this case, the nature of the conjecture's underlying monism—more fully than the available evidence will allow. Nevertheless, I suggest that the conjecture is too vague, positing only that there is *some* reality or substance more fundamental than either matter or mind but integrating the characteristics of both. The conjecture suggests that this fundamental reality exists in a "psychophysically neutral domain," in which neither position in space nor the passage of time is applicable.[43] Other than this, the conjecture does not describe the underlying reality any further or explain how or why the underlying reality integrates physical and mental aspects and then becomes physically effective in space-time. The conjecture does, however, speak of the underlying reality as if it were a stable, enduring existent—in traditional philosophical terms, a substance. This assumed image of a stable

43. Atmanspacher, "Pauli-Jung Conjecture," sec. 2. See also Atmanspacher and Fuchs, *Pauli-Jung Conjecture.*

underlying substrate fails to see the advantages of rejecting substance theories and embracing processive ones. Though the conjecture replaces Cartesian dualism with monism, it retains Descartes's image of reality as composed of some sort of substance-like stuff.

A final problem with the conjecture is that by making "meaning" the criterion for synchronicity, it implicitly elevates a feature of human experience to the status of a metaphysical principle. "Meaning" as Jung and Pauli viewed it is of obvious importance in human life, but it hardly applies to simpler life-forms or to events at the atomic level. I have argued that atomic-level processes tend to find intensity through the contrast produced by vibration and that each process is radically relational, tending to generate wholeness and increasing complexity. Synchronicity is an idea that pertains primarily to human experience. A new description of nature itself requires us to begin at a more elemental level.

In our next two chapters we consider the ideas of four philosophers whose ideas pertain directly to nature as a whole. All four philosophers advance ideas that represent a departure from the main streams of Western philosophy, especially philosophy during the past two centuries, which in the United States has focused largely on linguistic analysis. Therefore, prior to embarking on these chapters, it is well to ponder the following statement by Norwood Russell Hanson about the influence of quantum mechanics on contemporary philosophy:

> The difficulties and errors in . . . allowing one's expectation to dictate the results that one will accept will be obvious to every student of the history of thought.
>
> What alternative is there but to concede that the physical theory which tells us the truth, or a good part of the truth, about micronature can do so only if we accept *its* rules? Such rules as do structure quantum mechanics run clearly counter to any metaphysical preconceptions familiar to philosophers with a nineteenth-century outlook. Therefore it becomes a reasonable metaphysical possibility that nature is fundamentally indeterministic; that elementary particles are, ontologically, always in partially defined states; that they do not in any sense that is scientifically respectable and philosophically intelligible have both a precise position and an exact energy.
>
> The great physicists who spearheaded the spectacular advances of quantum mechanics reached philosophical conclusions concerning their own work far too hastily and without the care they lavished on the strictly scientific components of their

> research. But this should not blind us to the fundamental fact that quantum mechanics has changed the shape of the philosophical world for all of us.[44]

In chapter 8 I proposed that five philosophical conclusions can be gleaned from an analysis of quantum mechanics: (1) that nature is composed not of matter and energy but of processes; (2) that each process culminates in a conclusive state of definiteness before it can influence other processes; (3) that the processes found in nature occur not randomly or incoherently but in derivation from their relevant predecessors; and (4) that the causal influence of events on their successors is not merely physical but informational. Each event is said to consist of a dipolar integration of physical and conceptual elements, though in many physical processes the role of the conceptual pole may be negligible. I have also argued (5) that there is an outcome-oriented or teleological aspect to the ways in which abstract potentialities and physical influences are integrated. At the atomic level the purposive aspect is extraordinarily simple. This outcome-oriented aspect of atomic phenomena can be described in four ways. First, each atomic-level process is inherently future oriented or, in a sense, anticipatory. Second, each atomic-level process tends to seek intensity through the contrast produced by vibration. I have said that wavelike alternations between contrasting boundaries tend inherently to be engaging. Third, each atomic-level process is radically relational, so that the novel relations that are procured tend towards outcomes that display wholeness and increasing complexity. As the number of relations increases, systems become more complex. Fourth, the internal process of each occasion "perishes" (the wave function collapses) when the occasion acts upon its environment.

To some readers, these claims may seem audacious, if not in fact purely conjectural. These readers are asked to remember that all scientific claims and principles are based on assumptions about the nature of reality, whether those assumptions are stated clearly or presupposed tacitly, and that neither relativity nor quantum mechanics could have proceeded very far without the radical alteration of earlier assumptions about nature that had long been held to be self-evident. The five conclusions listed above are derived not from sheer speculation but from an honest appraisal of the facts of quantum mechanics as they are now understood.

44. Hanson, "Quantum Mechanics," 46, 48.

10

Process Philosophies and Nature: Heraclitus, Bergson, and James

We are considering a concept of nature in which the conventional notion of material substance is superseded by a systematic notion of process. I have interpreted special relativity and quantum mechanics in a way intended to support the claim that reality is founded not upon static substances but on relational processes. Fortunately we are not required to defend and expand this claim based only on arguments drawn directly from physics. Though it has typically been a minority perspective, process metaphysics has existed in philosophy for 2,500 years. The process philosophers of history contribute profound insights and convincing explanations related to the outlook on nature that I have proposed. We are not in a position to undertake a full analysis of the work of any philosopher. At this point I simply offer abridged descriptions of the ideas of four of the most important process thinkers: Heraclitus, Henri Bergson, William James, and Alfred North Whitehead. It is Whitehead who contributes most deeply to process metaphysics, and I will give fuller attention to his system of thought.

I. HERACLITUS OF EPHESUS

Heraclitus lived from approximately 535 to 475 BCE. He thus precedes Plato by a century and Aristotle by a century and a half. Our knowledge of Heraclitus's life is minimal. It is believed that he grew up in an aristocratic family but avoided positions of influence and prestige in favor of a life focused on philosophy. His writings are known as "fragments" because only isolated statements have been preserved. There are about a hundred and twenty of these fragments, and though they deal with various subjects, there is no doubt that their central theme is that the world is characterized by flux or process. Heraclitus is the author of the familiar proverb "You cannot step twice into the same river, for other waters are continually flowing on." He also said, "Everything flows and nothing abides," and, somewhat paradoxically, "It is in changing that things find repose."[1]

It may be asked what is so remarkable or distinctive about these statements. It seems obvious enough that the world is always changing. The question is rather whether these changes are superficial, or profound. Does the world of change point to something deeper—to some enduring, basic substances, or to some invariable principles or laws, or to both? Or is change itself the primary thing that endures? For Heraclitus, it is the latter assertion that is accepted. In an attempt to convey the idea that things change not only outwardly but in their very core, Heraclitus adopted the metaphors of fire and lightning:

- "There is exchange of all things for fire and of fire for all things, as there is of wares for gold and of gold for wares."
- "The phases of fire are craving and satiety."
- "This universe . . . always has been, is, and will be an ever-living fire, kindling itself by regular measures and going out by regular measures."
- "The thunderbolt pilots all things."[2]

These remarkable sayings reject underlying essences or substances and radicalize process as the foundation of nature.

It can be argued that this notion of "process all the way down" bears a remarkable conceptual resemblance to the discoveries of twentieth-century science. In Werner Heisenberg's examination of the relation

1. Wheelwright, *Heraclitus*, frag. 21, 20, 23.

2. Wheelwright, *Heraclitus*, frag. 28, 30, 29, 35. In the main, Wheelwright's fragment numbering follows the traditional numbering of Hermann Diels as revised by Walther Kranz in 1934. In this book, all further references to the fragments employ Wheelwright's numbering. The first modern compilation of the fragments of Heraclitus was created in 1817 by theologian Friederich Schleiermacher.

between physics and philosophy, though several ancient Greeks are mentioned, Heisenberg reserves special praise for Heraclitus:

> We may remark at this point that modern physics is in some way extremely near to the doctrines of Heraclitus. If we replace the words "fire" by the word "energy" we can almost repeat his statements word for word from our modern point of view. Energy is in fact the substance from which all elementary particles, all atoms and therefore all things, are made, and energy is that which moves.[3]

> In the philosophy of Democritus all atoms consist of the same substance if the word "substance" is to be applied here at all. The elementary particles in modern physics carry a mass in the same limited sense in which they have other properties. Since mass and energy are, according to the theory of relativity, essentially the same concepts, we may say that all elementary particles consist of energy. This could be interpreted as defining energy as the primary substance of the world. It has indeed the essential property belonging to the term "substance," that it is conserved. Therefore, it has been mentioned before that the views of modern physics are in this respect very close to those of Heraclitus if one interprets his element fire as meaning energy. Energy is in fact that which moves; it may be called the primary cause of all changes, and energy can be transformed into matter or heat or light. The strife between opposites in the philosophy of Heraclitus can be found in the strife between two different forms of energy.[4]

Heraclitus does not rely on didactic logic to defend the radical role of process. Instead, he blurts out this theme in aphorisms that are often described as "dark" or "riddling."[5] He writes, "It throws apart and then brings together again; it advances and retires." "Time is a child moving counters in a game; the [ruling] power is a child's."[6] There seems to be a deliberately enigmatic character to these words. Indeed, Heraclitus scholar Philip Wheelwright characterizes Heraclitus as "one of the

3. Heisenberg, *Physics and Philosophy*, 63.

4. Heisenberg, *Physics and Philosophy*, 70–71.

5. Wheelwright, *Heraclitus*, frag. 28, 23, 30, 31, 24.

6. Wheelwright, Heraclitus, frag. 31, 24. In fragment 24 I have taken the liberty of substituting "ruling" for the more customary translation of *basileia* as "royal," as suggested in William Harris's commentary on fragment 24 (Harris, *Heraclitus*). The Greek *basileuo* means "to rule" or "to control" (see Strong's 936).

subtlest, most impatient, and most paradoxical of philosophers."[7] To some extent the tenor of Heraclitus's writing may result merely from a personal bent towards subtlety and elusiveness. But I think there is another, more philosophical reason for his deliberate obscurity. Heraclitus recognized that the central claim of his philosophy—that the world of physical things is fundamentally not static but dynamic—would be deemed by most of his hearers to be unreasonable, contrary to everyday observation and in conflict with common sense. Obviously the world around us does change. But changes always appear to occur in a setting populated by existing objects—rocks, trees, birds, fish, utensils, water, air. As we have already noted, what we observe is that any changes that do take place always appear to pertain to *preexisting objects*. Objects may be moved from one place to another, and they may even be transmuted into some other type of material by, for example, being dissolved in water, burned in a fire, or exposed to a strong acid. But in all of these cases our very concept of change is anchored to our sense that *there are previously existing stable entities that are the subjects of that change*. Therefore, to say that change is more fundamentally real than the objects it pertains to is counterintuitive. Changes, we would say, have to happen to *some*thing.

And yet, Heraclitus proclaims, "The thunderbolt pilots all things"; "All things are fire"; "Everything flows, and nothing abides." Can Heraclitus actually be claiming that *change itself is what is really real*, even when referring to an object as solid as a boulder? This metaphysical riddle is what lies behind Heraclitus's deliberate obscurity. Heraclitus believed that the true nature of the world is inherently perplexing, for our senses and our minds rely intuitively on the assumption that the objects we observe are composed of matter or stable substance. Nevertheless, Heraclitus says, constant flux is the deeper truth. So he writes, "Nature loves to hide."[8] With this hiddenness of reality in mind, Heraclitus expressed his philosophy in phrases that are deliberately paradoxical. Today, we encounter this same paradoxical character in quantum mechanics' "weird" denial of the existence of the stable physical corpuscles posited by classical physics and by commonsense realism.

Though Heraclitus embraces a worldview based thoroughly on flux, he does not ask us to believe that this flux is simply blind, random, or chaotic. Process may be ceaseless, but it is characterized by order and

7. Wheelwright, *Heraclitus*, 12.

8. Wheelwright, *Heraclitus*, frag. 17.

pattern. Heraclitus therefore speaks of a Logos[9] that permeates and guides nature. "Wisdom is one—to know the intelligence that steers all things through all things."[10] From the transcendent perspective of this Logos, even in variation and contrast there is a kind of unity that emerges. "God is day and night, winter and summer, war and peace, satiety and want, but he undergoes transformations, just as [fire] when combined with incenses, is named according to the particular aroma which it gives off."[11] If there can be such a thing as "the viewpoint of the Logos"—which would be a radically transcendent, all-inclusive perspective—then from that perspective it is the *constant variation* of things-in-process that creates the *ultimate oneness* of the world. "Listening not to me but to the Logos, it is wise to acknowledge that all things are one."[12] Paradoxically, it is this very multiplicity that creates the reality of unity. "The bones connected by joints are at once a unitary whole and not a unitary whole. To be in agreement is to differ; the concordant is the discordant. From out of all the many particulars comes oneness, and out of oneness come all the many particulars."[13] But ordinary human perception is oblivious to this. "People do not understand how that which is at variance with itself agrees with itself. There is a harmony in the bending back, as in the case of the bow and the lyre."[14] "The hidden harmony is better than the obvious."[15]

An ironic and somewhat melancholy aspect characterizes not only Heraclitus's worldview but his life itself—for Heraclitus devoted himself to the articulation of insights and teachings that might serve as a light and a guide for others, and yet the quirky obscurity of his sayings has evidently hindered his philosophy from ever inspiring a noteworthy school of thought. The lack of any band of disciples or advocates, combined with

9. *Logos* is the Greek word for "word." For the Greeks, however, *logos* is a great deal more than simply a printed or spoken group of letters that can be looked up in the dictionary. The Greeks had the insight that our very use of language implies that there is something reasonable and consistent about the way the world works. *Logos* thus implies a rational principle of creation. This notion is found also in the first chapter of the Gospel of John, though John's understanding of the Logos is not identical to the concept of the ancient Greeks.

10. Wheelwright, *Heraclitus*, frag. 120.

11. Wheelwright, *Heraclitus*, frag. 121.

12. Wheelwright, *Heraclitus*, frag. 118.

13. Wheelwright, *Heraclitus*, frag. 112.

14. Wheelwright, *Heraclitus*, frag. 117.

15. Wheelwright, *Heraclitus*, frag. 116.

the relative difficulty in antiquity of creating and preserving written texts, has left us with his frustratingly meager corpus of loosely connected adages whose very label, "fragments," points to their jumbled, incomplete condition. This state of affairs helps to explain why it is that Heraclitus in our own time has been relegated to the periphery of Western philosophical studies, overshadowed among the classical Greeks by Socrates, Plato, and Aristotle and even by less well-known thinkers such as Pythagoras, Parmenides, Zeno, Diogenes, and Democritus.

Heraclitus is overdue for a fresh hearing. One could argue, as Heisenberg did, that major discoveries of the twentieth century in relativity physics, quantum mechanics, chaos theory, and field theory, and a growing awareness of the dynamic, relational, holistic quality of nature itself are grounds for a new and deeper appreciation of "the Dark Philosopher." In fact, the position taken in this book agrees with Heraclitus especially at three points. The first is that *it is more useful and more accurate to understand reality not in terms of unfeeling matter or static substances but in terms of relational processes.* The second is to *affirm the existence of a "Logos" or coordinating principle* that somehow channels information to events in such a way that more highly organized entities tend to emerge—in other words, that nature shares information in ways that foster the creation of larger and more complex wholes. The third is that *the functioning of what Heraclitus called "the Logos" can serve as a legitimate reference point for religious feeling and spiritual insight.* I suggest that it is still wise to believe, as Heraclitus did, that *whatever it is in nature that promotes creativity, wholeness, growth, and renewal is the ultimate basis for value, meaning, and hope, and is therefore worthy of being interpreted in religious terms.* In affirming this religious value, however, it is imperative to distinguish between Heraclitus's views and the conventional beliefs of monotheism. In the Western world, belief in a personal God is commonly, though not accurately, ascribed to *all* religion. Heraclitus's Logos is not much like the God of the Abrahamic faiths. According to Wheelwright,

> we must avoid the familiar Christian associations that are likely, even in disguised form, to cling to the word [God]. There is no supposition in Heraclitus' though of a single universal God who is at once personal, omniscient, and (despite all appearances to the contrary) deeply concerned about the ultimate destiny of mankind in general and, for better or worse, of each particular human individual. There is, to be sure, in Heraclitus' view, a

> unity and coherence that is somehow present as a hidden other aspect of the plurality and diversity of things that is seen everywhere, and he even goes so far as to describe that cosmic oneness as "wisdom" (Fr. 119) and "intelligence" (Fr. 120). Such words, however, express a metaphor intended to suggest the mysterious organizing power that pervades the universe as a whole, manifesting unlimited possibilities of comeback after defeat and of new creation after destruction and death. If we interpret the metaphor literally, we shall overstress the human analogy implied in the ideas of wisdom and intelligence, and thus shall fall into anthropomorphism.[16]

In Heraclitus we find an affirmation of three principles—namely, (a) that process, and not substance, is the basis of reality; (2) that in that process, information is utilized by the activity of Logos or reason; and (3) that the relation between nature and the Logos constitutes the basis for a religious worldview. These themes are considered more fully later in this chapter, and in subsequent chapters.

II. HENRI BERGSON

French philosopher Henri Bergson was born in Paris in 1859 and died there in 1941. His ancestors were mostly Jewish, and he received Jewish religious training as a child. As a student he displayed ability in science. When he was eighteen he was awarded a prize for solving a problem in mathematics; his solution was published in the journal *Nouvelles Annales de Mathématiques* in 1877. As he advanced into higher education he was forced to decide between the sciences and the humanities and, to the surprise of his teachers, he chose the latter. During this period he studied ancient philosophy, publishing a work on the materialist cosmology set forth by the Roman philosopher Lucretius in *De Rerum Natura*. He also completed an analysis of Aristotle's concept of place, which he appended to his dissertation. The dissertation itself was titled *Time and Free Will* and was published in 1889. According to Louis de Broglie, *Time and Free Will* "antedates by forty years the ideas of Niels Bohr and Werner Heisenberg on the physical interpretation of wave mechanics."[17] Of course, by saying this de Broglie cannot have meant that in 1889 Bergson anticipated the actual principles of quantum mechanics but only that the

16. Wheelwright, *Heraclitus*, 71–72.

17. De Broglie, "Concept of Contemporary Physics," 47.

philosophy found in *Time and Free Will* can readily be correlated to the ideas that were later advanced by Bohr and Heisenberg.

By the time Bergson completed his doctorate he had already developed a sympathetic interest in Darwin's theory of evolution and had read works by evolutionary philosopher Herbert Spencer (1820–1903). During this period "he also gave courses at the university in Clermont-Ferrand on the Pre-Socratics, and in particular Heraclitus,"[18] presumably because he recognized a significant similarity between Heraclitus's process philosophy and the processive, developmental thesis of evolution. Bergson went on to a distinguished career as a professor at the *Collège de France*. He lectured by invitation both in Europe and in the United States, formed a friendship with William James, received the *Grand-Croix de la Legion d'honneur*, and was awarded the Nobel Prize in Literature in 1927.

Bergson's philosophy is characterized by three central ideas: *intuition*, *duration*, and *creativity*. By *intuition* Bergson did not mean simply a passive state in which ideas happen to drift into the mind. Intuition for Bergson is a direct, unmediated awareness of our inner experience. Intuition remains thoughtful even while being receptive. For Bergson, the opposite of intuition is abstract intellectual analysis, which divides our perceptions into distinct, independently definable subunits.[19] Bergson notes that such intellectual, conceptual analysis has historically been the obligatory method of Western philosophy. The adoption of intuition as a contrasting philosophical methodology is a deliberate attempt to avoid such objectifying intellectualism. As a parallel, another enemy of metaphysical insight is language. It is the obvious goal of language to use words, and especially nouns and verbs, that have clear, fixed definitions. However, such well-defined, static terms conceal the inherently dynamic quality of the world they attempt to describe.

> Therein lies the ordinary function of ready-made concepts, those stations with which we mark out the path of becoming. But to seek to penetrate with them into the inmost nature of things, is to apply to the mobility of the real a method created in order to give stationary points of observation on it. It is to forget that, if metaphysics is possible, it can only be a laborious, and even painful, effort to remount the natural slope of the work of thought, in order to place oneself directly, by a kind of intellectual expansion, within the thing studied: in short, a

18. Wikipedia, "Henri Bergson."
19. Gunter, "Henri Bergson."

> passage from reality to concepts and no longer from concepts to reality. Is it astonishing that, like children trying to catch smoke by closing their hands, philosophers so often see the object they would grasp fly before them? It is in this way that many of the quarrels between the schools are perpetuated, each of them reproaching the others with having allowed the real to slip away.[20]

The more significant type of experience discovered by intuition is designated by Bergson as *duration*. In the same way that Bergson uses *intuition* to counter the reductionistic, objectifying consequences of traditional metaphysical reasoning, he uses the term *duration* as a way of countering the traditional notion of a durationless point-instant[21] and the accompanying assumption that time and experience are exhaustively divisible by using mathematical methods. Duration resists mathematical, clocklike description and exhibits fundamental spurts of creativity.[22] Bergsonian duration is not only an alternative to the mathematical division of time but a correction to the intellectual use of static categories in metaphysics. Bergson held that duration is *experiential*, encountered in our consciousness as a flowing state of awareness that is rhythmic and unsegmented. Duration is thus suggestive of the claim in Gestalt psychology that experience occurs in chunks that cannot be broken down for analysis without damaging the inherently holistic quality of the content. Bergsonian duration is also similar to what his predecessor, William James, called "the specious present," which James defined as "the short duration of which we are immediately and incessantly sensible."[23] James reasoned that when we see something move, our experience does not consist of a series of instants that must somehow be assembled; rather, we see the motion in a single, continuous packet of time. Therefore we do not say, "I saw it here, and then there, and then there," and so on; we say simply, "I saw it move." Bergson was well aware that theories of time do not ordinarily explain time in terms of human experience, but he continued to insist that our direct experience of duration reveals a

20. Bergson, *Introduction to Metaphysics*, 45.

21. A *point-instant* is "the smallest unit of space-time; a particular point along the continua of space and time" (Wiktionary, s.v. "point-instant," https://en.wiktionary.org/wiki/point-instant). This notion of a limit approaching zero is especially prominent in the use of calculus to describe changing motion and related physical phenomena. This concept is essential when calculus is used to describe rates of change in the physical sciences.

22. Gunter, "Henri Bergson."

23. See Wikipedia, "Specious Present." See James, *Principles of Psychology*, 609.

crucial metaphysical truth. His goal was the development of "a philosophy which sees in duration the very stuff of reality."[24]

The third element of Bergson's philosophy is creativity. Having been educated during a time when Darwinism was a topic of worldwide debate, Bergson found it important to discover a connection between biological evolution and metaphysical philosophy. The process-oriented doctrine of a universal creativity that is present throughout nature is probably Bergson's most well-known concept. This creativity, which he called the *élan vital* or "vital force," is immanent in all organisms. In *Creative Evolution*, Bergson describes spontaneous morphogenesis (the generation of form or self-organization) in living organisms. As we have seen, Darwin's theory regards biological adaptation as an accidental process in which less successful variants are weeded out by the comparative failure of species to reproduce. Bergson argues that this idea of purely random variations fails to explain biological adaptation. An organ such as the eye of a vertebrate is a functional whole made up of interacting parts. It is reasonable to think that if a few or even only one of those parts randomly varied independently of the others, the functioning of the entire organ would, in most cases, be impaired or disabled. But, since evolution actually does occur, it is necessary to accept the fact that in successive mutations all the parts of the organs in a surviving species must have varied *in concert*; otherwise, over time, the organs and their interactions would become irreparably dysfunctional. For Bergson, it is utterly implausible to claim that the parts-whole synchronization that we do observe is the end result of nothing more than random rearrangements.[25]

Moreover, Bergson said, Darwinism does not explain why living things have evolved in the direction of greater and greater complexity. Superficially it appears that organs and entire species are simply reproduced as copies from one generation to the next, with any variations being so minor that physical, genetic copying generally seems adequate to explain the result. About this, Bergson asks,

> But can an organic structure be likened to an imprint [i.e., a copy]? We have already called attention to the ambiguity of the term "adaptation." The gradual complication of a form which is being better and better adapted to the mold of outward circumstances is one thing, the increasingly complex structure of an instrument which derives more and more advantage from

24. Bergson, *Creative Evolution*, 296.

25. This argument is stated in Goudge, "Bergson, Henri," 292.

> the circumstances is another. In the former case, the matter merely receives an imprint; in the second, it reacts positively, it solves a problem. Obviously it is this second sense of the word "adapt" that is used when one says that the eye has become better and better adapted to the influence of light. But one passes more or less unconsciously from this sense to the other, and a purely mechanistic biology will strive to make the *passive* adaptation of an inert matter, which submits to the influence of its environment, mean the same as the *active* adaptation of an organism which derives from this influence an advantage it can appropriate.[26]

Bergson holds that the problems pertaining to a mechanistic interpretation of biology can be solved only by recognizing the functionality of a creative agency other than random variation. This creative impulse, the élan vital, must be in effect in order to produce the holistic adaptations and the complexification that we do observe in evolving mutations. It may seem that Bergson's élan vital is nothing more than a renewed version of vitalism, the eighteenth- and nineteenth-century doctrine that life depends on a peculiar vital principle that is distinct from physical and chemical factors. Bergson, however, viewed his own outlook not as a new version of vitalism but as an alternative to it, inasmuch as the élan vital operates not only in living organisms but throughout all of nature, and not in contrast to physical factors but in integration with them.

Bergson's ideas deserve much more attention than we can give them here. For present purposes it is sufficient to say that Bergson's philosophy generally concurs with and expands upon the main ideas that I am proposing: that nature is not fundamentally mechanistic but dynamic and relational; that nature embodies a creative principle in which holism and complexity evolve spontaneously; and that a primordial tendency towards purposiveness is naturally associated with organismic wholeness. Certainly in any analysis of types of philosophy, Bergson would be included among the process philosophers.

There are differences, however, between Bergson's approach and the approach developed in this book. Though, in the end, my conclusions are very much like Bergson's, his work is grounded in biology; the method of this book begins with twentieth-century physics. In a deliberate effort to avoid objective, intellectual modes of analysis, Bergson relies on "intuition." I cite the laboratory discoveries, mathematical aspects, and

26. Bergson, *Creative Evolution*, 78–79; emphasis original.

conceptual principles of special relativity and quantum mechanics. In the passage just quoted, Bergson illustrates the difference between "passive" and "active" adaptation by referring to the goal-achieving tendencies of biological evolution; I base the principle of active adaptation on the role of mathematical information in quantum processes so as to introduce nonmechanistic potentialities in the context of quantum uncertainty and superposition, thereby resulting in novel outcomes. Because of these differences between Bergson's approach and mine, though I remain appreciative of his ideas, I do not appeal extensively to his philosophy in the remainder of this book.

III. WILLIAM JAMES

William James was born in New York City in 1842 and died in New Hampshire in 1910. Though he traveled extensively in Europe, knew several languages, and maintained contact with numerous scholars overseas, he was thoroughly a New Englander and spent most of his career teaching at Harvard. Today he is known as one of the founders of modern psychology, having taught the first college course called "psychology" ever offered in the United States. He was and still is viewed as a prominent philosopher, one of the creators and the foremost proponent of the peculiarly American school of thought known as pragmatism. Also, though he espoused no specific religious tradition, James devoted intensive study to religious experiences and beliefs. Though James was naturally active, inquisitive, and sociable, he also suffered debilitating episodes of what today would be called clinical depression.

At his core James was an avid lover of experience itself. Robert D. Richardson opens his biography of James by recounting James's experience of the great San Francisco earthquake. In 1906 James happened to be serving as a visiting professor at Stanford, and thus he and his wife, Alice, were living in Palo Alto when the earthquake struck the Bay Area on the morning of April 18. At 5:12 a.m. a foreshock was felt throughout the area. After about twenty seconds the earthquake itself began. Massive upheavals punctuated the continuous, violent shaking, which lasted for forty-five to sixty seconds. The quake was felt from southern Oregon all the way to Los Angeles and inland as far as mid-Nevada. The US Geological Survey has determined that the shifting fault line reached all the way from central Oregon to a location near Monterrey Bay, a distance of 296 miles.[27]

As the quake occurred people reacted by running outside, whether to observe the cause of the cataclysm or simply to escape from buildings that were being shaken apart. Not surprisingly, most people were panic-stricken, while others were reduced to a state of silent shock. But not James. "James's first unthinking response to the quake was, he tells us, one of 'glee,' 'admiration,' 'delight,' and 'welcome.' He felt, he said, no sense of fear whatever. 'Go it,' I almost cried aloud, 'and go it stronger,' James recalled." He ran to his wife's room. Alice was not injured and also was feeling no fear. Clearly, as Richardson observes, "James possessed what has been called a 'great experiencing nature'; he was astonishingly, even alarmingly, open to new experiences."[28]

James's inborn fascination with "lived experience" was reinforced by the empirical orientation of his studies. His undergraduate work, begun in 1861, was in chemistry, which in James's time was even more laboratory-dependent than it is today. In 1865 James joined an expedition to Brazil led by of one of his teachers, biologist Louis Agassiz. Coping with conditions that today would be deemed unacceptably dangerous, the group traveled up the Amazon collecting specimens. Agassiz, an opponent of Darwin's theory, was legendary for requiring his students to gather copious samples and to study those samples in meticulous detail. These scholarly habits were imparted to James. James's graduate degree was, interestingly, an MD, granted at Harvard in 1869. This was at a time when medical studies focused primarily on anatomy and physiology, since fields such as microbiology, epidemiology, pharmacology, and genetics were still in their infancy. The fact that, professionally, James moved gradually from chemistry and medicine to psychology and finally to philosophy is a testimony not only to his intellectual scope, persistence, and originality but to his tireless desire to apply academic inquiry to the practical issues of life.

There are many dimensions and nuances to William James's philosophical ideas. At present we can consider these ideas only briefly. I will focus on the two concepts that are most relevant to the purposes of this book. These are *the pragmatic theory of truth* and *radical empiricism*. Historically, pragmatism appeared before radical empiricism, but

27. USGS, "San Francisco Earthquake." See also USGS, "1906 Rupture."

28. Richardson, *William James*, 3.

conceptually the two are so mutually implicated that they can be presented in either order.

Stated simply—or simplistically—the pragmatic theory of truth asserts that truth is based on usefulness—that is, according to a belief's effectiveness in clarifying our circumstances and guiding our actions. James said, "Ideas become true just so far as they help us to get into satisfactory relations with other parts of our experience."[29] Broadly, then, in the pragmatic theory, truth is associated with the practical consequences of specific assertions or convictions. It may appear at first that this definition of truth amounts to little more than an absolutization of one's personal circumstances and wishes. For example, since it leads to being in "satisfactory relations with other parts of my experience," I may pragmatically choose to believe that my present situation in life—employment, personal relationships, health, and so on—is quite satisfactory as it is at present. However, if, in accord with the pragmatic theory of truth, I do adopt this belief, it would appear that this belief blinds me to potential improvements such as finding more meaningful work, pursuing deeper intimacy in my relationships, and scheduling regular medical checkups. The pragmatic theory of truth thus serves to encourage wishful thinking and circumvent the commonsense notion of reality testing. James's answer to this objection is that his notion of "other parts of our experience" entailed more than just a superficial review of our immediate circumstances. It entailed a constant curiosity regarding the larger world, including all of its potential risks and rewards. In other words, James's theory of truth required a broad attempt to verify testable conditions.

This need to rely on the verification of testable conditions in order to see whether the beliefs in our minds "correspond" with facts in the real world is, actually, the basis of the theory of truth commonly encountered among most people—the *correspondence theory*. The correspondence theory thus presupposes a constant bridging of the apparent gap between the inner mind and the outer world. Thus the theory relies on an intuitive trust in the validity of reality testing. In philosophy the other classic theory of truth is *the coherence theory*. The coherence theory in its historical form does not merely emphasize the interrelatedness (the "coherence") of all true propositions; it affirms that truth (or perhaps Truth) is, by its very nature, one monistic reality, an expression of something like an Absolute Mind or an innate rationality that lies behind the unity of the

29. Flew, *Dictionary of Philosophy*, s.v. "Pragmatism."

natural world.[30] In our attempt to recognize the significance of James's pragmatic theory of truth, it is necessary to understand the shortcomings that pertain to the correspondence theory and to the coherence theory.

I have said that the correspondence theory is based on what we call "reality testing." The correspondence theory takes it for granted that there is a clear division between our mental perceptions and ideas, on the one hand, and the actual, physical world, on the other. This differentiation between inner experience and the outer world, or between subjective experience and objective fact, is not merely an invention created by academic theorizing; we are forced to deal with this distinction from the time of our childhood. Youngsters display a manifest tendency to confuse imagination with physical fact. Whether it is about the tooth fairy who exchanges money for extracted teeth or the monster who hides in the bedroom closet, children easily confuse fantasy with reality, and parents understand that one of the basic duties of parenting is to teach children how to distinguish between the two. This act of distinguishing between fact and imagination is at the heart of the correspondence theory of truth. The correspondence theory is essential not only for children but for adults. For example, if, in a case of mistaken identity, the police arrest you for robbing a bank yesterday, it is a simple matter to explain that during the robbery you were attending an activity at church where you were observed by several dozen witnesses. Such observations of the objective world *correspond* with your claim and corroborate your innocence. Our whole lives are filled with experiences in which we "check and see" whether something is true.

But the correspondence theory is not as obviously applicable as it may seem. The central difficulty with the theory is that we have many experiences that are not like your observable attendance at a public event. The correspondence theory is especially problematical when it attempts to deal with broader questions related to matters such as personal commitments and social values, moral and ethical decisions, esthetic assessments and preferences, issues of meaning and interpretation, spiritual experiences and religious convictions, and truths that are symbolic or figurative rather than literal.[31] *Romeo and Juliet*, *Huckleberry*

30. Flew, *Dictionary of Philosophy*, s.vv. "Coherence Theory of Truth," "Idealism." See also White, "Coherence Theory of Truth," 130–33.

31. I have already noted, in chapter 5, that even "literal" language is actually symbolic. Even statements about concrete objects rely on the fact that common nouns are abstractions that may be applied to similar objects in varying circumstances.

Finn, *The Grapes of Wrath*, and classical myths and fables all convey a type of truth that cannot be verified or interpreted simply by establishing a correspondence with the observation of specific facts. The heart of the problem with the correspondence theory lies in its assumption that we are able to perceive an external world that is simply objective and real. In the example of mistaken identity, reality is easy to establish, but there are many situations—some of them vitally important—when the nature of external reality is the very thing that is being questioned. James wrote, "Objective evidence and certitude are doubtless very fine ideals to play with, but where on this moonlit and dream-visited planet are they found?"[32] James realized that our perception of the objective world is more subjective than we commonly assume. In fact, as George Berkeley pointed out, mental content is actually the *only* thing that we can consciously experience, and this mental characteristic includes our sensory impressions of physical objects. We know that, actually, physical objects are mostly empty space, occupied only scantily by subatomic particles. Furthermore, even our knowledge of those particles can lay claim only to a simplified account of their reality, for we can detect only certain of those particles' properties, and even those properties are not observed directly but are inferred from various laboratory procedures and calculations. Besides, even laboratory observations and calculations are forms of subjective experience. Because of special relativity we now know, as James could not have known, that even measurements that were once thought to be absolutely objective vary depending on the frame of reference of the observer. Moreover, quantum mechanics posits the existence of physically significant processes that are *innately* unobservable, such as superposition, entanglement, and quantum tunneling. The correspondence theory applies to phenomena such as these only indirectly, and perhaps not at all.

If James recognized the weaknesses of the correspondence theory, he had even less patience for the coherence theory, with its idea that truth is a monolithic network of mutually implicated propositions. James was possessed by a strong aversion to the idea of a "block universe," as he called it, in which everything fits together in a tidy system of concepts. Though it is desirable that varying truth claims should cohere with one another, James found no evidence that any systematic coherence could be affirmed as a universal principle. James observed that experiences

32. James, "Will to Believe," 45.

actually happen in terms of specifics, not in terms of generalities or universals. The pragmatic theory of truth recognizes that all truth claims are made within specific contexts by specific individuals or groups who are motivated by specific needs and goals and under the sway of specific assumptions. Thus, because the objectivity of an external world that can be observed impartially is a failed criterion, and because all truth claims are contextual, the pragmatic theory of truth shifts its emphasis from correspondence to the applicable usefulness of propositions in specific situations. Reality testing is retained within the pragmatic theory not because it leads to invincible certitude but simply because of its demonstrable pragmatic value.[33]

We must now recognize that the centrality of *contexts* in the pragmatic theory of truth leads to the centrality of *lived experience* in philosophy. The meaning of "context" points to the presence of actual persons in actual situations having actual needs and doing actual deeds. This connection between the pragmatic theory of truth and actual experience leads directly to James's radical empiricism. In James's philosophy, the notion of experience is broadened dramatically. Where the correspondence theory reduces imagination and the life of the mind to a secondary or derivative status, the pragmatic theory of truth recognizes the importance of mental life itself—its practical impact—as an essential form of experience. James famously wrote,

> To be radical, an empiricism must neither admit into its constructions any element that is not directly experienced, nor exclude from them any element that is directly experienced. For such a philosophy, *the relations that connect experiences must themselves be experienced relations, and any kind of relation experienced must be accounted as "real" as anything else in the system.*[34]

James held that our emotions, desires, dreams, yearnings, aversions, and personal convictions *are* real and important because they are actual experiences and have actual consequences. This does not mean that your dream about a unicorn requires that unicorns must really exist somewhere. Naturally James acknowledges that imagination and physical

33. For a more thorough analysis of the nature of truth than can be included here, see Wheeler, *Religion*, ch. 6 (189–275). For an incisive critique of the correspondence theory, see Crosby, *Specter of the Absurd*, 174–85. My perspective on truth is very similar to Wheeler's and Crosby's.

34. James, "World of Pure Experience," 42; emphasis original.

existence are not the same thing. But the fact that dreams are imaginary is not a defect; in important ways, the imaginary aspect is actually an advantage. In psychotherapy, for example, the interpretation of dreams allows the therapist and the client to glimpse the needs and tendencies of the psyche when it is unfettered by the physical and social restraints of our waking experience. When insightfully interpreted, such dreams can be life-changing. In principle, then, an actual truth can often be discovered even when the path to that truth begins in imagination. Beyond the narrow setting of dream interpretation, the functioning of imagination is essential. Imagination is what allows us to ponder possibilities as we plan for the future, to assess options as we reach a decision, to discover our own feelings and priorities when we make a commitment, or simply to engage in creative meditation. For James, such imaginative yearning, ruminating, and deciding are at the very heart of human existence.

James's intuition that experience is fundamentally real leads unavoidably to the question of whether experience can be attributed not only to human beings but to nature as a whole. James was reluctant to embrace panpsychism—with good reason, in my view—but something like panpsychism seems to lurk in his virtually metaphysical endorsement of experience. According to James's student and most well-known biographer, Ralph Barton Perry,

> He [James] had always been attracted to panpsychism, and had for brief periods succumbed to its charms. It provided an interpretation of physical nature that was realistic without being materialistic. It was also a convenient auxiliary to radical empiricism: imputing to every entity an inner experience of its own, and thus providing in experiential terms for regions of value falling outside the experience of man. But on the other hand, panpsychism was a variety of substantialism,—it gave things a core,—whereas James had been reared on the empiricist doctrine that a thing is the sum of its appearances. James was disinclined to accept any unbridgeable chasms; and panpsychism, despite its name, introduced a profound dualism—that, namely, between the psychical inwardness of things and their physical outwardness. . . . It was a seductive friend, but not a vigorous ally.[35]

35. Perry, *Philosophy and Psychology*, 394–95. Regarding James's hesitancy about panpsychism, see also Seigfried, *James's Radical Reconstruction*, 238.

In chapters 11 and 12 I discuss the problem created by what Perry calls the "profound dualism" introduced by panpsychism. At present we can only note that James himself never converted his radical empiricism into a description of reality—that is, into an explicit system of metaphysics.[36] Though, as we have seen, he stated forthrightly that an empirical philosophy ought to focus entirely on experience, he typically avoided any plainspoken assertion that experience is what truly exists and that it is *all* that exists. James dodged the intellectual tension that this avoidance implies by calling pragmatism "merely a method."[37] Nevertheless James's philosophy still stands as a timeless summons to philosophy to rise above its frequent preoccupations with technicalities and the analysis of language to an honest encounter with the vexing enigmas, existential hurdles, and cumulative meaning—if there is any—of life's experience in its wholeness.

One of my daughters lived for several years in another part of the United States. A few summers ago she sent an email including some photos of a sunset she was enjoying from the balcony of her second-floor apartment. She remarked on the beauty of the sky, and then wrote, "There's a special mystery in these moments. I'm convinced that. . . . I'm trailing off now, I apologize. But Dad, is it beautiful because we perceive it to be beautiful, or is that a human projection of beauty? Is it beautiful on its own? What beauty does anything have, if beauty is just a human perspective?"

In response to this I wrote a rather long-winded reply in which I mentioned Aristotle, Newton, Comte, Dewey, and several others. I also included the following account of an experience of my own:

> When I was in high school my friends and I would occasionally sit out on the curb by our house on summer nights and talk about lighthearted subjects such as the existence of God and the meaning of life. I remember arguing that the existence of beauty in nature points to the work of a benevolent Creator. One of my friends quickly replied, "But that's just all in your mind. The flowers and birds aren't beautiful in themselves; we just think of them that way." This argument, which makes the world seem so empty and so dead, discouraged me for a while. But studying in

36. Evidently James was poised to write about his "metaphysic, the systematic general treatise on philosophy" (James, *Some Problems of Philosophy*, 200), but died before he could do it. For a moving account of this, see Seigfried, *James's Radical Reconstruction*, 394–96.

37. Seigfried, *James's Radical Reconstruction*, 282.

> seminary made me realize that whether beauty is in the world ("nature") or in our own minds, there is still beauty in nature, because our minds themselves are products of natural evolution. It's obvious if you think about it that the human brain is just as much a result of natural processes as is a blooming rose—but most people don't connect those dots.
>
> The bottom line: If experience is real, then certainly beauty is real, whether it exists in the flower, in the beholder of the flower, or in both.[38]

Clearly, my daughter's question does not pertain only to beauty and the experience of the esthetic. It pertains to the inescapable subjectivity of every single experience—whether fulfillment or frustration, pleasure or agony, intense purpose or aimless uncertainty, generous service or selfish gratification, high aspiration or dull repetition, creativity and growth or diminishment and failure. From the exuberant joy of children at play to the unspeakable horror of the concentration camps, life is composed of experiences. Are these experiences real? Or does physics require us to demote them to a second-class, or even illusory, status? I ask, are not our *experiences* even *more* real than the colliding corpuscles that have for so long taken center stage in our discussions about reality and nature?

My daughter has never studied philosophy. Her note illustrates that ordinary people face this question about the significance of their own experience in a world that, we are told, is actually constructed of purposeless particles and unfeeling forces. James saw that, with its division between the inner mind and outer objectivity, the correspondence theory of truth reinforces this pernicious cleavage between the outer world of facts and the inner world of meaning. The pragmatic theory of truth and the doctrine of radical empiricism were his efforts to overcome that cleavage. Those efforts arrived at a point where James was faced with a choice between traditional materialism and panpsychism. He saw the advantages of the latter, but died too soon to gain a perspective on the nature of physical reality that would have enabled him to formulate a more plausible, more nuanced revision of panpsychism. In other words, he lacked special relativity and quantum mechanics. That lack did not pertain to Alfred North Whitehead. It is to his philosophy that we now turn.

38. Email correspondence between Cathy Conner and the author, June 18, 2021.

11

Nature and Process Philosophies: Whitehead

There are physicists who have wanted to deny the relevance of quantum mechanics to any particular description of reality. The concepts involved in quantum theory have seemed so inexplicable and self-contradictory that some authorities have deemed it judicious simply to dismiss any attempt at interpretation. In a 2023 article in *Scientific American*, Australian scientist Chris Ferrie argues that the philosophical interpretation of quantum physics leads only to confusion.

> Quantum physics says nothing about how the world *is*. Instead, quantum physics only describes the experiments we do to test our theories of how the world works—it gives us probabilities for the outcomes that may happen in an experiment. The compulsion to interpret quantum physics concepts as prescriptions for physical reality derives from the unfortunate way we traditionally teach physics.[1]

This attitude is nothing new. In fact it was expressed by Niels Bohr himself, who, as we have noted, declared, "There is no quantum world. There is only an abstract quantum mechanical description. It is wrong to think that the task of physics is to find out how Nature is."[2] Bohr made this statement during the early years of quantum theorizing, and he may

1. Ferrie, "Quantum Entanglement."

2. Petersen, "Philosophy of Niels Bohr," 12. I previously quoted this statement in chapter 8.

perhaps be excused for wanting to avoid premature speculation about the meaning of the discoveries that were being made. But in our own day we must admit that Ferrie's recommendation seems unnecessary and, in fact, extreme. Human beings have a vital interest in distinguishing the imaginary from the real, and any experiment that produces consistent results may prompt legitimate questions about the nature of reality. It is important to know that Thomson's plum pudding model of the atom is wrong and that the discoveries of Rutherford and Bohr led to fundamental improvements. In my opinion, the claim that quantum physics tells us nothing about "how the world is" is both anti-intellectual and obstructive—anti-intellectual because it rejects, in advance, *any* application of philosophy to quantum physics, no matter how interesting or useful that philosophy might have turned out to be; and obstructive because it obstructs the possibility that the picture of nature revealed by quantum theory can make a constructive contribution to the evolving worldviews not only of intellectuals but of ordinary people. Despite Ferrie's demurrals, *all* persons do make functioning assumptions about the nature of reality, whether or not those ideas have been carefully examined. Most people today tend to accept a vague form of naïve realism in which the world is composed of physical forces and materials, nature is "law abiding," and the appearance of objects shows them as they really are. Is such a worldview helpful? Or healthy? Or—dare I utter the word—*accurate*? If Ferrie's advice is followed, this adapted naïve realism, deprived of quantum theory's insights about uncertainty, entanglement, and so on, would be the worldview we are stuck with. If adopted consistently, the attitude that scientific theories tell us "nothing about how the world is" would reduce the sum of human knowledge to a very large set of circumstantially determined rules of thumb.

In this chapter I discuss the philosophy of the British mathematician Alfred North Whitehead, who developed a philosophy of nature based on the special theory of relativity and quantum mechanics, which in Whitehead's time were recent discoveries.[3] I conclude that Whitehead's philosophy provides not only an improved interpretation of

3. Two studies that deal at some length with the relation between quantum mechanics and Whitehead's philosophy are Epperson, *Quantum Mechanics* and Hättich, *Quantum Processes*. See also Eastman and Keeton, *Physics and Whitehead*. In this third book, I think the chapters by Shimon Malin and Henry Stapp are of interest. However, all three of these books are more appropriate for readers who are already familiar with quantum science.

twentieth-century physics but a basis for a more integrative, holistic view of nature that, I argue, is very direly needed in today's world.

Whitehead was born in 1861 in Ramsgate, Kent, a seaside town on the English Channel about eighty miles east of London. His father was an Anglican minister. When he was fourteen Whitehead was sent to Sherborne, a public school in southwest England. One of Whitehead's intellectual gifts must have been a facility in foreign languages. He reported drolly, "We read Latin and Greek. . . . They were not foreign languages; they were just Latin and Greek. Nothing of importance in the way of ideas could be presented in any other way. Thus we read the New Testament in Greek."[4] However, Whitehead's primary interest was not languages but math. In 1880 Whitehead entered Trinity College at Cambridge with a scholarship in mathematics. Four years later he graduated with a dissertation on James Clerk Maxwell's *A Treatise on Electricity and Magnetism*, a pioneering two-volume work that had been published only eleven years earlier. In that same year Whitehead was elected as a fellow at Cambridge.

In 1890 Whitehead married Evelyn Wade. The Whiteheads became ardent supporters of women's right to vote; in 1907 Whitehead became chairman of the Cambridge branch of the national Men's League for Women's Suffrage.[5] Price indicates that the Whiteheads were "thick with liberal politics," sometimes to the point of public disapproval. With typical irenic spirit, Whitehead said, "It was exciting work. . . . Rotten eggs and oranges were effective party weapons and I have often been covered with them. But they were indications of vigour, rather than of bad feeling."[6] Whitehead taught mathematics at Cambridge until 1910. He had been one of Bertrand Russell's teachers, and he and Russell coauthored the groundbreaking three-volume *Principia Mathematica*, pertaining to theoretical mathematics and symbolic logic. In 1910 the Whiteheads moved to London, where he lectured at University College and was appointed in 1914 as professor of applied mathematics at the Imperial College of Science and Technology. In London Whitehead focused on ideas regarding the philosophy of education, the philosophy of science, and physics. In 1920 Whitehead published *Concept of Nature*, based on a series of lectures critiquing common assumptions about motion, matter and its attributes, and space and time or "space-time." Einstein's theories are cited.

4. From a conversation reported in Price, *Dialogues*, 5.

5. Lowe, *Alfred North Whitehead*, 1:314.

6. Price, *Dialogues*, 8.

Next, in 1924, at an age when many people contemplate retirement, Whitehead was invited to Harvard to become a professor of philosophy.[7] Newly arrived in America, in the fall of 1924 Whitehead was invited to deliver the Lowell Lectures, a series of public talks given in Boston on various popular and technical subjects and funded by a bequest of John Lowell, who had died in 1836. Whitehead later said that he wrote the lectures "at the rate of one a week as wanted for delivery."[8] This feat is remarkable and shows that Whitehead had already been thinking intently for some time about the ideas involved. Whitehead's Lowell Lectures became the basis for his first major philosophical work, *Science and the Modern World*, published in 1925.

Science and the Modern World is a very important book in its own right, but nowadays it tends to be overshadowed by Whitehead's other major works: *Adventures of Ideas*, *Modes of Thought*, and, most of all, *Process and Reality*,[9] which is the longest and is usually regarded as the definitive expression of Whitehead's philosophy. Whenever *Process and Reality* seems to diverge from *Science and the Modern World*, scholars tend to assume that *Process and Reality* overrules *Science and the Modern World* because it was written later. Moreover, both by training and by personal inclination, most scholars tend to prefer precise definitions and systematic coherence—and that is what *Process and Reality* offers, more than any of Whitehead's other books. But, in my view, Whitehead actually wrote *Process and Reality* not as a definitive conclusion or a final product but more nearly as a pragmatic attempt to formulate a single, coherent philosophy of nature. The subtitle of the *Process and Reality* is actually "an essay in cosmology," not "an essay in metaphysics"—although since Whitehead's philosophy is a form of naturalism that makes no allowance for a distinct realm of ideas or spirit (*Geist*), cosmology and metaphysics amount to nearly the same thing. Whitehead viewed the formulation of metaphysical principles as an ongoing process of practical suitability, as "an imaginative experiment" whose success "is always to be tested by the applicability of its results beyond the restricted locus from

7. Editors of the Encyclopædia Britannica et al., "Career in the United States."

8. Price, *Dialogues*, 149.

9. Whitehead, *Adventures of Ideas* (1933); *Modes of Thought* (1938); and *Process and Reality* (1927). *Science and the Modern World*, *Adventures of Ideas*, and *Modes of Thought* are currently published in paperback by The Free Press, a division of the Macmillan Company. The standard edition of *Process and Reality* is now the corrected edition, edited by David Ray Griffin and Donald Sherburne. See bibliography.

which it originated."[10] He himself noted that "the greatest metaphysician" regarding depth of insight is often "the poorest systematic thinker,"[11] that philosophy itself is based finally not on logic or reasoning but on the intuitive interpretation of evidence,[12] and that "a learned orthodoxy suppresses adventure."[13] (This comment, however, has not prevented a learned orthodoxy from springing up around Whitehead's ideas.) The heart of the problem in philosophy, and to some extent in science, too, is that language by its very nature is incapable of the kind of accuracy and insight that the words seem to convey. In *Process and Reality* itself Whitehead wrote,

> Philosophers can never hope finally to formulate these metaphysical first principles. Weakness of insight and deficiencies of language stand in the way inexorably. Words and phrases must be stretched towards a generality foreign to their ordinary usage; and however such elements of language be stabilized as technicalities, they remain metaphors mutely appealing for an imaginative leap.[14]

In *Science and the Modern World* there is a very important statement that has been persistently overlooked by many of Whitehead's interpreters.[15] Referring to his own "philosophy of organism," Whitehead wrote,

> It is equally possible to arrive at this organic conception of the world if we start from the fundamental notions of modern physics, instead of, as above, from psychology and physiology. In fact by reason of my own studies in mathematics and mathematical physics, I did in fact arrive at my convictions in this way. Mathematical physics presumes in the first place an electromagnetic field of activity pervading space and time. The

10. Whitehead, *Process and Reality*, 5.

11. Whitehead, *Adventures of Ideas*, 166, referring to Plato.

12. Whitehead, *Modes of Thought*, 48–49.

13. Whitehead, *Adventures of Ideas*, 277.

14. Whitehead, *Process and Reality*, 4.

15. I have attended philosophical and theological conferences at which self-avowed Whiteheadians declared publicly that Whitehead was not familiar with quantum mechanics, despite the fact that *Science and the Modern World* contains an entire chapter on "The Quantum Theory" (ch. 8) and also a chapter on "Relativity" (ch. 7). These erroneous statements can perhaps be blamed on the fact that Whitehead's philosophy can be described and applied without any extended reference to physics. There is also the likelihood that most professional philosophers and theologians deem relativity and quantum theory to be largely irrelevant to their own field of study.

> laws which condition this field are nothing else than the conditions observed by the general activity of the flux of the world, as it individualises itself in the events.[16]

Whitehead does not belabor this point about the source of his ideas, and I have wondered whether his redundant use of the phrase "in fact" points to his characteristic discomfort about citing his own personal experiences as a philosophical example. In any case, the origins of Whitehead's philosophy in mathematical physics are of decisive importance to our use of his philosophy in this book.[17]

In *Science and the Modern World*, in a chapter entitled "Mathematics," Whitehead turns his attention explicitly to quantum mechanics. Remarkably, this was written in a year when Bohr, Schrödinger, Pauli, de Broglie, Dirac, and others were themselves still debating the meaning quantum theory. Whitehead writes,

> At present physics is troubled by the quantum theory. . . . One of the most helpful lines of explanation is to assume that an electron does not continuously traverse its path in space. The alternative notion as to its mode of existence is that it appears at a series of discrete positions in space which it occupies for successive durations of time. . . . So far the question is purely one for mathematics and physical science to settle between them, on the basis of mathematical calculations and physical observations.
>
> This discontinuous existence in space, thus assigned to electrons, is very unlike the continuous existence of material entities which we habitually assume as obvious. . . . Accordingly if this explanation is allowed, we have to revise all our notions of the ultimate character of material existence.
>
> There is no difficulty in explaining the paradox, if we consent to apply to the apparently steady undifferentiated endurance of matter the same principles as those now accepted for sound and light. . . . [In this case,] we shall conceive each primordial element as a vibratory ebb and flow of an underlying energy, or activity. . . . This system, forming the primordial element, is nothing at any instant. It requires its whole period in which to manifest itself. . . . The path in space of such a vibratory

16. Whitehead, *Science and the Modern World*, 152–53 (ch. 9).

17. Theologian Catherine Keller should be credited for recognizing that Whitehead "became a philosopher in order to connect the radical new insights of Einstein's relativity theory and quantum indeterminacy to our living sense of value" (Keller, *On the Mystery*, 23).

> entity—where the entity is *constituted by* the vibrations—must be represented by a series of detached positions in space, analogously to the automobile which is found at successive milestones and at nowhere between.
>
> We first must ask whether there is any evidence to associate the quantum theory with vibration. This question is immediately answered in the affirmative. It seems, therefore, that the hypothesis of essentially vibratory existence is the most hopeful way of explaining the paradox of the discontinuous orbit.[18]
>
> [A] new problem is now placed before philosophers and physicists, if we entertain the hypothesis that the ultimate elements of matter are in their essence vibratory. . . . We have already got rid of the matter with its appearance of undifferentiated endurance. Apart from some metaphysical compulsion, there is no reason to provide another more subtle stuff to take the place of the matter which has just been explained away. The field is now open for the introduction of some new doctrine of organism which may take the place of the materialism with which, since the seventeenth century, science has saddled philosophy. It must be remembered that the physicists' energy is obviously an abstraction. The concrete fact, which is the organism, must be a complete expression of the character of a real occurrence.[19]

Again, it seems quite striking that Whitehead included these ideas in a series of public lectures given in the fall of 1924.

Whitehead uses the word *organism* not to suggest that the entities in question are living creatures but as a metaphor reflecting (1) the innate wholeness or unity of a wave, including the presence of identifiable but non-isolable parts; (2) the fact that the entity itself is a distinct unit, having a birth and a termination; (3) the fact that the individual events are intrinsically in relation, for each event gives rise to its successor(s); and (4) the whole event cannot be divorced from the field in which it occurs. Like an organism, a vibration has parts that cannot be isolated or extracted from the whole, it is dynamic rather than static, and it is inherently in relation. Of singular importance is Whitehead's emphasis that "the entity is *constituted by* the vibrations." This anticipates Whitehead's

18. Note that at this time the term *orbital* had not yet been used in physics and chemistry. It was introduced in 1932 by physicist Robert S. Mulliken (Wikipedia, "Robert S. Mulliken"). Note that Robert A. Millikan (1868–1953) and Robert S. Mulliken (1896–1986) are two different physicists.

19. Whitehead, *Science and the Modern World*, 34–36; emphasis original.

later statement that "the process itself is the actuality, and requires no antecedent static cabinet."[20]

I will now elaborate on several key elements in Whitehead's philosophy that correspond more or less directly to the philosophical principles that we have derived from special relativity and quantum mechanics. The gist of the argument is this: There are several notable similarities or points of compatibility between twentieth-century physics and Whitehead's philosophy. One is the notion of universal relatedness. Many quantum theorists regard the universe itself as a unified quantum system. In quantum theory there is no truly empty space, for the quantum field, with its constant potential for birthing virtual particles, parallels Whitehead's idea of a universal creative advance in which all events are implicated. Indeed, the continual appearance and disappearance of virtual particles resembles the ontological status of Whitehead's actual occasions. A second compatibility exists between the quantum notion of superposition and Whitehead's notion of concrescing occasions that achieve "satisfaction"; reality takes place as definiteness (the collapse of the wave function) is achieved.[21] A third compatibility is implied between the centrality of mathematics in quantum mechanics (an atom is most accurately depicted as a wave function; the Pauli exclusion principle is essentially mathematical) and Whitehead's notion that each actual occasion is a dipolar unity between the mental and physical poles of each occasion; the presence of the mental pole allows for the ingression of mathematics into physical events. A fourth compatibility exists between Einstein's demonstration that space and time are not absolute receptacles in the Platonic sense and Whitehead's doctrine of space-time as a mode of relation. This list could be expanded at length. In each of the six topics below, there is a direct association between concepts pertaining to physics and the basic tenets of Whitehead's philosophy.

I. THE ACTUAL OCCASION, OR ACTUAL ENTITY

In quantum mechanics we have learned that the atomic realm is populated not by tiny material corpuscles but by events that exist in a state of uncertainty until they are disturbed, whether by experimental probing or by some other means. The state of uncertainty itself has a dynamic or

20. Whitehead, *Adventures of Ideas*, 276.

21. On this point, see Malin, "Whitehead's Philosophy," 80–81.

processive quality associated with its relevance to a mathematical wave function and to the simultaneous presence, in superposition, of various potentialities or pending outcomes. Whitehead used the term "actual occasion" or "actual entity" to refer to these events, saying that actual entities are "the final real things of which the world is made up."[22] The single actual entity in Whitehead's philosophy very much resembles the virtual particles that spontaneously arise in quantum fields.

Whitehead further stipulated an "ontological principle," which holds that actual occasions or groups of occasions are the only functioning causes in nature; in other words, principles, abstractions, patterns, and "laws of nature" may be very useful as descriptions, but they themselves do not cause events to occur. For example, we often say that the law of gravity caused an object to fall, but in fact it is the mutual attraction exerted by two masses that causes the motion of falling. Gravity itself is a descriptive term, not a force that somehow exists throughout the universe as an independently existing reality even apart from the presence of matter. In other words, if there are no masses, there is no gravity.[23] This dependence on specific actualities expresses the attitude behind Whitehead's ontological principal.

II. GENETIC AND MORPHOLOGICAL ANALYSIS

In quantum mechanics there are two basic ontological states: (a) the state of uncertainty stated by Heisenberg's principle, reflected in the Schrödinger equation's interpretation in probabilistic terms, and associated with the phenomenon of superposition; and (b) physical definiteness, as measured in experiments and observed in the properties of macroscopic objects. (See the section above in chapter 8 regarding the second principle derived from quantum mechanics.) *These two states in quantum physics (superposition and ordinary physical definiteness) parallel Whitehead's distinction between the genetic analysis and the*

22. Whitehead, *Process and Reality*, 18.

23. In quantum theory, all forces are thought to be mediated by particles; the particle proposed regarding gravity is the graviton. The Higgs boson has been said to contribute to the existence of mass but is not viewed as the cause of gravity. As of this writing, the significance of the Higgs particle is still being debated. Whitehead developed his own theory of gravitation, though that theory is not usually viewed as a major component of his philosophy. See Wikipedia, "Whitehead's Theory of Gravitation."

morphological analysis of actual occasions.[24] *Genetic analysis* considers the actual occasion as it develops or "concresces" (Whitehead's term) internally—that is, in a state when potential outcomes are still pending. The event is still becoming whatever it will finally become. The phases of concrescence involve potentials, not actuals. Because the process does not involve interactions between parts that are already actual or real, the genetic process does not take place in time. It is a nontemporal process. These same comments can, in general, be applied to the ontology of quantum superposition.

Morphological analysis considers the actual occasion as a determinate object in relation to other objects. Morphological analysis more or less parallels the analysis of relations found in classical physics, but with important modifications. All observable physical cause-and-effect relations pertain mainly to morphological analysis. Genetic analysis is by definition not observable because there are not yet any final outcomes that could be observed. Therefore genetic analysis must be undertaken by inferential reasoning—but it should be noted that inferential reasoning is not peculiar to Whitehead's philosophy but is at the heart of both theoretical and experimental physics. For example, inferences are drawn from the direct observation of traces in a cloud chamber.[25] Genetic and morphological analysis reflect the reality of two fundamental types of process or "fluency"—namely, the internal process of a particular instance of concrescence, and the external, observable transition from past to present to future events as events are related in space-time.[26]

Regarding the possible outcomes that result from superposition, our primary and sometimes sole source of knowledge is to rely on the specialized mathematics of quantum mechanics.[27] This dependence on math makes sense because in a superposed state the potential outcomes themselves are still abstractions, and mathematics employs abstract methods. Seeking to broaden the applicability of the idea of superposition to a general philosophical concept that is applicable throughout nature, Whitehead went into considerable detail about the internal formation of a single actual occasion. As noted, Whitehead called this process of self-formation "concrescence." *Concrescence* is an unfortunate term if it suggests the stable material of concrete; what Whitehead had in

24. Whitehead, *Process and Reality*, 66, 219–20.

25. Wikipedia, "Cloud Chamber."

26. Whitehead, *Process and Reality*, 66.

27. Heisenberg, *Philosophical Problems*, 56–57.

mind was rather the *process of hardening*—that is, of becoming definite. Metaphysically, the internal process of concrescence is common to all actual occasions throughout the cosmos, a feature of the fields and the energy-events from which all natural phenomena are derived.

Whitehead proposed that concrescence is initiated by an "initial aim" (in Heisenberg's wording, an innate "tendency")[28] and continues by including and integrating information gleaned from the occasions of the past. In Whitehead's theory, the initial aim is conceptual (scientists might prefer the word *mathematical*), but concrescence must also include physical influences, and those are gained from the occasion's inherited environment. The actual occasion integrates physical and conceptual elements by assessing or "grading" conceptual and physical influences in terms of relevance to possible outcomes. Some may feel that this language is excessively humanized or psychologized. I discuss this criticism below. At present, the point is that *the gathering of information is a spontaneous feature of every actual occasion*. This "gathering of information" is the basis in Whitehead's philosophy of the connectedness or radical relationality of nature.

As anecdotal evidence of this relatedness, we can consider the gravitational influence that operates slowly at enormous distances between galaxies. Other examples are the James Webb Space Telescope's ability to observe astronomical objects that are thirty-five billion light-years from Earth,[29] and the effects of quantum nonlocality, which have now been shown to convey information instantly across great distances.[30] In my opinion, in Whitehead's philosophy, entanglement's instantaneous transmission of information can be explained by assuming that the information conveyed is purely conceptual (mathematical) and therefore not subject to special relativity's prohibition of faster-than-light velocities pertaining to physical entities.[31] This explanation, of course, depends on

28. Heisenberg, *Physics and Philosophy*, 53. See also 147–48,

29. Irving, "James Webb."

30. This is the claim of Nadeau and Kafatos, *Non-Local Universe*. However, faster-than-light modes of transmission are still being debated among physicists. See Greenstein, *Quantum Strangeness*, ch. 12; Berardelli, "Quantum Physics Gets 'Spooky'"; and Wolchover, "Experiment Reaffirms Quantum Weirdness."

31. I am no physicist, but it seems to me that the Gisin experiment demonstrates conclusively that entangled objects transmit information at faster-than-light velocities. Not all physicists agree with this. The original objections to faster-than-light speeds raised by Einstein, Podolsky, and Rosen in 1935 (see their paper online at Einstein et al., "Quantum-Mechanical Description") sometimes seem to have inspired an orthodox

the postulate that atomic events traffic in information that is not only physical but conceptual or mathematical. I discuss this further in section 7 below. In any case, the faster-than-light transmission of information is not the primary feature of quantum nonlocality but rather the relatedness of actual occasions.

In Whitehead's genetic analysis of actual occasions, the universal relatedness of things—established by the transmission of information—is conveyed by a process that Whitehead called *prehension*. Prehension refers essentially to the activity of spontaneously gathering data from the occasion's environment. When questioned, Whitehead said that by "prehension" he meant an act of *grasping*.[32] As everyday examples resembling prehension we can think of the way an infant spontaneously grasps objects that are placed in her hand or the way the tendril of a growing bean plant spontaneously wraps itself around a trellis. It is crucial to understand that a prehension is not a thing that can be reified, like a message that is sent from one entity to another. A prehension is an action, an activity, carried out by the concrescing occasion itself.

In English the word *grasping* can also signify *comprehension* or *understanding*, and Whitehead stipulated that there are two basic types of prehension—physical prehension and conceptual prehension (sometimes called "mental" prehensions).[33] If purely mathematical information (say, regarding a spin of +½ or −½) is what is transmitted in quantum entanglement, then this explains why such information is not limited to the speed of light, since this limitation pertains to objects with mass or to physical forces (gravity, light, magnetism). It also means that entanglement provides direct empirical evidence for Whitehead's notion of mental prehensions and for his doctrine that each actual occasion has a mental pole.

In the phases of concrescence, mental and physical prehensions are not kept separate but are typically transformed into what Whitehead

---

dogma that faster-than-light speeds are simply impossible. See Wikipedia, "Quantum Entanglement," which includes the statement, "However, all interpretations agree that entanglement produces correlation between the measurements, and that the mutual information between the entangled particles can be exploited, but that any transmission of information at faster-than-light speeds is impossible." For a more thorough discussion see Ball, *Beyond Weird*, 160–95.

32. Whitehead's explanation of prehension as "grasping" was reported frequently by my teacher, Harvey Potthoff, who was a student of Whitehead's during Whitehead's final year of teaching at Harvard, 1936–1937.

33. Whitehead, *Process and Reality*, part 3.

calls "hybrid prehensions"—integrated prehensions in which both physical and conceptual qualities are present. Virtually all prehensions that are experienced in human consciousness are hybrid prehensions of this type. This is actually an important point, and I discuss it in chapter 12.

Whitehead's theory of prehensions is thus the answer to the questions about information that I posed in chapter 3 as to whether information is physical or conceptual. Whitehead's answer is that it is both, either separately or in hybrid integration. In chapter 3 I proposed that information can be viewed in three ways: as physical, as abstract[34] or nonphysical, or as a combination of the two. In chapter 3 I used the word *conceptual* as it was used in the realist-nominalist debate to refer to information that connects universals with physical nature. The term *conceptualism* in medieval philosophy is the nearest equivalent to "hybrid prehension" in Whitehead's philosophy. Whitehead uses the term "conceptual" prehensions to refer to information that is purely nonphysical. The word *conceptual* in this context is not limited to consciously entertained images or ideas. For Whitehead, "conceptual prehensions" are simply prehensions of data that are nonconcrete and not tied to some specific object or situation. From this point onwards I refer to conceptual prehensions as Whitehead does, to refer to information that is ideational and not tied to specific physical events.

The combined form of information that I had in mind in chapter 3 is explained philosophically by Whitehead's notion of hybrid prehensions. The functioning of purely conceptual prehensions includes mathematical information. The inclusion of mathematical information as conceptual prehension provides Whitehead's explanation of the seemingly mysterious role that mathematics plays in nature (see the discussion of the fourth principle gained from quantum mechanics in chapter 8, including the comments made by Richard Feynman and Eugene Wigner). In Whitehead's genetic analysis of the process of concrescence, the concrescing occasion's various prehensions are compared and contrasted until some definite culmination is achieved. In events at the atomic level, the phases involving comparison and contrast are minimal so that generally the outcome simply reproduces the physical outcomes achieved by

34. The word *abstract* has its own intellectual history. The root meaning in Latin means to "draw out" or "pull out," which suggests that abstractions are extracted from specific physical circumstances—which they may be—but this is not what I intend. I mean information that is nonconcrete and immaterial, and not derived from specific circumstances. Since the current chapters refer extensively to physics, most of the examples of abstract information are mathematical.

the occasion's relevant predecessors. For example, one electronic occasion yields a subsequent electronic occasion. Whitehead's description of the "higher phases" of concrescence, which entail more complex forms of comparison and contrast, is intended to explain the novelty, originality, and intelligence that are exemplified in behavior of more complex organisms.

The theory of prehensions is not merely a theory about information. It is also the basis of Whitehead's whole idea of causality. Purely physical prehensions convey efficient causes (mechanical or Newtonian causes), and conceptual prehensions convey the influence of abstractions, understood in quantum theory as mathematical potentials or probabilities. Though pure physical prehensions tend towards near certainty in outcome, in the end it is the concrescing occasion itself that is the cause (*causa sui*). Because of the nondeterministic influence of conceptual prehensions, absolute determinism is avoided.

One advantage of the conceptual content of mental and hybrid prehensions is that the concepts involved may be used to concatenate many prehensions with a common aspect into the prehension of a single object. Whitehead referred to this process as "the Category of Transmutation."[35] The specific quality or trait is abstracted from a large number of prehensions into the prehension of a single society of occasions (for example, a solid object).

Each actual occasion culminates in a phase that Whitehead called "satisfaction." Satisfaction may seem to imply the pleasant feelings that attend a sense of completion, but it can also signify a state in which certain conditions have simply been met. Thus we speak of contractual obligations that have been "satisfied." In Whitehead's philosophy, "satisfaction" is not a psychological term but a metaphysical one. Satisfaction is the final, physically determinate state of the actual occasion. If satisfaction is never achieved, there is no physical conclusion, and no actual occasion ever takes place. In Whitehead's philosophy, satisfaction is a parallel to the collapse of the wave function in quantum theory. Emphatically, Whitehead is not suggesting that all occasions culminate in some state of ideal completion or fulfillment. Very often the outcome of the final phase of concrescence fails conspicuously to achieve the originating initial aim, whether because of conflicts involving mutual incompatibilities or for other reasons.

35. Whitehead, *Process and Reality*, 101–2, 251.

III. NONTEMPORAL PROCESS

In the genetic analysis of actual occasions or actual entities, Whitehead stipulates that *the phases are nontemporal.* In our intellectual analysis of the phases they may seem to be temporal or even described as "successive" phases, but in actuality they are not. The basic reasoning behind this is that the phases are not separable entities that relate to one another but potentialities that are not actual. Since the phases are not preexisting things, there can be no spatiotemporal relations between them.

The notion of nontemporal process is counterintuitive. Again, however, Whitehead is simply making philosophical use of what were, at that time, recent discoveries in physics. By the theory of special relativity, we know that time is not a universal receptacle in which all processes happen, but rather a mode of relationship between existing objects. Whitehead's phases of concrescence are not progressing periods of time but coexisting potentialities. As is the case with electromagnetic radiation, from the frame of reference of an external observer, the completed actual occasion may very well have taken a finite amount of time, but from within there is no experience of time because there are no actual objects that could have temporal relationships. At the outset, we must understand that the prehensions that are involved in the phases of concrescence, and the phases themselves, are not independently existing objects. Time, in Whitehead's view and in Leibniz's, is a mode of relation, a way of distinguishing the before-or-after status of various events. Prehensions and the phases of concrescence are activities being carried out by the actual occasion that is, itself, still in a state of potential. The phases are indeterminate, not actual things that can be put in order temporally, logically, or any other way. The phases pertain to *potentials,* not to actuality. Our very concept of order is based on our waking experience—of objects, of events, or of concepts. But there is a sense in which the whole point of concrescence is that it is not restricted by these ordinary rules of factual existence. In this, concrescence very much resembles quantum superposition, in which conflicting states coexist. Insisting that the phases must be in some reasonable order is a bit like insisting that dreams should conform to the conditions of waking experience.

To some persons this state of affairs has seemed paradoxical if not in fact unacceptable, and the nontemporal status of the phases has been debated by Whitehead scholars. This discussion is somewhat technical, but it is important to clarify the issue here in order to prevent ongoing misunderstanding. In his 1966 classic *A Key to Whitehead's "Process and Reality"* Donald Sherburne wrote,

> [Whitehead] provides a genetic analysis of the satisfaction of actual entities—that is, [he] analyzes the phases of concrescence. The difficulty involves the relation of the genetic process to physical time. Whitehead incorporates the relativity theory of modern physics (discussed [by Whitehead] in Chapter Five, Section II [of *Process and Reality*]) into the basic principles of his system. This theory entails the idea that there is no absolute time as a sort of container within which actual entities become; rather, time is an abstraction from the succession of actual entities. Concrescence is not in time; rather, time is in concrescence in the sense of being an abstraction from actual entities. But if this is the case, then it seems strange to speak of one phase of concrescence as prior to another when the passage from phase to phase is not in physical time. This is the difficulty.
>
> William Christian (*An Interpretation of Whitehead's Metaphysics*, Yale, 1959, pp. 80–81) asks what sort of priority this priority of one phase to another could be, concluding that it is not temporal priority, logical priority, part-whole priority nor dialectical priority, but must be priority *sui generis* [of its own kind].
>
> It may seem to many that this suggestion is ad hoc and unsatisfactory. But a better suggestion is hard to find. This difficulty does not admit of easy solution and Whitehead scholars increasingly are going to have to puzzle over it, for the sequence of phases of concrescence is too central to the system to be abandoned, and yet difficult to incorporate into the system.[36]

More recently, physicist-philosopher Frank Hättich has joined this discussion, commenting that "unfortunately the nature of this genetic order is not clear from Whitehead's writings."[37] Hättich attempts to solve the problem by suggesting that "the genetical order of phases in a concrescence can be explicated by reason of the outcome range of qualities of each phase: *a phase is genetically earlier than another if (if and only if) the definiteness of the latter is strictly contained in that of the former.*"[38] Hättich seems to say that if, among several possibilities, one possibility is eventually included in a subsequent phase, it is *that* possibility that may

36. Sherburne, *Key*, 38.

37. Hättich, *Quantum Processes*, 9.

38. Hättich, *Quantum Processes*, 10; emphasis original.

be said to be genetically earlier. However, upon analysis it appears that Hättich's proposed solution is merely a variant of Christian's "logical order," which has already been deemed unhelpful. In Hättich's proposal, the several possibilities that did include the one that was eventually adopted would still in some way have to precede the process of adopting it. In my opinion, Hättich's proposal muddies the water more than it clarifies it. I suggest a different approach entirely.

In the *Critique of Pure Reason* (1781/1787), Kant presents four "antinomies"—four pairs of assertions in which each pair, though reasonable and plausible by the intellectual standards of Kant's time, are mutually contradictory. In other words, reason seems to be at odds with itself.[39] In another section, Kant argues that all human experience occurs within the context of space and time and that, since it is impossible for us to compare or contrast our own spatiotemporal state to any other unknown (and unexperienceable) mode of existence, we have absolutely no basis for assuming that any conclusions we may claim to reach regarding spatiotemporal existence are anything more than arbitrary attempts to expand our own experience into a description of ultimate reality.[40] In other words, human thought is inherently limited, both intellectually and contextually. Kant concluded that final metaphysical conclusions—systems of thought that attempt to explain reality—are unattainable.

However, Kant regarded metaphysics as a completed scheme of coherent ideas that were, at least in theory, unassailable. In contrast, in the tradition of Heraclitus, Bergson, James, Whitehead, and others, my view is that metaphysics is a practical endeavor, never perfectible but nevertheless of vital importance. As I have already noted, all human beings make assumptions about the nature of reality, whether or not those assumptions are consciously adopted or thoughtfully adapted. These assumptions contribute in decisive ways both to our daily behavior and to our overall view of life. The overall purpose of this book is to offer a more integrative worldview in which our era's chronic conflicts between the sciences and the humanities, between facts and values, between daily routines and spiritual vision, can be overcome. Strange as it may seem, our current discussion about the sequentiality of the phases of concrescence—which is really also about quantum indeterminacy and the origins of goal-oriented outcomes in nature—is of decisive importance, for

39. Kant, *Critique of Pure Reason*, 1.2.2.2.2 ("The Antimony of Pure Reason").

40. Kant, *Critique of Pure Reason*, 1.1 ("Transcendental Aesthetic").

the way in which nature traffics in information—in Whitehead's terms, the prehensions that are envisaged through the phases of concrescence—is the main doorway by which we move beyond the obstinate claim that nature is merely a reductionistic system of physical forces and mechanistic collisions.

I cite the *Critique of Pure Reason* because I think Kant's comments about antinomies and space-time apply, in a general way, to William Christian's list of "temporal priority, logical priority, part-whole priority, [and] dialectical priority."[41] The point is that *Christian's types of priority are derived from temporal human thought and practice, not from the relativity physics that Sherburne does mention.* At this point, it seems that Sherburne fails to see the important philosophical implications of special relativity. Having deemed Christian's list comprehensive and then found it ill-suited to Whitehead's notion of nontemporal process, Christian and Sherburne conclude that Whitehead's phases must manifest an ad hoc, *sui generis* ordering. The words *sui generis* insinuate that Whitehead invented the idea of nontemporal process as an ad hoc expedient to fill a gap in his system. Sherburne is typically one of Whitehead's most astute interpreters, but he mentions Whitehead's reference to the theory of relativity only in passing. Perhaps he fails to suspend his book's focus on *Process and Reality* long enough to return to *Science and the Modern World* and to Whitehead's exploratory consideration of Einstein's ideas.

In chapter 8 I provided the example of radiant energy traveling from Proxima Centauri to the earth. From our own frame of reference, the light waves or photons require about 4.25 years to reach us, but at the speed of light the distance traveled is foreshortened to zero, and the time required is likewise reduced to zero. The waves involved, at least in our way of understanding them, have a peak and a trough, each aspect of the wave following its predecessor and heralding its successor, *and yet internally the entire process is nontemporal.* This is one of the so-called paradoxes of relativity. For light itself, the zeroes representing distance and time are not a matter of philosophical theory; they are the mathematical result of the Lorentz transform, which by now has been verified experimentally in ways too numerous to mention. Relativistically, there is no single correct frame of reference. Both "time = 0" and "time = 4.25 years" are accurate. This multiple interpretation of accuracy (or "accuracies") can be fleshed out more fully by recurring to James's pragmatic theory of truth, but that

41. Sherburne, *Key*, 38.

discussion must be postponed until chapter 14. My point at present is that anyone who has difficulties with Whitehead's nontemporal phases must also have difficulties with the special theory of relativity and, in fact, with nature itself. Whitehead's own comment in *Science and the Modern World* that his philosophy was inspired by mathematical physics is pivotal at this point. It would in fact be surprising if Whitehead's ideas did *not* reflect some of the "paradoxical" aspects of relativity and the "weirdness" of quantum mechanics. Whitehead himself noted that the phases of concrescence manifest a type of wholeness that resists logic's typical method of analysis—which Bergson pointedly rejected—which breaks things down into smaller parts.[42] The whole notion of nontemporal process is essential to the worldview found in this book because it is the precondition of the indeterminacy, the self-determination, and the teleology or axiology that, I am arguing, are at the heart of nature.

IV. SERIALLY ORDERED ROUTES OF OCCASIONS

Whitehead held that the entities we typically think of as particles are, ontologically, self-sustaining repetitions of serially ordered events. At the beginning of this section on Whitehead's thought, I quoted several paragraphs from *Science and the Modern World*. To repeat: "One of the most helpful lines of explanation is to assume that an electron does not continuously traverse its path in space. The alternative notion as to its mode of existence is that it appears as *a series of discrete positions in space which it occupies for successive durations of time*."[43] These words make it indisputable that Whitehead's notion of serial order, which pertains to his entire explanation of enduring objects at the macroscopic level, was inspired by quantum mechanics.

42. Whitehead wrote, "On the other hand, Hume's train of thought unwittingly emphasizes 'process.' His very scepticism is nothing but the discovery that there is something in the world which cannot be expressed in analytic propositions. Hume discovered that 'We murder to dissect.' He did not say this, because he belonged to the mid-eighteenth century; and so left the remark to Wordsworth. But, in effect, Hume discovered that an actual entity is at once a process, and is atomic; so that in no sense is it the sum of its parts. Hume proclaimed the bankruptcy of morphology" (Whitehead, *Process and Reality*, 140). I am suggesting that in their critique of Whitehead's notion of nontemporal process, Christian and Sherburne have attempted to force morphological analysis onto a metaphysical insight pertaining to which that method of analysis is inapplicable.

43. Whitehead, *Science and the Modern World*, 34; emphasis added.

In quantum theory itself, when the notion of serially ordered events (a "chain of events") replaces the traditional Newtonian notion of "solid, massy, hard, impenetrable, movable particles,"[44] most of the allegedly "weird" quantum phenomena become easily understandable. For example, quantum tunneling may be understood simply as the occurrence of a successor event with new spatial coordinates. The enigmatic quality of wave–particle complementarity is obviated because, as Papatheodorou and Hiley put it, "an entity which is both a wave and a particle is conceivable, not as a peculiar type of substance moving in an *a priori* existing space-time, but as an unfolding *process* with all of its energy transactions occurring at those specific centers of its activity which otherwise comprise the points of the co-forming space-time pattern of a wave."[45]

Essentially, in Whitehead's view, fundamental particles, and ultimately all of physical reality, may be likened to the schedule of a traveling group of performers. Fans of a certain musical group can look at the group's schedule and see a list of locations and dates. Perhaps the group is appearing in Montreal, followed by Philadelphia, Chicago, St. Louis, and Kansas City. Naturally the group pays certain persons to manage the complicated details of their travel arrangements, but fans give no thought to that process. The purpose of the tour is the performances, and the group's followers are interested only in time and location. Clearly there is a kind of continuity and wholeness that pertains to the whole tour. This type of continuity and wholeness (but minus the intervening travel arrangements) characterizes Whitehead's understanding of the endurance of particles. An "enduring" particle is viewed as a series of discrete events.

If we now imagine the behavior of a specific electron in an atom, quantum mechanics holds that it may transition from one orbital to another but that physical acceleration, deceleration, and an intervening trajectory are forbidden. Rapid acceleration and deceleration—especially at the virtually instantaneous rates of change needed—would require the input of additional energy that is not available. Some physicists, ironically including even Erwin Schrödinger, have found this so-called quantum jump implausible. But it is imperative to see that *the conceptual difficulty pertaining to the "jump" is actually not about the physical discontinuity (the "jump") itself but with the reluctance of most interpreters of quantum*

44. Newton, *Opticks* 3, query 31 (375–76).

45. Papatheodorou and Hiley, "Process, Temporality, and Space-Time," 272. At the time when this article was written the authors were members of the Theoretical Physics Research Unit at Birkbeck College in the University of London.

mechanics to jettison their lingering loyalty to the image of tiny chunks of matter. What is needed instead is an ontology of processive, relational events that are serially organized. The "jump" arises from the fact that, by definition, quantum numbers do not allow for fractional values derived by mathematical division. Whitehead's notion of sequentially ordered events derives directly from this integer-dependent character of quantum mechanics. The intervals between quantum values correspond to the specific performances in the analogy of the touring performers. The replacement of enduring corpuscles with serially ordered events easily accounts conceptually for phenomena such as orbital transitions.

Beyond individual particles, Whitehead developed a detailed theory of what he called "societies of occasions." Some societies such as particles are organized serially or episodically. These would include single particles. However, societies are also commonly organized spatially as enduring groups; such societies include molecules, mineral structures, and macroscopic objects. Whitehead described various levels of organization, ranging from uncomplicated aggregates such as the sand on a beach to the complexity of living organisms. I describe the theory of societies briefly in chapter 12, as it is crucial to understanding the natural emergence of living creatures and human social relationships. The theory of societies, involving a "presiding occasion," is also necessary to formulating an answer to the hard problem of consciousness.

V. MASS-ENERGY EQUIVALENCE

Whitehead's philosophy makes use also of the relativistic understanding that matter is a form of energy. Einstein's celebrated equation

$$E = mc^2$$

represents not merely a peculiar calculation pertaining to nuclear reactors and bombs but a universal principle of dynamic interchange observed throughout nature. In chapter 3 I noted that even a spring that has been coiled tightly contains slightly more mass than a relaxed spring. This is because potential energy has been added to the bonds between the molecules in the spring. Conversely, during a medical PET scan, positrons collide with electrons and mutually annihilate one another, creating two gamma rays of energy that are emitted in opposite directions.

This is a direct conversion of matter into energy. Such interchanges between energy and matter seem a great deal more natural if the mass involved is conceived not as a self-identical enduring substance but as a self-renewing series of dynamic, processive events.

VI. DIPOLAR UNITY

In our consideration of the views of Pauli and Jung (chapter 9) we encountered the idea of a "psycho-physically neutral reality" in which the physical and the psychical are united, and we noted that this underlying neutral reality was proposed by Jung and Pauli as the basis of the ultimate unity of mental and physical phenomena. I explained this psycho-physically neutral reality as an example of what philosophers of mind refer to as a "dual-aspect ontology," "dual-aspect monism," or a "neutral monism."[46] But then I noted that there are drawbacks to this terminology. In chapter 3 I observed that the philosophical term *monism* is meant to emphasize that nature is not divided into a dualism between material reality and psychical or mental reality but that there is only a single basis of existence. As for the word *neutral*, it suggests that neither matter nor mind is preeminent; neither is the original source from which the other is derived. I then stated that "neutral monism" is an ill-chosen phrase because *monism* implies that reality has a permanent, unitary character, and *neutral* suggests an ineffectual, disengaged state of affairs. I suggested that what is needed instead is a concept of nature that draws attention to the ubiquity of process and relationality and that also acknowledges the existence of abstract possibility, in the form of mathematical information, that is dynamically involved in all natural processes.

Instead of "neutral monism," Whitehead's characteristic way of speaking of this integrative, underlying reality is to use the word *dipolar*. The term can be applied in many fields, but it is perhaps most commonly used in chemistry and physics. *Dipolar* is a well-chosen word because it suggests the nature of a magnet or magnetism, in which it is impossible to separate the north pole from the south. If you cut a bar magnet in half, the result is two magnets, each with its own north and south pole. Other metaphors might be the two sides of a single coin or directional pairs such as left and right. In chemistry and physics there are dipolar phenomena related to positive and negative electrical charges, but in this

46. See Wikipedia, "Double-Aspect Theory."

case the two are separable because the charges are typically carried by negative electrons and positive protons that can be and often are separated. In contrast, magnetism provides an excellent illustration of Heraclitus's maxim that "all things are one"—a unity in which the contrast of opposites is precisely what creates the reality of the unity. In physics this is illustrated by the fact that even unitary particles such as the electron have a "dipole moment"[47] in which the magnetic field is aligned in diametrically opposite directions. The particle's very unity is fundamental to the manifestation of its bidirectional magnetic field.

Obviously, Whitehead's use of *dipolar* is not about physical magnetism. It is a metaphor meant to illustrate the ontological unity-in-contrast of the physical pole and the conceptual pole, which coinhere in each concrescing actual occasion. In physics, this unity of the physical and conceptual is observed in superposition and quantum entanglement and in the inherent indeterminacy or state of potentiality that characterizes superposition's ontological status.

Superposition is not, as it may at first seem, merely an esoteric concept required by quantum theory. Superposition is an unseen but actual phenomenon in ordinary events. In recent years, superposition has become more well-known because it is at the core of quantum computing, a technology that is currently undergoing intensive development. In quantum computing, in addition to the data of ones and zeroes, a third state—the state of superposition—is also available, and the state of superposition itself is not limited to two conditions.[48] Quantum computing depends on the existence of a quantum state in which, ontologically, physical influences and abstract, mathematical possibilities are mutually implicated. It is this mutual implication of the physical and the conceptual that Whitehead signifies by using the term *dipolar*.

Superposition is understood as a state in which several potential outcomes are simultaneously present. Since these potential outcomes are not all achievable, and since by their nature they are conceptual rather than physically actual, the final outcomes are inherently uncertain. In chapter 5 I offered the illustration of two persons who were riding in a car and considering a choice between two restaurants. They come to a T-intersection, which forces them to make a decision. However, there is

47. The word *moment* does not imply that the magnetism is fleeting; it refers to the mathematical method of calculating the field strength at a specific point in space and a specific instant in time.

48. Mathas, "Basics of Quantum Computing."

a third possibility—namely, that they fail to decide, and thus remain in an indeterminate state. This indeterminacy or uncertainty exists because the various possibilities are, in their minds, still represented by *abstract possibilities* and not by physical conditions that already exist. In physics there are not two personal discussants but simply indeterminate potentials. In quantum superposition this indeterminacy is associated with Heisenberg's original uncertainty principle of 1927.

Physicists describe the possibilities present in quantum processes mathematically. This dipolar, mutually implicated integration of the conceptual and the physical is at the heart of superposition, and it is the core of Whitehead's conception of concrescence. The prehensions being integrated during concrescence may be conceptual, they may be physical, or they may be a hybrid combination of the two, but in toto there is always some degree of dipolarity. The word *dipolar* may be "only" a metaphor, but in fact metaphors and models are indispensable in particle physics.

VII. QUANTUM UNCERTAINTY AND FINAL CAUSATION

It hardly seems surprising that new mathematical information is constantly being created by nature. In a lake, the weight of the water increases when there is rain. Whenever a tree falls, there is a transfer of energy that can be quantified, at least approximately. Such processes create information of a de facto character. In many circumstances, *descriptive* information can be used to *predict* outcomes. This is its main use in, for example, weather forecasting and in real estate appraisal—but even beavers anticipate that if a certain branch is placed at a certain place, their dam will be strengthened, and bees communicate predictive information regarding the location of pollen-laden flowers. In all of these instances the information *describes* physical conditions, but it does not necessarily *change* them.

In certain atomic processes, however, the role of mathematics is in another category. The possible outcomes represented by the abstract or mathematical information—*conceptual prehensions*, in Whitehead's terms—do not merely describe; they *influence*. They function as options or tendencies. They place parameters around the possibilities and function as guides or lures. The fact that there is quantum uncertainty indicates that the purely mathematical or conceptual information is not mechanistically deterministic; it functions as one possibility among others. It is the event itself that determines the outcome, *causa sui.*

Another way of looking at this situation is to acknowledge that there is a mutually causal relation between the specific electrons in an atom and the organismic wholeness that pertains to the atom in its entirety. In entanglement and in the application of the Pauli exclusion principle, the individual parts create a holistic system in which the system in turn influences the behavior of the parts. We considered the holistic character of atomic structure in chapter 9. This general principle is the basis for the emergence of complexification and organism in nature. Whitehead writes,

> The concrete enduring entities are organisms, so that the plan of the *whole* influences the very characters of the various subordinate organisms which enter into it. . . . Thus an electron within a living body is different from an electron outside it, by reason of the plan of the body. The electron blindly runs either within or without the body, but it runs within the body in accordance with its character within the body; that is to say, in accordance with the general plan of the body, and this plan includes the mental state. But the principle of modification is perfectly general throughout nature, and represents no property peculiar to living bodies. . . . [This] doctrine involves the abandonment of the traditional scientific materialism, and the substitution of an alternative doctrine of organism.[49]

Whitehead mentions not only the physically complex condition of the body but its mental state. It is implicit in this statement that purposiveness or teleology is implicated. Materialists and reductionists cringe when the word *teleology* is introduced, but this cringing is partly due to the fact that teleology and the Greek root word, *telos*, have unjustifiably been imbued with examples drawn from conscious human decision-making. For Aristotle, *physis* (nature) itself has an embedded directionality, and a purposive artisan is simply attempting to imitate the telos of nature.[50] In chapter 9 I noted that in the New Testament *telos* can signify the final outcome or culmination of any event whether or not it is caused deliberately,[51] and I stated that in this book "final cause" and "teleology" do not refer only to conscious choices but to physical tendencies derived

49. Whitehead, *Science and the Modern World*, 79–80.

50. Peters, *Greek Philosophical Terms*, s.v. "télos" (192).

51. See Souter, *Pocket Lexicon*, s.v. "telos."

from abstract information, such as the mathematical probabilities signified by the Schrödinger equation.

Because symbols and goals can become more complex over time, organisms emerge and tend to evolve into more complex forms. Atoms produce simple molecules and then more complex molecules, and organisms eventually appear. If evolution really could be reduced to the formula "survival of the fittest," then granite boulders and stagnant oceans would have the advantage. But life struggles onward towards complexity and enjoyment. As Bergson observed, the difficulties posed by a mechanistic interpretation of biology can be solved only by accepting the effects of a creative directionality influencing the effects of random variations. I will say once again that it is time to acknowledge that physicalism is little more than an intellectualized version of naïve realism, formed in the early experiences of childhood and then gradually reinforced by layers of positivistic ideology and by the impressive mathematics pioneered by Newton and Leibniz. Naturally, in a great many settings, Newtonian physics is absolutely indispensable, and it will remain so. But a physicalist-reductionistic philosophy does not make for an adequate philosophy of nature, or of life. In fact, it is an obstruction.

I repeat that in this book we are searching for a more integrative, life-affirming, unifying worldview—a worldview that affirms freewill and human responsibility, that seeks to interrelate scientific thinking with humanistic thinking, and that finds in the natural world not merely instrumental value to be exploited but intrinsic value to be grateful for and to preserve. In this chapter I have introduced some of the basic ideas of Alfred North Whitehead as a unique resource in the development of this worldview.

Unfortunately Whitehead's philosophy is not universally recognized as such a resource. His ideas have too often been presented as a system of ready-made abstractions that then seem excessively speculative and unempirical—a result that is often unappealing, especially to scientists. To counter this depiction, and because it contains chapters specifically about relativity and quantum mechanics, I have focused on *Science and the Modern World* rather than on *Process and Reality*. However, Whitehead has been neglected not only by scientists but by professional philosophers. To most philosophers Whitehead's work seems unrelated to the major themes of twentieth- and twenty-first-century philosophy—though, as

Lisa Landoe Hedrick has shown,[52] this perception is flawed.[53] A third difficulty faced by Whitehead's philosophy is that, frankly, his ideas have too often been oversimplified by fussy philosophers who, for some reason, seem eager to move beyond description and analysis in order to reveal Whitehead's alleged shortcomings. Bertrand Russell, in his sometimes untactful way, commented that "a stupid man's report of what a clever man says is never accurate, because he unconsciously translates what he hears into something that he can understand."[54] Whitehead's critics are certainly not stupid, but, just the same, a good many of them have charged flaws to Whitehead's philosophy that were more nearly attributable to gaps in their own understanding.[55]

A final reason why Whitehead's philosophy is viewed with suspicion is that he unabashedly included a concept of God in his system. God is not a popular component in cosmology, either among scientists or philosophers. The problem is worsened because even among Whitehead's supporters, his idea of God is interpreted in widely varying ways. I have not introduced Whitehead's God-concept in this book because, within the scope of our discussion, it adds an unnecessary and page-consuming complication. In my view, Whitehead's God functions as something like a universal field that supplies not physical forces but mathematical-conceptual goals to each actual occasion. Whitehead called this goal the

52. Hedrick, *Whitehead and the Pittsburg School.*

53. A friend who is an instructor of philosophy at a local college reports that when she suggests a course in Whitehead's philosophy, the members of the philosophy department simply refuse to discuss it. This response is evidently not unusual.

54. Russell, *History of Western Philosophy*, 83.

55. For example, R. G. Collingwood claims that Whitehead had "never read Aristotle's *Metaphysics* for himself," despite the fact that Aristotle's philosophy is cited over thirty times in *Process and Reality*, and it happens that Whitehead was president of the Aristotelian Society in London for the term 1922–1923 (Collingwood, *Idea of Nature*, 170). Terrence Deacon attempts to summarize Whitehead's philosophy in three pages, wrongly stating that Whitehead combines "micro homunculi to make a macro homunculus" so that "raging rivers and storm clouds should exhibit at least as much mentality and intentional causal power as human brains" (Deacon, *Incomplete Nature*, 77–79). Deacon evidently lacks a basic understanding of Whitehead's explanation of differing types of structure in enduring objects, and especially the difference between a "nexus" and a "society" (see Whitehead, *Process and Reality*, 83–129). Even Gary Dorrien, normally an incisive analyst, claims that Whitehead was opposed to the idea of entropy and that "Whitehead's principle that individual organisms move toward complexity by virtue of their creative urge to live and flourish contradicted the second law of thermodynamics" (Dorrien, *American Liberal Theology*, 521). Whitehead recognizes entropy explicitly in his discussion of why organisms require food (Whitehead, *Process and Reality*, 105).

"initial aim" and held that it is automatically a part of the ontological structure of each actuality, reflecting not a single, universal purpose for the cosmos but whatever outcome tends to promote intensity and complexity for each occasion in its own context.[56] Some philosophers and theologians ascribe a quasi-personal character to Whitehead's God, but I think this is a mistake. My views on this subject have been published elsewhere.[57]

There is of course the opposite mistake of accepting Whitehead's philosophy too uncritically and viewing its ideas as simple verity, in the spirit of the correspondence theory of truth. I have noted already that Whitehead himself would not have been comfortable with this approach, and I have tried to avoid that approach here. Despite this, however, I do affirm the superlative discernment and remarkable applicability that characterize Whitehead's work. I am in agreement with John Cobb when he affirms, "For myself I am persuaded that he ranks with Plato, Aristotle, and Kant as one of the greatest creative thinkers of all time. I regret that from the use I will make of his work in this book the reader is unlikely to receive any adequate sense of Whitehead's genius."[58]

We have now concluded our consideration of relativity and quantum mechanics and are ready to apply the principles thus discovered to the macroscopic world, including an understanding of natural evolution, life, and human experience.

56. See Whitehead, *Process and Reality*, 67 (end of the first full paragraph) and 224–25 (paragraph beginning "The ground, or origin of the concrescent process," and especially 245); Whitehead, *Adventures of Ideas*, 130, and 168 (paragraph beginning "Plato has definite reasons"); Whitehead, *Modes of Thought*, 102–3; Whitehead, *Science and the Modern World*, ch. 11; and Whitehead, *Religion in the Making*, ch. 3 ("Body and Spirit").

57. Conner, "Functional-Empirical Approach"; Conner, "Whitehead the Naturalist"; Conner, "Plight of a Theoretical Deity"; and Conner, "Beyond Emergentism."

58. Cobb, *Christian Natural Theology*, 16. Among Whitehead scholars, Cobb has been perhaps the ablest interpreter and the most effective advocate.

12

Nature, Organism, and Wholeness

SPECIAL RELATIVITY AND QUANTUM mechanics have led us to an ontology of processes. In this ontology, nature at the atomic level is composed not of particles set in motion but of events. Each event is derived from the environment of its past and is comprised internally of a dipolar integration of physical and conceptual factors. The event seeks intensity obtained through patterns of vibration, and the vibration itself is regarded as the core reality; in Whitehead's words, "the process itself is the actuality, and requires no antecedent static cabinet."[1] Events are inherently relational and tend to form enduring patterns or structures based on their relations. As William James says, the relations themselves are as real as the entities that are related. Each event tends towards an outcome that is partly determined by its past but is, ontologically, self-caused. The non-mechanistic role of abstract information in the outcome of events allows for the observed indeterminacy at the atomic level and is the vehicle for the inclusion of final causes or goal-oriented results in the natural world.

The phenomenology of the primordial process may be summarized as follows: (1) All events in the universe are produced within the context of fields (electric, magnetic, gravitational, and so on).[2] (2) Fields include mathematical information pertaining to relations between specific events. Fields are thus the physical phenomenon underlying the radical

1. Whitehead, *Adventures of Ideas*, 276.

2. In physics every particle is associated with its own type of field. The field is a matrix of potential and actual relations. In this book fields are said to produce not solid, material particles but events. Particles result when a particular type of event is reproduced episodically (or serially) through time.

relationality of all events. (3) The events that populate the field are characterized by vibration and by contrast within wholeness. Vibration and contrast entail not merely static physical endurance but an informational aspect. *Each event must therefore be viewed as a dipolar integration of physical and conceptual information.* (4) The events are inherently relational, deriving their initial conditions from the existing world and contributing to the new world that is produced by their own outcome. (5) Each event seeks intensity, relation, and some type of concrete definiteness in its outcome. Quantum uncertainty illustrates the fact that the outcome is not predetermined; it is the event itself that makes the final determination. Its aim or telos involves other events, and this aim tends to create novel complexity and organismic wholeness. Nature at the most primordial level tends to use information to create novel and more holistic, more organismic outcomes.

In the self-caused aspect of the event itself we observe the creative aspect of the processes of nature. Though nature displays many processes that are blindly repetitious, wholeness and goal-driven novelty still emerge. I have argued that the self-creative function of the event cannot be understood apart from its use of an abstract or conceptual element—what Aristotle called a final cause, a telos, a goal—that allows each event to seek an outcome that has not been entirely determined in advance by preexisting conditions. However, these goals do not often resemble the conscious aspirations of human beings. At the quantum level they are simply physical culminations in which vibration and relationality predominate.

The basic premise of this chapter is that this rudimentary, self-creative element of goal-formation at the atomic level is the ontological basis for the increasingly sophisticated manifestations of purposiveness, sensibility, and meaning that we observe emerging in the natural evolution of organisms, beginning with single cells and gradually producing plants and animals. In this book I assume that it is not necessary to present a step-by-step explanation of the ways in which final causation at the atomic level gradually complexifies and eventually produces conscious intelligence. Whitehead offers an explanation in considerable detail.[3] It should be intuitively obvious that final causation at the atomic level is, metaphysically, the basis for final causation at the level of consciousness, intelligence, and reason. In chapters 4 through 11 I have relied on special

3. Whitehead, *Process and Reality*, 39–280.

relativity and quantum mechanics to illustrate the importance of abstract or mathematical information in nature and to reveal the essential aspect of goal-seeking or telos in the way which that information is embodied in specific events.

Admittedly there is no guarantee that an ontological analysis based on relativity and quantum mechanics has produced an accurate depiction of nature. All of our metaphysical theories are just that: theories. But certainly, theories that accommodate themselves to the discoveries of twentieth-century physics are a vast improvement over the materialistic, mechanistic worldviews that have prevailed for the past three hundred years. I am now arguing that this newer worldview has some important advantages beyond a mere reinterpretation of physics. By integrating physical and conceptual influences, the worldview presented in this book provides a metaphysical basis for emergent social structures among living organisms, including human beings. The process-relational worldview thus tends to overcome the current separation between the sciences and the humanities and even between science and religion. As we shall see, the new worldview solves the "hard problem" of consciousness. Also, it depicts nature as being "alive" rather than "lifeless."[4] By doing that, our ontology based on relativity and quantum physics provides a metaphysical basis for the intrinsic value of nature, both in its parts and in its wholeness. This represents an enormous improvement over our existing tendency to lapse into piecemeal, human-centered motives for environmentalism and ethical behavior. As the ground of value, nature itself becomes sacred, and our ethical treatment of other human beings may be understood as a part of the whole web of life.

In chapters 4 through 10 I focused on phenomena at the atomic level. This was necessary because of our concentration on physics. In the present chapter we redirect our attention to the larger structures that emerge from atoms when they are gathered or organized in various ways. As human beings we have a natural interest in the macroscopic objects and materials that are essential to our own experience. But these objects and materials are composed of extremely large numbers of atoms. For example, in a single twelve-ounce beverage container there are roughly 1.14×10^{26} (114 septillion) molecules of water.[5] The water molecules in a container are not organized in any lasting pattern; we would call them

4. Whitehead, *Modes of Thought*, chs. 7 and 8.

5. Author's calculation.

a simple aggregate. Fortunately, however, the atoms—or, in Whitehead's terms, actual occasions—that populate the world are in many instances organized in ways that are more than mere aggregates. It is these more complex patterns of organization that make it possible for macroscopic objects to function as single items or as wholes. Thus, as we consider these various types of organization, the question of wholeness and holism becomes important.

There are still numerous opponents to the idea of holism. In chapter 2 I described the views of physicist Ethan Siegel, who argues that "the Universe really is 100% reductionist in nature," and that "the whole isn't greater than the sum of its parts; that's a flaw in our thinking. Nonreductionism requires magic, not merely science."[6] Another opponent of holism is physicist Sabine Hossenfelder, who asserts that

> if you really want to understand what an organism does and how it does it, you will look for an explanation on the level of constituents. . . . If we want to really understand something, we look for a reductionist theory. Why? Because we know from experience that reductionist theories have more explanatory power. . . . Indeed, the whole history of science until now has been a success story of reductionism.[7]

Hossenfelder concludes, "I cannot think of any scientific fact that is better established than that the properties of the constituents of a system determine how the system works."[8] Hossenfelder links the claim that all phenomena can be reduced to the interactions of the parts of the system to an emphasis that the parts simply follow the laws of nature. Thus all phenomena can be explained in terms of physical cause and effect. Not surprisingly, as an accompaniment to reductionism, Hossenfelder denies the reality of free will,[9] arguing with considerable passion and conviction that experiences such as passion and conviction are nothing more than a deceptive by-product of neurological processes that are strictly physical.

There are several reasons why Hossenfelder's reductionistic physicalism fails, but the most obvious one is that her argument commits

6. Siegel, "Reductionist in Nature."

7. Hossenfelder, "What Is Reductionism?"

8. Hossenfelder, "What Is Reductionism?"

9. Hossenfelder has a video on YouTube explaining why "free will is incompatible with the [currently known] laws of nature" and why the idea is "nonsense" anyway (Hossenfelder, "You Don't Have Free Will," 1:21, 3:50).

the fallacy of *petitio principii.* The premises of her argument are, first, that all complex systems are simply combinations of simpler parts that are exclusively physical and, second, that the laws of nature determine the behavior of all these physical parts throughout the universe. Obviously these premises contain reductionistic physicalism not as a logical deduction but as an initial assumption, for the beginning assertion that *all* events are solely physical rules out in advance the possibility that any events can be caused by nonphysical factors such as concepts, goals, or purposes. If *all* X is caused solely by Y, then *no* X can be caused by Z. Moreover, though the efficacy of the laws of nature is a key principle in Hossenfelder's argument, she makes no attempt to explain how those laws were created or how it can be that seemingly abstract principles ("laws") can exercise a determinative physical influence on particular objects. Responding specifically to Hossenfelder's statement "that things are made of smaller things, and if you know what the small things do, then you can tell what the large things do," Don Howard writes,

> But deep down at the quantum level, that is not so, because when two or more systems become entangled with one another through an interaction, the postinteraction state of the joint system cannot be written as a product of separate states for the subsystems. In other words, $\psi 12 \neq \psi 1 \otimes \psi 2$. There is no more scientifically well-established example of holism—the idea that the whole is more than the sum of its parts—than that.[10]

In quantum entanglement, entangled particles cannot be understood as two independent entities, each with its own isolable properties.

> The entanglement of two quantum particles is a holistic effect. All the intrinsic properties of the two particles, such as electrical charge, together with all their extrinsic properties, such as position, still do not determine the state of the two-particle system. The whole is more than the sum of its parts. The atomistic picture of the world, in which everything is determined by the properties of the most elementary building blocks and how they are related in spacetime, breaks down. Instead of considering particles primary and entanglement secondary, perhaps we should think about it the other way round.[11]

10. Howard, "Physicist Gets Philosophical." Hossenfelder's preceding idea is quoted in this article. Howard teaches at Notre Dame University in Indiana.

11. Kuhlmann, "Physicists Debate."

Physicists such as Feynman, Wigner, Pauli, Heisenberg, and Bell found mystery and wonder in the role of mathematics in nature. For Hossenfelder, the applicability of mathematics seems to be merely an expedient assumption.

It is an empirical fact that holism exists in nature. *Wholeness*, as distinguished from holism, may be said to occur whenever a group of things functions together as a unit to produce a result that could not have been produced by the individual members of the group. As Howard indicates, in contrast to the wholeness of a mere aggregate, *holism* exists especially when the well-known formula "the whole is greater than the sum of its parts" can be observed in nature. Understood this way, holism is not a new idea; it was discussed, for example, by Parmenides, Empedocles, Plato, and Aristotle.[12] Compelling empirical evidence for this holistic, teleological character is found in the gradual complexification that we observe in natural evolution. In this context the word *evolution* can be applied not only to biological organisms but to the gradual complexification displayed by the cosmos in the formation of solar systems and planets composed of increasingly complex molecules and, at least in the case of our own planet, living organisms, including our own experiences of thinking, feeling, and longing.

Obviously, in the interpretation of nature we have no basis for granting paradigmatic status to human experience. Human consciousness counts as part of nature, but it is misleading to use human experience as a typical example of the utilization of information in natural processes. If we begin with human experience, there is a tendency to assume mistakenly that all purposive intelligence is a product of conscious thought. To understand how final causation is present throughout nature, it is more instructive to begin with the simplest examples we can find.

In this chapter, I discuss four types of wholeness. The first, which displays only a modest level of holism, may be called an *aggregate*, by which I mean an unorganized group of similar components. Examples would be a pile of gravel, the water droplets in a cloud, or an underground layer of petroleum in an oil field. The second type of wholeness can be referred to as a *structured aggregate*. Snowflakes, quartz crystals,

12. Peters, *Greek Philosophical Terms*, s.v. "holon" (84–87). In English the word *holism* is associated especially with the book *Holism and Evolution*, written and published in 1926 by Jan C. Smuts (1870–1950). In addition to his interest in philosophy, Smuts was a military commander and was the prime minister of the Union of South Africa (1919–1924, 1939–1948).

and molecules of clay are examples. The aggregate is bound together by patterned physical structures. The third type consists of entities whose wholeness is necessary in order to serve some larger purpose. Such entities would include the leaves on a tree, the eyes of an eagle, or, as a subcategory, human artifacts such as a flint arrowhead or a cooking utensil. The fourth type consists of living organisms, which display a palpable wholeness that may be observed in the experience of individual subjectivity.

(1) Simple aggregates occur in nature largely because of natural processes acting on their constituent parts. Coal seams occur underground because the carbon they contain was once a part of large forests that died and were subsequently buried by geologic processes. Water organizes itself into lakes and seas because water runs downhill and collects in low places. The water exhibits modestly holistic qualities such as the formation of rivers, large waves, and glaciers that act in unitary ways to cause phenomena such as erosion and flooding. An avalanche is an aggregate of rocks and soil that, acting in a specific complex event, carves out a single path through a mountain forest. High- and low-pressure atmospheric fronts act in a unitary way to create extensive weather patterns. As a whole, aggregates themselves display efficient causation rather than final causation. To the extent that any final causality is involved at all, it exists only in a rudimentary way in the maintenance of the atomic orbitals that function in each of the aggregate's atoms. Whitehead referred to such aggregates as *nexus*. (Actually *nexūs*, with a macron, is the plural form.)[13]

(2) Structured aggregates illustrate a higher degree of unity. The clay mineral kaolinite consists of alternating, sandwich-like layers of silica tetrahedra and alumina octahedra;[14] this is usually depicted graphically as a relatively complicated Tinkertoy-like structure. Kaolinite molecules tend to preserve and enlarge their structure by spontaneously but repetitiously bonding with specific atoms or with other clay molecules to form extended sheets. Thus there is a level of self-organization. The molecular sheets in their wholeness have macroscopic properties, such as

13. Whitehead's explanation of enduring objects or "societies of occasions" is found in chapter 3 of *Process and Reality* (83–109) and continuing intermittently through page 218. Whitehead's term *nexus* refers to disordered aggregates. He used the term *society* to apply to collections of objects that are "ordered among themselves" according to some "defining characteristic" (*Process and Reality*, 89). The second, third, and fourth types of wholeness that I describe here are, generally, examples of Whitehead's "societies of occasions."

14. For a graphic diagram, see Whittaker, "Beevers Crystal Structure."

the tendency to absorb water between the layers. This is what causes the layers to swell apart, resulting in expansion that can cause cracks and the structural destabilization of buildings that are constructed on clay-containing soils. The regular pattern of the tetrahedra and octahedra is caused by the quantum mechanical properties of the chemical bonds between atoms of aluminum, oxygen, hydrogen, and silicon.[15] Thus in the kaolinite molecule we observe an information-based morphology and an attendant characteristic of holism. In chapter 11 I argued that the quantum structure of a single atom exemplifies holism.

(3) A third type of wholeness pertains to objects that are not complete organisms but that do serve some definable purpose, though in nature the purposes usually are not formed or implemented deliberately by any conscious being. I have already mentioned the leaves on a tree and the eyes of an eagle. The individual cells in a living body could also be cited. Such wholes have an internal structure of their own and manifest organismic unity. When surgeons transplant a heart or a liver, the indispensable condition is that the organismic structure of the transplanted organ is sufficiently similar to the structure of the original to serve as a replacement. Human inventions may be viewed as a subcategory of this type of holism. An automobile is a very complicated machine, and even some of its parts are very complicated, but its primary purpose—the transport of passengers and goods—is unitary and easily definable. However in an automobile this purpose is not self-created but, rather, is imparted by its human manufacturer.[16] The same thing is true pertaining to all human inventions, including even computer programs that convincingly simulate the deliberate responses of human beings. The reductionist's claim that the holism of a leaf or an eagle's eye can easily be explained away by dividing such wholes into their smallest parts is based on an entrenched refusal to acknowledge the ontological status of the purpose or purposes of the item in question. Such purposes are arbitrarily disallowed in advance by the physicalist premises of the reductionist ideology.

(4) The fourth type of holism is found in entities that experience subjective unity. Living organisms are the most obvious example. In such organisms there is a sense of selfhood. Again, this selfhood is not

15. See Wikipedia, "Kaolinite."

16. Whitehead stipulated that a true society is self-sustaining (Whitehead, *Process and Reality*, 89). Arrowheads and automobiles are not self-sustaining—unless we consider that their eminent usefulness prompts their human creators to maintain them as long as it is feasible to do so.

necessarily experienced consciously. It is discernible, rather, in the self-sustaining, self-fulfilling responses of the organism within the context of its environment. The phenomenon of selfhood thus includes plants such as trees and carrots and even single-celled bacteria. The argument of this book is that such responses have been made possible in nature by the element of final causation or teleology that exists even at the atomic level—for even atoms and electrons experience subjectivity to the extent that they preserve themselves by receiving and utilizing information in ways that are outcome oriented and nondeterministic. Such final causation depends on the availability of conceptual or mathematical information as an innate aspect of the ontology of quantum events, or what Whitehead called actual occasions. As nature evolves, the goal-oriented aspect of an organism's behavior tends to become more complex and innovative as the increasingly intricate physical structure of the organism allows it to do so—and yet the individual atom itself is a candidate for this fourth type of holism. The pattern of orbitals, each with its own uniquely identifiable electrons, illustrates the holistic quality of the fourth type and bestows a type of unity on the atom. In fact this is why we call it an "atom"—an undivided reality. Thus the atom, which is the elemental component of all the enduring objects of our daily experience, illustrates that there is overlap in all four types of wholeness. Aggregates, for example, are typically composed of atoms whose structure is more complex than the minimal structure of the total aggregate. My point is that holism and organismic structure exist even at the fundamental levels of nature.

In chapter 9 I briefly described the amazing intelligence of bees, trees, slime molds, and tomato plants. We will now consider several examples of intelligence and purposiveness in the kinds of living things that have long been regarded as insentient. Our observation of plants, of simpler animals, and of the functioning of our own bodies ought to convince us immediately that purposes existed in nature long before there were human minds or our own capacity for consciousness. The reluctance of Western intellectuals to acknowledge this purposiveness may be blamed, first, on religious traditions that place human beings in a category superior to animals and plants (it was widely believed that only human beings have souls) and, following the Enlightenment, on the emergence of scientific materialism, with its general skepticism regarding the ontological status of consciousness and purpose, especially in animals and plants.

There is a growing acknowledgment that a type of purposive intelligence is present in plants. Botanists Paco Calvo, Monica Gagliano, Gustavo M. Souza, and Anthony Trewavas maintain that "the drive to survive is a biological universal. Intelligent behaviour is usually recognized when individual organisms including plants, in the face of fiercely competitive or adverse, real-world circumstances, change their behaviour to improve their probability of survival."[17] The authors describe the relationship of intelligence to adaptability and emphasize the need to recognize such intelligence in plants, showing it to be goal-directed and purposeful. Intelligence, understood this way, is frequently reported even in single cells and microbes. (In chapter 2 I described the seeming intelligence of tiny "DNA replication machines" that are constantly at work in the human body.) Numerous examples of plant intelligence are reported by Evans:

> Sometimes the plant summons a parasite that attacks the insect larvae instead of eggs. After cabbage butterflies lay eggs on a black mustard plant, the plant produces a volatile compound that attracts a parasitic wasp to feeding caterpillars. This plant can even detect what sort of insect eggs have been laid. If a black mustard plant detects egg laying by a butterfly whose larvae adore the taste of black mustard leaves, the plant goes into defense mode. If the eggs are deposited by a species whose larvae are likely to move on without doing much damage, the mustard plant doesn't waste energy building up its defenses.
>
> The Scots pine has a particularly remarkable response to egg laying by sawflies. Exactly three days after eggs are laid, the pine needles produce a chemical that summons an insect that parasitizes sawfly eggs. Why wait three days? Because that is when the insect predator is most successful at parasitizing the eggs.[18]

The African acacia tree defends itself from being overgrazed by producing tannin-C, a chemical that combines with proteins in the animal's gut to form a poison that is lethal to kudus and other antelopes who eat its leaves. Interestingly, the tree does not produce the tannins until after ten or fifteen minutes of being grazed. Then the tree also releases the chemical ethylene, which spreads by air to other trees and activates their defense systems.[19] Walnut trees produce a toxin called juglone that

17. Calvo et al., "Plants Are Intelligent."
18. Evans, "Science to the Grower."
19. Flowers, "Incredible Killer Tree."

impairs or kills competing trees and other plants. This effectively allows new walnut trees to germinate and grow more quickly.[20]

To ensure more effective methods of pollination, many species of orchids have developed ways of relying on certain insects to spread their pollen. Some orchids offer a nectar that is appealing especially to a specific type of insect; the advantage is that these insects then seek out only flowers of that orchid. This distributes the orchid's pollen more efficiently since the insects then search only for flowers of that species of orchid. Some orchids use methods of deception. Bee orchid flowers display a large lip-shaped petal that looks like a female of a species of bee. Male bees attempt to mate with the flower, inadvertently gathering pollen. The bee then looks for other females, finds other orchids, and pollen is transmitted. Hammer orchids use a trap mechanism to attach pollen to insects. The middle petal of the hammer orchid resembles a female wasp, and it also produces a perfume that lures male wasps. As the male struggles to mate with the flower, the petal of the flower swings over and flips the male wasp upside down onto the stigma precisely onto the location on the wasp's body where the pollen will later be most effectively dispersed.[21]

Michael Pollan reports that it is common for plants to display goal-oriented responses to their environment. Bean plants waste no time and energy blindly groping about for a pole to grow on, for they somehow sense the location of a pole if one is already present. Pea plants sense both the presence of above-ground structures that will support their tendrils and below-ground structures in which their roots can grow, sending information back and forth between tendrils and roots in order to coordinate plant growth.[22] When sage brush leaves are eaten or clipped, the sage plant sends chemical warning signals to its neighbors. The mimosa or "sensitive plant" folds its leaves protectively when touched or jostled, but after repeated disturbances the plant learns to distinguish between impacts that are potentially harmful and those that are not.[23]

However, despite these and many other examples, Pollan finds that conventionally minded scientists react with considerable animus when

20. Government of Ontario, "Walnut Toxicity."

21. Brewer, "Sneaky Orchids."

22. Guerra et al., "Asymmetrical Distribution."

23. Pollan, "Intelligent Plant."

assessing researches into "plant intelligence"[24] or "plant neurobiology." He writes,

> Many plant scientists have pushed back hard against the nascent field, beginning with a tart, dismissive letter. . . . Lincoln Taiz, an emeritus professor of plant physiology at U.C. Santa Cruz and one of the signers of the . . . letter, told me . . . [that] "the mechanisms [in plants] are quite different from those of true nervous systems." Taiz says that the writings of the plant neurobiologists suffer from "over-interpretation of data, teleology, anthropomorphizing, philosophizing, and wild speculations." He is confident that eventually the plant behaviors we can't yet account for will be explained by the action of chemical or electrical pathways, without recourse to "animism." Clifford Slayman, a professor of cellular and molecular physiology at Yale, who also signed the . . . letter . . . was even more blunt. "'Plant intelligence' is a foolish distraction, not a new paradigm," he wrote in a recent e-mail. Slayman has referred to the letter [signed by himself, Taiz, and others] as "the last serious confrontation between the scientific community and the nuthouse on these issues."[25]

It is obvious that this exasperated antagonism to the idea that purposiveness exists in plants is based not simply on science but on the same doctrinaire physicalism that we have witnessed in other fields of study. Biologists who peremptorily dismiss plant intelligence do so without recognizing that their ostensibly objective scientific beliefs are trapped in an intellectual cage based on their own physicalist metaphysical ideology. In evidence of this, note how unthinkingly Taiz adopts a false dichotomy between "chemical or electrical pathways" (representing physicalism) and "animism" (representing the "nuthouse"). But actually, of course, the intellectual options are not limited to this simplistic choice between materialism and spiritualism.

As we noted in chapter 11, an alternative to this false dichotomy is found in the outlook that nature is composed not of material particles but of dynamic processes, and that these processes integrate physical

24. Regarding the term *intelligence*, Pollan notes that "most definitions of intelligence fall into one of two categories. The first is worded so that intelligence requires a brain; the definition refers to intrinsic mental qualities such as reason, judgment, and abstract thought. The second category, less brain-bound and metaphysical, stresses behavior, defining intelligence as the ability to respond in optimal ways to the challenges presented by one's environment and circumstances. Not surprisingly, the plant neurobiologists jump into this second camp" (Pollan, "Intelligent Plant").

25. Pollan, "Intelligent Plant."

information with abstract or conceptual information. This dipolar integration of the physical and the conceptual obviates the criticisms of biologists who dispute the presence of teleology in plants simply because plants lack the detectable morphology of a "true nervous system." The functionality of abstract possibilities and goals exists in the ontological structure of each actual occasion—the most rudimentary components of nature—and therefore need not be restricted to a macroscopically visible morphology. It is the very nature of actual occasions to prehend their predecessors—that is, to utilize information. The reception and assessment of information, which is the subject of Whitehead's theory of prehensions, is the basis of how each actual occasion forms itself. He writes,

> Each actuality is essentially bipolar, physical and mental, and the physical inheritance is essentially accompanied by a conceptual reaction partly conformed to it, and partly introductory of a relevant novel contrast, but always introducing emphasis, valuation, and purpose. The integration of the physical and mental side into a unity of experience is a self-formation which is a process of concrescence, and which by the principle of objective immortality characterizes the creativity which transcends it. So though mentality is non-spatial, mentality is always a reaction from, and integration with, physical experience which is spatial.[26]

When Whitehead says that the "physical inheritance" prompts a "conceptual reaction" that is "partly conformed" and "partly introductory of a relevant novel contrast," he means that the mental prehension is partly faithful to the physical prehension that prompted it but also partly an original creation that serves to suggest an outcome or response that is as yet undetermined. This "introduction of a relevant novel contrast" is the basis for the differing states that are simultaneously present in a state of quantum superposition.

Obviously in an example from physics no central nervous system is necessary, or even possible. The point at present is that nature does not have to wait for a brain or some other neurological morphology to appear in order for information to be used in a meaningful way. Whitehead, who referred to his own system as "the philosophy of organism,"

26. Whitehead, *Process and Reality*, 108. By "each actuality" Whitehead means each *actual entity*, not each aggregate of entities or each macroscopic object. "Objective immortality" refers not to the subjective immortality posited by, for example, the orthodox Christian belief in life after death but to the ongoing influence of events throughout the web of nature, even after the events themselves have ended.

constructed an elaborate theory of types of social structure pertaining not only to groups of living beings but to virtually all enduring objects that manifest self-produced patterns of organization.[27] As previously noted, he used the term *nexus* to refer to a mere aggregate. Other objects, both living and nonliving, are organized in various ways. Some societies are democratic, with little or no need for hierarchy. This pertains, for example, to many species that live in herds, flocks, or colonies. Other societies depend on the guidance supplied by an identifiable source of coordination or leadership. An animal's body itself is a society of occasions in which subsocieties such as individual cells and various organs may function democratically but in which there is a "presiding occasion" that, Whitehead guessed, is moving about in the interstices of the brain.[28] The presiding occasion's influence is sufficient to guide the organism's behavior, but the presiding occasion's hegemony is certainly not absolute. We observe the effects of the presiding occasion when the creature responds as a whole to stimuli.

In any case, the utilization of abstract information is primordial in nature, and therefore in organisms such as plants, information can be shared in ways that do not require a central nervous system. A remarkable example of this is the slime mold, which clearly has no central nervous system but has been shown to navigate its way to food sources through complicated mazes. In fact, in one experiment in which bits of edible material were placed on a flat surface in predetermined locations resembling a map of the Tokyo area, the slime mold effectually made paths to pieces of food in a way that produced a more efficient diagram of the Tokyo regional subway system. The behavior of slime molds alone should be astonishing enough to convince any skeptic concerning the presence of purposiveness in nature.[29]

If the goal-oriented use of conceptual information is detectable in plants, it is, at least for human observers, even more detectable in animals. Copepods—tiny animals that live in drops of both sea water and fresh water—are the most plentiful animals on earth.[30] Though related to crustaceans such as lobsters and crawfish, copepods are much smaller, ranging from about 0.2 to 1.7 millimeters (less than one-tenth of an inch)

27. Whitehead, *Process and Reality*, 83–129 (part 2, chs. 3 and 4).

28. Whitehead, *Process and Reality*, 105–9.

29. Ito, "Slime Holds Intelligence Key." See also BBC Earth Science, "Can Slime Mould Solve Mazes?" and WIRED, "Blob Solves Mazes."

30. Rothenberg Gritz, "Drop of Seawater."

in length. The little animals go through six larval and six juvenile stages between egg and adulthood and grow a new pair of legs at each stage. Though they do have a brain and a nervous system, these neurological structures are tiny in comparison with those of most other animals. Nevertheless copepods are continually engaged in obtaining food, detecting and fleeing from predators, and finding a mate.[31]

Though most animal species possess nothing like human self-awareness and intelligence, even the simplest animals engage routinely in elementary mental acts of assessment. One summer afternoon while sitting on our patio I noticed a roly-poly or "pill bug" (*Armadillidium vulgare*) crawling across the concrete. The little creature, which is somewhat smaller than a pinto bean, came to the edge of a steep curb where it would have fallen into a puddle of water if it had continued. However, the bug stopped at the edge and remained motionless for almost a full minute, seeming to peer over the brink and analyze its circumstances. Finally it backed up, turned around, and crawled away to safety. It is this type of nonconscious intelligence, and not human thinking, that best illustrates the existence of purposiveness in nature.

Such goal-oriented uses of information are ubiquitous in animals. Salmon return to the waters where they were born. This prevents them from swimming to a location that would not be suitable for spawning. Biologists think salmon find their way by sensing the earth's magnetic field and then relying on their sense of smell to locate their home stream.[32] In Africa, colonies of termites construct large, porous chimneys that use passing breezes to force fresh air into their living quarters, which are underground and would become uninhabitable without an air supply.[33] In New Guinea, male bowerbirds use twigs and other materials to assemble a door-like structure on the ground; they then decorate the structure with brightly colored objects. This little edifice is intended to gain the attention of a potential mate. Female bowerbirds assess the male's creation as the chief criterion in selecting a partner, and the females seem to have very exacting tastes.[34] The arctic tern migrates annually all the way from the Arctic to the Antarctic for a total distance of over eighteen thousand miles. Individual birds do not simply depend on visible landmarks but are able to vary their route depending on weather conditions and the

31. Yen et al., "Sensory-Motor Systems of Copepods."

32. USGS, "How Do Salmon Know."

33. PBS, "Incredible Termite Mound."

34. Wikipedia, "Bowerbird."

availability of food sources.[35] Monarch butterflies migrate from Mexico to Canada; no single butterfly lives long enough to complete the route, so the total migratory pattern is somehow passed from one generation to the next.[36] Dolphins hunt in organized packs and use strategy to trap their prey.[37] Several types of birds can count,[38] and crows can count out loud—that is, while vocalizing.[39] Crows, elephants, dolphins, and chimpanzees seem to "hold funerals" or to engage in other ritual-like behavior associated with mourning their dead.[40] Dogs are regularly reported to have acted deliberately and with foresight in order to save the lives of human beings; this includes one German Shepherd who knocked aside a rifle as a woman was pointing it at her chest in an attempted suicide.[41]

Despite examples such as these, many scientists still remain cautious about the idea that animals experience consciousness. On April 19, 2024, a distinguished group of scholars signed "The New York Declaration on Animal Consciousness," affirming that consciousness *may* be attributed to animals other than humans. What seems remarkable about this document is not what it says but that it says it so tentatively ("evidence indicates at least *a realistic possibility* of conscious experience in all vertebrates"[42]). It is a good guess that anyone who has ever spent much time around animals would view the allegation that animals experience consciousness simply as a common-sense assertion. Reluctance among scholars to attribute mental awareness to animals can be traced back to Descartes, who denied that animals possess souls or consciousness.[43] The

35. Ramroop and West, "To the Ends of the Earth."

36. Reppert and de Roode, "Demystifying Monarch Butterfly Migration."

37. NOAA Fisheries, "Common Bottlenose Dolphin."

38. Ditz and Nieder, "Neurons Selective."

39. Starr, "Count Out Loud."

40. Weisberger, "Why Crows Hold Funerals." The author attempts to explain away any elements of grief or ritual by claiming that crows attend to their dead merely "to identify predators and potential threats," but this survival-of-the-fittest reductionism seems strained. Undoubtedly animals do experience loss and sadness.

41. Sieczkowski, "Loyal Dog."

42. New York University, "Animal Consciousness"; emphasis added. See also Bush, "Scientists Push New Paradigm," and Ghosh, "Are Animals Conscious?" I cite these articles not to support the idea that animals experience consciousness but to illustrate the incredible tentativeness of many intellectuals to accept the idea that animals "may be" conscious.

43. Williams, "Descartes, René," 354.

Cartesian separation of mind and matter as an *ontological* dichotomy bestows the ability to reason only upon humans.

My chief criticism of recent declarations about animal consciousness is that they assume that consciousness or sentience is a clearly defined, substance-like phenomenon—that is, that it either *is* present or it *isn't*, thus failing to understand that there are vitally important modes of mental functioning that are not fully "conscious" but that biologically must have existed in living organisms prior to the development of consciousness as we humans now experience it. One reason why it is so difficult for scientists to affirm the presence of consciousness in simpler organisms is that they tend to reason backwards from their own experience of human self-awareness and then find it difficult to ascribe this presupposed human paradigm to, say, fish or insects. This, as I have said, is the implicit legacy of Descartes.

We commonly explain the purposive activities of animals by citing instinct, but the word *instinct* is grossly misused at this point. We human beings have an unfortunate habit of assuming that when we invent a word to refer to some phenomenon, we have also *explained* that phenomenon—but description and explanation are two different things. Dictionary definitions of *instinct* are not explanatory; they are merely descriptive. *The American Heritage Dictionary* defines instinct as "an inborn pattern of behavior that is characteristic of a species and is often a response to specific environmental stimuli." This definition does not even attempt to show how a "pattern of behavior" can be "inborn"; it simply states that it is. *The Century Dictionary* goes somewhat further by defining instinct as "a special innate propensity, in any organized being, but more especially in the lower animals, producing effects which appear to be those of reason and knowledge, but which transcend the general intelligence or experience of the creature; the sagacity of brutes."[44] This definition acknowledges the question, but does not actually answer it, as to how animal behaviors can seem to be based on "reason and knowledge" despite the fact that those behaviors "transcend the general intelligence or experience of the creature." So, how *does* the animal know what to do? When someone "explains" knowingly that an animal "just does that by instinct," this is not a genuine explanation but more nearly a tautology—a matter of word definition. In situations like this, the word *instinct* has been uttered as a putative explanation, but there is no

44. Definitions from both dictionaries can be found at Wordnik Online Dictionary, s.v. "Instinct," https://www.wordnik.com/words/instinct.

genuine elucidation of the process by which the animal envisions and then undertakes the behavior in question. This use of *instinct* amounts to little more than an intellectual dodge.

In this book the basis of instinct is explained in terms of three philosophical doctrines: (1) nature at its core is composed not of inert substances but of events; (2) by their own act of self-constitution, these events spontaneously receive and transmit information not only physically but abstractly or conceptually; and (3) a goal-oriented or teleological influence is manifest in the way that this information is implemented. For example, when the termite colony produces a ventilation chimney, it is not necessary for the requisite construction plan to be taught to and utilized by each termite as if each one were a conscious human student; rather, information is shared spontaneously throughout the colony as a whole. This does not require the type of conscious understanding that we associate with human thinking. Once we replace the idea that colliding particles are the ultimate basis of reality with the idea that information-transmitting events are the reality, it is much easier to see how the termite colony responds as a whole to the project of chimney building. Partly, the urge towards chimney building is built into their DNA, but the coding is not purely physical; it must contain conceptual or abstract elements or else the construction project could not be adapted to the unique demands of specific locations. In Whiteheadian terms, the serial route of occasions that guide the behavior of each individual termite prehends the "lures for feeling" that contribute to chimney construction.

This may sound suspiciously like panpsychism or mental telepathy, but it is not. When the electrons in an atom share their quantum numbers or quantum states with one another, this is not telepathy or panpsychism. It is simply a direct transmission of data that may variously be described as "mathematical" or "abstract" or "conceptual." Readers are asked to recognize that, whether or not we find this idea easy to accept, this transmission of data is the way that nature operates. If we find this assertion unbelievable, it is not because this type of data transfer is innately inconceivable but because, especially if we are scientific-minded, we have been indoctrinated with the belief that information can be shared only by means of a physical intermediary—in other words, only by local rather than by nonlocal means. Physicalism has been our entrenched intellectual habit ever since the time of Newton, and before. But we can now see that the transmission of conceptual information among termites is, in principle, not unlike the transmission of mathematical information among the electronic events that form a system of orbitals according to the requirements of the Pauli exclusion principle.

Throughout nature, an aim or final cause is necessarily involved in order to explain the fact that each actual occasion is the ultimate reason for its own outcome. I am not claiming that *all* events are freely self-determined. Macroscopic events involving aggregates—such as the behavior of gases, the acceleration of falling bodies, and the energy released by certain chemical reactions—are, for all practical purposes, mechanistic and deterministic. But on the level of single quantum systems, if events were not self-caused, each event would of necessity be caused externally by preexisting conditions, and nature would be deterministic—but determinism is not what we observe in quantum events. It is also not what we observe throughout nature, for the nondeterministic implementation of nonphysical, abstract information is observable in the behavior of all holistic systems. Thus each termite mound is adapted to its own location, each bowerbird deliberately builds a different bower, each monarch butterfly selects its own milkweed, and each salmon swims to its own birthplace. Even the lowly roly-poly might not have turned and headed towards safety; it might have continued forward and plummeted over the curb into the water. Final causation allows for individual differences in a way that utter determinism does not.

It is simply preposterous to claim, as many scientific-minded people still do, that behaviors such as the ones just described can be explained away by a recourse to physicalist reductionism. In his book *D'Alembert's Dream*, which was first published in 1782, the French philosopher Denis Diderot (1713–1784) imagines a conversation between himself and the physicist d'Alembert. Though Diderot appreciated the advances represented by Newtonian physics, which at that time were being widely touted, he nevertheless ridiculed attempts by European intellectuals to reduce life to the workings of mechanistic forces. In *D'Alembert's Dream* Diderot reminds d'Alembert of the way that the original contents of an egg spontaneously evolve into flesh, a beak, wings, eyes, and feet. Finally the egg hatches.

> Now the wall is breached and the bird emerges, walks, flies, feels pain, runs away, comes back again, complains, suffers, loves, desires, enjoys, it experiences all your affections and does all the things you do. And will you maintain, with Descartes, that it is an imitating machine pure and simple? Why, even little children will laugh at you, and philosophers will answer that if it is a

> machine you are one too! . . . Just listen to your own arguments and you will feel how pitiful they are. You will come to feel that by refusing to entertain a simple hypothesis that explains everything—sensitivity as a property common to all matter or as a result of the organization of matter—you are flying in the face of common sense and plunging into a chasm of mysteries, contradictions and absurdities.[45]

Diderot makes his point convincingly. However, it will be objected that his phrase "sensitivity as a property common to all matter" leads directly to panpsychism and that panpsychism is no more believable—perhaps it is even less believable—than reductionistic physicalism. I tend to agree with this critique and have said several times in the preceding chapters that this book does not endorse panpsychism. We have now reached a point at which this statement must be explained.

Psyche is the Greek word for soul. It has been incorporated into modern English as the basis for our word *psychology*, with all of psychology's rich applications and implications pertaining to human experience. Psychology encompasses complex topics such as assorted theories of personality, theories about learning, the distinction between conscious and unconscious experience, various definitions of neurosis and psychosis, the interpretation of dreams, differing approaches to psychotherapy and behavior modification, the diagnosis of mental illness, and so on. Therefore, when applied to nature itself, the term *panpsychism* cannot help but encumber physical existence with the onus of embodying the complexities and subtleties of human behavior. This, I submit, is simply an incongruous notion. Insofar as panpsychism envisions *human* experience as a model that is followed throughout nature, panpsychism is not merely too humanocentric and personalistic; it is grandiose, for it takes human experience as the definitive example for the rest of the world. This problem becomes even more conspicuous for those panpsychists who propose that the world itself has a soul, for there is no clear evidence to support this claim. Thus, even when various qualifications are stated, panpsychism claims too much—and because it claims too much, panpsychism inspires widespread skepticism among many people who are not predisposed to interpret the world in personalistic terms.

What we need is a more nuanced doctrine of what can be called *panexperientialism* or *pansubjectivism*. Even these terms, however, must

45. Diderot, *D'Alembert's Dream*, 166–67, quoted in Prigogine and Stengers, *Order Out of Chaos*, 80–81.

be carefully defined. There are only certain types of entities that may be said to possess subjectivity or to have experience. Earlier in this chapter I distinguished four types of enduring objects: (1) aggregates, (2) structured aggregates, (3) organs or other organismic structures that serve one or more purposes in a larger context, and (4) entities that experience true subjective unity. We now ask in what sense, if any, these objects can be said to have experience or to possess subjectivity. As an answer we can immediately suggest that the more holistic an object is, the more likely it is that the entity may be understood as an experiencing subject. Using the criterion of holism, it is clear that objects such as rocks and clouds do not have subjective experience. Rocks and clouds have already been noted as examples of aggregates that are not good examples of holism. Holistic entities that may accurately be said to possess subjectivity include atoms, one-celled organisms, plants, and animals, though entities such as these certainly do not experience self-awareness in the way that human beings do.

As human beings we naturally tend to view experience and subjectivity in terms of our own thoughts, feelings, motives, and actions. But I am not attributing human experiences such as these to nature in general. Rather, as a stipulative definition I am proposing that we *redefine experience and subjectivity* so that these terms may be applied to *any holistic or organismic entity that responds to its environment in a unitary, goal-oriented way*. We have observed that even slime mold behaves with a kind of intelligence, if by "intelligence" we mean a characteristic ability to respond creatively and constructively to the environment. In other words, the entity's response entails not only a sensitivity to its circumstances but also an assessment of the possible outcomes of various actions. For most people, it is unbelievable that plants and insects think thoughts or make conscious decisions—but it is not unbelievable that they respond to "fiercely competitive or adverse, real-world circumstances"[46] by purposely and consistently changing their behavior to improve their probability of survival. A tree, for example, does not think, but it does behave as an organismic unit, sending its roots in the direction of water, turning its leaves to receive sunlight, manufacturing cellulose by photosynthesis, and adding annual growth rings to its trunk.

Because of the way the theory of evolution was presented by Darwin and others, today we tend to assume that the principal goal sought

46. Calvo et al., "Plants Are Intelligent."

by living organisms is *survival.* Obviously, an organism whose actions did not contribute to self-preservation would soon die, and in many behaviors such as finding food and avoiding predators, survival is the foremost result. However, I now suggest that in the experience of the organism itself the more motivating goal is not survival but, actually, an immediate, intuitive enjoyment or sense of satisfaction pertaining to consequences that are more immediate.

For example, when the male bowerbird constructs a wicket and the female accepts it, the long-term result is likely to be the reproduction of the species; but an awareness of that eventual result is based not on logical reasoning on the part of the bowerbirds but on our own, human observation of the outcomes of bowerbird behaviors that we witness over a span of time. It seems very unlikely that the birds themselves have any understanding of the connection between bower building and the survival of their species. It is more probable that they are drawn forward by the lure of feelings of enjoyment and satisfaction that are associated with certain imminent factors such as the attractiveness of the bower itself and feelings of pleasure in finding and interacting with a mate.

The truth is that we human beings ourselves live more of our lives in this way than we admit, making decisions on the basis of feelings of satisfaction that are anticipated, and only afterwards explaining those decisions as a product of rational deliberation. I am now suggesting that even plants and microbes experience a means of assessing their environment and responding in ways that fulfill appropriate goals. Their behavior does not include conscious deliberation, but functionally it amounts to a decision nevertheless and therefore counts as a form of experience. By this nonconscious, nonrational type of experience, nature produces the complexity, innovation, and wholeness that are conspicuous in the world around us. Possibilities are assessed and acted upon. This requires the reception and utilization of information. The information is not only physical, pertaining to existing conditions, but abstract or conceptual, serving as a means of assessing potential outcomes.

In fact, the purposive use of information is found not only in living organisms but in holistic types of what has historically been referred to as "inanimate matter." In quantum physics, phenomena such as entanglement, nonlocality, and the formation of quantum systems qua systems all point to the reality of what it means to be an experiencing subject at the atomic level. Each atom is a subject that has its own history and its own experience, not in the sense of having thoughts or emotions but in the sense of receiving information as a unit and acting as a unit, a whole—in bonding with other atoms, for example. From this example it is obvious that there are coexistent layers of subjectivity and layers of experience throughout nature. In the same way, in the human body each cell has its own history and experience, and the human being as a whole has her or his own history and experience. Redefining subjectivity and experience in ways that apply to plants, microbes, and atoms is invaluable as we seek to understand the basis of human sentience and consciousness. Human self-awareness may be viewed as a natural sequel to simpler forms of experience that have already emerged in nature. The existence of mind in human beings is thus not an anomaly but rather a more complex example of nature's general tendency to traffic in information.

THE HARD PROBLEM OF CONSCIOUSNESS

It is not a primary purpose of this book to solve the mind–body problem or the hard problem of consciousness, but it will nevertheless be helpful now to discuss these topics as a way of clarifying the concept of experience that I am advancing. I presented the hard problem of consciousness in the closing pages of chapter 2. I noted that the phrase "hard problem" was first coined by philosopher David Chalmers in his paper "Facing Up to the Problem of Consciousness" in 1995.[47] Chalmers asked how the transmission of an enormous multitude of signals between the neurons in the brain can result in the consolidated, unitary phenomenon that we experience as consciousness or self-awareness. Another way of stating the problem is to ask how thoughts and feelings ("qualia") can arise from a preexisting matrix of nonsentient matter. During the years since Chalmers's initial article, the hard problem has not dwindled away but has instead overflowed beyond the boundaries of philosophical specialty and been taken up by scholars in various fields, including even a few physicists.[48]

It is beyond the scope of this book to undertake even a short survey of the answers that have been suggested. Instead, I will state directly that, in the philosophical orientation being defended here, the "hard problem" of consciousness is, to put it frankly, a fake problem. It is fake in the

47. Chalmers, "Facing Up to the Problem."

48. For an early example, see Penrose, *Emperor's New Mind*. Concerning the current state of the debate, see Horgan, "25-Year-Old Bet."

sense that is based on erroneous premises and is therefore an attempt to solve a problem that is more nearly imagined than real. As commonly set forth, the hard problem does at least acknowledge the authenticity of our mental life—that our thoughts, feelings, and desires do somehow exist and are not merely epiphenomenal. In this, the statement of the hard problem does go beyond uncompromising physicalism. But then, unfortunately, the statement of the problem proceeds by contrasting our conscious experiences to what is presumed to be the solely physical basis of the biology of the brain. Thus, though Descartes has not even been mentioned, the premises on which the hard problem is based are distinctly Cartesian. The doctrine of mental substance, which is the natural consequence of Descartes's foundational *cogito* ("I think"), is posited in the hard problem's description of the subjective quality of consciousness. But the hard problem then introduces a Cartesian conundrum by presupposing a physicalist understanding of the brain's constitutive neurons. The hard problem is "hard"—in fact, as framed, it is unsolvable—because it can be solved only by explicitly rejecting the metaphysics of physicalism on which its ideas about brain physiology continue implicitly to be based. Regrettably, however, the majority of writers on the subject of the mind–body problem seem to be gripped by an aversion to any genuinely original metaphysics.

Whitehead presented a nuanced analysis of the body-mind problem a century ago,[49] but his approach has generally been neglected,[50] often for no other reason than that Whitehead's philosophy is said to be complicated—with an implication that any careful analysis of it can always be postponed until later—or because Whitehead's thought has mistakenly been identified with panpsychism,[51] and then panpsychism

49. Whitehead, *Process and Reality*, part 2, ch. 3, esp. secs. 9–11.

50. A typical example is the Wikipedia article on the mind–body problem (Wikipedia, "Mind–Body Problem") in which Whitehead is mentioned only in a single sentence and no attempt is made to summarize his ideas. In this article David Griffin is credited with promoting the term *panexperientialism*, but when I was a student at Claremont (1975–1976) *panexperientialism* was in general use and not viewed as the brainchild of any specific scholar.

51. Even many of Whitehead's supporters identify him as a panpsychist, but this description is mistaken. Whitehead never used the term, and though his use of the word *feeling* may suggest panpsychism, Whitehead redefines "feeling" to refer to the sensitivity of every actual entity to the factors in its environment and to the elements in its own phases of concrescence. The assertion that Whitehead was a "pansubjectivist" rather than a panpsychist is now widely attributed to the scholar David Griffin, but the statement that "Whitehead was not a panpsychist" was first made in my hearing

has been peremptorily dismissed.[52] Whitehead may be said to endorse "panexperientialism" or "pansubjectivism" but not "panpsychism," and especially not if panpsychism is understood as the belief that consciousness or rationality is present throughout nature. Whitehead wrote, "The principle that I am adopting is that consciousness presupposes experience, and not experience consciousness. It [consciousness] is a special element in the subjective form of some feelings. Thus an actual entity may, or may not, be conscious of some part of its experience."[53] "Mental activity is one of the modes of feeling belonging to all actual entities in some degree, but only amounting to conscious intellectuality in some actual entities."[54] As Marjorie Suchocki explains, "Consciousness is itself the narrowing of focus, the creation of a foreground and a background by lifting 'this' rather than 'that' into significance."[55] Suchocki is saying that consciousness is simply the means by which we ignore some aspects of our current experience in order to focus our attention on other aspects. (This point has already been made in chapter 3.) She writes, "In any one moment, far more impinges upon the self than can be cognated. Even less can be carried over into the continuing self in the form of significant memory."[56] As vital as consciousness is, consciousness points beyond itself to a deeper type of intentionality—a core selfhood—that is *not* conscious, but which coordinates or controls consciousness, and which has been investigated thoroughly by Freud, Jung, and other depth psychologists. Suchocki writes,

> For example, an adult remembers childhood not in its detail, but in terms of an overall retention of feeling that may be positive or negative in tone, with some of the occasions that contributed to that type of feeling available for remembrance at any time. But by far most of the events of childhood are no longer immediately accessible to memory. The remembered past of childhood is like the experience of crossing a brook on stepping stones—the stones upon which one steps may be the only ones seen, but

---

by John B. Cobb in a classroom lecture, "Advanced Seminar in Process Studies," at Claremont Graduate School, in fall 1975. This preceded the more recent attributions of the idea that Whitehead "was not a panpsychist" to Griffin.

52. An example of a superficial, uninformed dismissal of Whitehead's philosophy is provided by Deacon, *Incomplete Nature*, 78–79.

53. Whitehead, *Process and Reality*, 53.

54. Whitehead, *Process and Reality*, 56.

55. Suchocki, *Fall to Violence*, 38.

56. Suchocki, *Fall to Violence*, 38.

> they are there because of the whole rocky bottom of the brook. There are many stones, hidden from the surface, and their presence supports the stones upon which one steps. One remembers a few events from childhood, but there were many, many more that are no longer easily available to conscious memory.[57]

On an empirical level, it is indisputable that, as Whitehead said, in creatures with brains there is a *dominant occasion* or a *presiding occasion* that provides a level of informed guidance and control for the organism as a whole. Without such a presiding occasion the organism would not be able to act in any unitary way—but organisms *do* act in unitary ways, as selves. (For example, you are now reading this book.) Though the phrase "presiding occasion" (or "serially ordered route of occasions") may sound speculative, the functioning of this presiding occasion is obvious whenever any creature acts as a single being with an identifiable purpose. The bottom line is that *the experience of this presiding occasion is associated with what we recognize as consciousness*. But the experience of the presiding occasion is not always conscious; in fact, frequently it is not, as when you are asleep or when you act out of habit without being fully aware of your actions.

The consciousness associated with the route of presiding occasions is not found exclusively in human beings but in numerous species of animals, occurring in the forms of experience that are best suited to the circumstances and purposes of the species in question. We may assume that consciousness coevolved in animal life concurrently with vision, hearing, and other senses, as a coordinated way of informing the organism's responses to changes in its environment. We human beings assume that our modes of intelligence are superior—and in some cases, they are—but plentiful examples can be cited of ways in which animals cope with circumstances that leave us humans perplexed, oblivious, or helpless. We should also note that the type of consciousness observed in animals would be of no use in plants since plants lack the mobility and physical dexterity of animals. Consciousness evolved in order to direct the behavior of creatures who possess a significant freedom of movement.

The upshot of all this is that if in our metaphysics we dispense with insensate material particles and instead affirm that nature is composed of serially ordered events whose innate nature is to integrate both physical and conceptual elements, the unbridgeable division between physiology

57. Suchocki, *Fall to Violence*, 38.

and consciousness never arises. In a process-oriented, organismic philosophy, consciousness is not viewed as a peculiar phenomenon that has somehow sprung into existence, complete and fully formed, in the lives of human beings. Consciousness has instead evolved naturally from the complexification of much simpler elements in nature—but elements that, even in their most primordial form, evince the presence of experience. But finally, there certainly is no reason for claiming that consciousness is ubiquitous in nature.

A final observation regarding consciousness is that the contents of consciousness consist entirely of *symbols*. By *symbol* I do not mean a graphical representation on a page or a metaphorical interpretation found in art or writing. By *symbol* I mean any mental representation that has been constructed by the mind, whether as a result of neurological signals from the body or as the product of sheer imagination originating in the cells of the brain.

Consider, for example, the way an infant learns to recognize the faces of her parents. We know that the parents' faces are actually composed of cells and that the cells, in turn, are composed of various types of molecules. The faces and other objects that we see are, in the way that they appear, constructed from millions of energy events in our retinas by the joint operation of our eyes, our optic nerve, and brain's occipital lobe. Memory is also required, because our minds constructively compare present sensations with images stored in memory, thus making the world recognizable and intelligible. It is these constructed images that an infant uses as she or he learns to recognize faces and to interpret experience generally, and that I am referring to as symbols. In the present discussion, the word *symbol* means more or less the same thing as the word *qualia*.

Symbols are not only visual but auditory, aromatic, gustatory, tactile, and visceral. In fact, symbols are used also to refer to emotions, values, purposes, questions, convictions, and social mores. Indeed, the fact that symbols can be adapted by the imagination constitutes one of their chief advantages. The ability to create imaginative symbols represents not an unfortunate loss of contact with reality but an essential component of nature's bid towards creativity and novelty. In 1938 Whitehead wrote, "Importance generates interest. Interest leads to discrimination. In this way, interest is increased; and the two factors, interest and discrimination, stimulate each other. Finally consciousness develops, gradually and fitfully; and it becomes another agent of stimulation."[58] Note Whitehead's insight that the creative process does not originate in consciousness, and that consciousness becomes a factor only when interest is strong enough.

The subjective aspect of experience need not be a reason to downgrade experience's epistemological value. In chapter 10 I discussed the philosophy of William James and his radically empiricist belief that experience in itself is real. James's approach, and Whitehead's, remind us that the role of symbols in our mental life is not simply a source of private mental invention but an aspect of reality—a means not only of accurately representing[59] our existing circumstances but a path to imagination, creativity, problem-solving, personal meaning, and social progress.

58. Whitehead, *Modes of Thought*, 31.

59. The notion that our conscious experience of the "external world" consists of "representations" is known as *representationalism* and has a complex history in Western philosophy. A full discussion of this topic is beyond the scope of this book, but I discuss the issues briefly in chapter 13.

13

Experience, Symbols, and Creativity

The preceding chapter concluded with the assertion that all of our conscious experiences rely on symbols. I stated that by *symbols* I do not mean only visualizable shapes or signs but any mental means of identifying the content of an experience.[1] For example, the sweetness of sugar may be regarded as a symbol. The sweetness is derived from the chemical composition of the sugar molecule, but *sweetness itself* is a qualia or an experience that may be understood as a symbol or, as it is sometimes called, a construct. Sugar really *is* sweet in a way that salt is not—but sweetness *as a quality* exists as a personal experience and not in the sugar itself. This definition of symbol is much broader than the more common idea that a symbol is simply a visual image.

The mind uses symbols as a way of representing the contents of experience, both conscious and unconscious.[2] When we associate symbolism with the general contents of experience, it becomes evident that symbols are used not only by human beings but by simpler organisms. Symbolism is not limited to human consciousness. The role of symbols is evident in much simpler natural phenomena. When a bowerbird constructs a bower in an attempt to attract a mate, it is relying on symbols. When a trout distinguishes between an edible insect and an inedible particle of debris, it is relying on symbols. When bees work together

1. The meaning of the word *symbols* in this chapter closely resembles the meaning of *qualia*. In general, my analysis of symbolism follows Whitehead's. See Whitehead, *Symbolism*.

2. It is evident that many forms of experience are unconscious. See Bargh, "Unconscious Thought and Perception."

to construct a hive, they are relying on symbols. Symbolism, as defined here, is at work in any setting where nature identifies and reproduces patterns qua patterns, whether they are patterns of physical structure or patterns of behavior. In the present chapter, however, the focus is on human modes of perception.

A primary function of symbols is the consolidation and simplification of large amounts of neurological data. The perception of a red book on a table is initiated when millions of photons strike the retina of the eye, but what we see is not millions of individual red flashes. We see a single red area that experience has taught us to identify as a book.[3] Our minds constantly consolidate massive amounts of data by using symbols to represent specific objects, sensations, and feelings. The amalgamating aspect of symbols is especially evident when the mind integrates a multitude of differing nerve impulses into a novel form of experience. For example, a mixture of tiny, evenly interspersed red dots and blue dots on a printed page appears to us as the color purple. The dots are not purple; they are either red or blue. The perception of the color purple does depend on the existence of an objective source—the dots—but the purple color is constructed by the mind as an act of blending and consolidation. In this type of example—and in most conscious experiences—both objectivity and subjectivity are necessary. We can say that the experience is dipolar, integrating an objective pole (depending on the dots) with a subjective pole (depending on mental synthesis).

The constructive ability of the mind is also evident in the representation of objects or ideas that do not exist outside of the mind. Solving a problem in mathematics, reading a novel, and planning a dinner menu all involve the constructive imagination of objects or situations that have no actual physical existence. Relatively simple physical activities such as speaking, singing, or playing a sport require constructive imagination. We guide our muscles in ways that, we imagine, will successfully sing a certain note, slice an onion, or return a serve. By an act of imagination we

3. Mental operations allow multiple perceptions of the same quality to be unified into a perception of a single object that has a single property or quality throughout. Whitehead called this prehension of unity out of multiplicity "the Category of Transmutation" and regarded it as a basic metaphysical category, operative not only in the minds of animals with larger brains but at the level of molecules and atoms (Whitehead, *Process and Reality*, 61–65, 250–54). Because each actual entity has a mental pole, each actuality has at least some small capacity to identify a common characteristic found in multiple past occasions and consolidate or objectify those data into the prehension of a single object. This is what Whitehead meant by transmutation.

try to anticipate the trajectory of a pitched ball as we swing our bat. In all of these cases, the mind makes use of what I am referring to as *symbols*.

However, the mind's continual reliance on symbols has not always been greeted with joyous appreciation. To many people it is troubling to think that our perception of the world is so thoroughly dependent upon mental construction. We have an understandable need to believe that what our perceptions convey is not merely symbolic but factual. As children we learn that it is essential to distinguish between imaginary things (for example, that our teddy bear can talk) and real things (that our teddy bear is actually a stuffed piece of fabric). This distinction becomes even more important as we grow older. Especially, in our perception of the world it is essential to be able to correct errors. A frequent example is that for centuries most people believed the earth is flat.[4] Eventually, however, numerous empirical considerations—including calculations by astronomers and the voyages of explorers—led to the recognition that it is spherical. The point is that there must be something about the way we perceive things that allows for reality testing. Though perceptions led to the original error of believing that the earth is flat, further perceptions, when more critically interpreted, revealed the error. Experimental observation is the basis not only of all our sciences but of all forms of practical knowledge. But, the question now being asked is, if all of our conscious experiences consist of symbols, how can we be assured that the symbols are accurate? Science itself has shown us that ordinary modes of perception do not always lead to accurate depictions of physical reality. We have already noted that the seemingly solid objects of our daily experience are actually composed of large numbers of particles flying at high speeds through what is mostly empty space.

In academic philosophy, this discrepancy between macroscopic appearance and microscopic functionality is investigated under the heading "philosophy of perception." Since the early twentieth century, the discussion of perception has mostly centered on a debate between two schools of thought: *direct realism* and *indirect realism*. Another position, philosophical idealism, emphasizing that all perceptions exist in the mind as *ideas*, was prevalent at the beginning of the nineteenth century but has gradually diminished in influence.[5] We considered the doctrine

4. But even in antiquity, thinkers such as Pythagoras, Eratosthenes, and Aristotle reasoned that the world is spherical. See Chodos, "Eratosthenes Measures the Earth."

5. Hirst, "Realism," 77. In chapter 9 I cited the philosophical idealism of George Berkeley (1685–1753). See also Wikipedia, "Direct and Indirect Realism."

of realism in chapter 9 as an aspect of the thought of physicist Bernard d'Espagnat, who contrasted realism with instrumentalism. D'Espagnat defined realism as the belief that the properties of material objects and physical systems exist whether or not they are observed by human beings. "Instrumentalism," on the other hand, makes *observability* an essential aspect of truth-finding, and therefore instrumentalism surprisingly resembles the orientation of idealism, which focuses on the existence (or at least the possibility) of mental content.

The theory of realism generally assumes that nature is composed of material objects that are located at definite positions in space, where they are able to be observed or tested.[6] In its idea of an objective material world, realism leads to the notion that, at least in principle, it should be possible to have a "God's-eye view" of the world—that is, a view that would be unbiased, factual, comprehensive, and free of errors. In contemporary philosophy *direct* realism is the opinion that perception is a direct awareness of external reality—a straightforward encounter by one or more of the senses with the object being observed.[7] *Indirect* realism distinguishes between the external objects themselves and the private sensa that are the mental results of brain processes caused by the effects of the external objects on the sense organs.[8] In the words of Willard Van Orman Quine, "Impressed with the fact that we know external things only mediately through our senses, philosophers from Berkeley onward have undertaken to strip away the physicalistic conjectures and bare the sense data. Yet even as we try to recapture the data, in all their innocence of interpretation, we find ourselves depending upon sidelong glances into natural science."[9] Science, in other words, seems to be needed as an intermediary so that we can explain the connection between the objects perceived and their appearance as we perceive them. Science tells us, for example, that the light waves are focused by the lenses of our eyes onto our retinas, which then send nerve impulses to the visual cortex, and so on. One concept related to this insight is the notion of "*the myth of the given*," associated with the philosophy of Wilfrid Sellars (1912–1989).

6. Whitehead disputed the commonsense idea that physical objects simply endure in a well-defined location in space, calling this idea "the fallacy of simple location." He also disputed the notion that the objects we perceive endure from moment to moment through time because they consist of unchanging material substance, calling this "the fallacy of misplaced concreteness" (Whitehead, *Symbolism*, 38–39).

7. Whitehead, *Symbolism*, 78.

8. Whitehead, *Symbolism*, 80.

9. Quine, *Word and Object*, 1–2.

Sellars denied that there can be any truly objective basis—a "given" or a foundation—for our statements, as if they could derive their meaning directly from confrontation with a particular kind of object or experience. Sellars is thus anti-foundationalist in his theories of concepts, knowledge, and truth.[10]

A third stance, *critical realism*, is the view that some of our sense data accurately represent certain external conditions—such as whether a door is open or closed, or the location of an approaching vehicle—while other sense data, though causally related to external conditions, are more nearly subjective creations, as when a combination of blue and red dots is perceived as purple. In other words, critical realism holds that we perceive the real world but that a process of critical, quasi-scientific measurement or analysis is required to distinguish between perceptions that are nearly literal and those that are more highly symbolic. In this book critical realism is largely accepted, but with crucial adaptations.[11]

First, it must be recognized that the so-called external world is not filled with well-defined objects or predetermined conditions that, even in theory, can be perceived or measured unambiguously. If, as was long assumed, reality did consist of uniformly extended space that was populated with enduring material substances, it would be possible at least theoretically, as Laplace claimed, to create an accurate picture of the world simply by making precise measurements. In contrast, the view being defended here is that the world is populated not with material objects but with *events* that are constantly coming into definiteness and then leaving a wake of influence. In field theory, every atomic event exists as a potential influence in the coming-to-be of future events. Because specific events always arise in the context of one or more fields, it is inaccurate to speak of an "external world" with the traditional implication that each event can most clearly be understood as a "thing in itself" with properties of its own and modes of influence that can be reduced to physical mechanism. Instead, all events are considered to take place in an interrelated field of weblike influence. Events do not merely *have* relationships; rather, relations with past events really enter into the formation of present events, so that there is a sense in which events *are* their relationships.

In the philosophy embraced in this book, the originating point of influence is not a static particle sitting in space. Rather, it is an event

10. DeVries and Sachs, "Wilfrid Sellars," sec. 4 ("Epistemology").

11. As I mentioned in chapter 9, Ian Barbour discusses critical realism and other modes of interpreting human perception in *Myths, Models, and Paradigms*, 34–38.

that has been completed and may or may not be succeeded by a direct descendant. Furthermore, influences are not entirely deterministic, and influence is necessarily received not in general but from the peculiar standpoint of each particular subsequent event as it becomes whatever it does become. Physical reality consists of a constant assemblage of events in which final causes are just as real as efficient causes, and there is an element of self-causation or subjectivity even in subatomic events. For these reasons there can be no perception from a standpoint that is universal or absolutely objective. Consider an example.

Let us revisit the behavior of a muon, which I mentioned in chapter 4, that travels at high velocity from the upper atmosphere toward the surface of the earth. From the frame of reference of an earthbound observer, the muon's clock "slows down" as a result of relativistic time dilation. From the muon's frame of reference, its clock runs at the same rate as usual, but the distance it travels is shortened as a result of the relativistic contraction of space in the direction of motion. Which times and distances are correct—the observer's, or the muon's? They both are! There is no universal clock or fixed point in space from which objective measurements can be made. This example from special relativity does not pertain directly to human perception, but it is valuable as an analogy. The point is that something like perspective or interpretation (involving a "frame of reference") is not peculiar to human mentality but is, rather, ubiquitous throughout nature.

In addition to examples from special relativity, we must also recognize the quantum mechanical principle that the physical world we perceive is permeated with degrees of freedom. We remember Heisenberg's uncertainty principle and recognize that the positions of electrons around an atomic nucleus must be depicted not as a result of predetermined orbits but as clouds of various orbitals described by probability waves. Thus when we speak of critical realism, we must understand that no amount of critical error correction can ever establish a single, perfected picture of the world—for flux, uncertainty, and perspectivity are at the heart of nature itself. Even in theory, there is no objective God's-eye view of reality.[12]

Beyond these examples from physics, the idea that human beings can legitimately seek one objective or factual depiction of reality is also undermined by the essentially ambiguous nature of the *symbols* that are

12. For a fuller critique of the idea that there can be a single correct view of the world, see Crosby, *Specter of the Absurd*, 137–49.

our only means of perception or description. Consider the sentence "The neighbor's dog has jumped over the fence into our backyard and chewed up the children's toys." The sentence is populated not with high-level abstractions but with references to simple physical items and activities. The sentence's meaning seems abundantly clear. However, the words of the sentence arouse in our minds various symbols—dog, jumping, fence, backyard, children's toys, and chewing. We should notice that, though the sentence is about specifics, *the sentence itself is meaningful only because each symbol can be applied in more than one context to an entire category of things*—dogs, jumping, fences, and so on. Thus, though it may seem ironic, the applicability of each symbol to a specific thing depends upon its ability to refer to a whole class of things. This statement leads us to see that in language, and indeed in nonlinguistic experience also, there is a type of dialogue, a constant back-and-forth, between specificity and generality. This back-and-forth requirement implies that there is a type of impreciseness built into language and in all symbols, since the symbol—dog, for example—is a generalization, an abstraction, in which specific details are left undefined. In contrast, an actual dog necessarily exists in terms of specifics—either black, white, brown, or tan, young or old, female or male, weighing not twenty pounds but twenty-five, and so forth.

This leads to another source of uncertainty. The meaning of any sentence—and the meaning of all symbols—will also vary because of the unique perspective of the person speaking and of the person or persons who are listening. Ultimately, words and symbols are always interpreted by specific individuals. Returning to the example of the neighbor's dog, perhaps the speaker is an irate husband who happens to dislike dogs and who is speaking to his wife, who happens to be fond of dogs, and especially of the dog next door. Or perhaps instead the speaker is a teenager who has been hired to mow the lawn and who is now on the phone, describing the backyard scene to a friend. It is easy to see that, depending on the *Sitz im Leben* of the speaker and of the listener, the symbols that come to mind will be associated with widely varying thoughts, memories, comparisons, and feelings. Furthermore, symbols frequently take on new connotations in the context of a group or a collective. Indeed, there can scarcely be any experience of symbols that is entirely free of social implications. The meanings of symbols may not be entirely subjective and unrestricted but neither are their meanings ever entirely objective and exact.

Because of (1) the inherent limits of perspective, (2) the inevitability of linguistic impreciseness, and (3) the inherent differentiation between internal experience and the external world, some philosophers have suggested that we ought simply to give up altogether on the idea that language can function as an adequate description of reality. For example, philosopher Richard Rorty (1931–2007) argued that we ought to stop worrying about the reliability of language as an accurate description of the objective world.

> Rorty rejected the long-held idea that correct internal representations of objects in the outside world are a necessary prerequisite for knowledge. Rorty argued instead that knowledge is an internal and linguistic affair; knowledge relates only to our own language. Rorty argues that language is made up of vocabularies that are temporary and historical, and concludes that "since vocabularies are made by human beings, so are truths." The acceptance of the preceding arguments leads to what Rorty calls "ironism"; a state of mind where people are completely aware that their knowledge is dependent on their time and place in history, and are therefore somewhat detached from their own beliefs.[13]

Rorty, like many other philosophers, observes that a long-standing dichotomy has been presumed to exist between the external world, which is factual and objective, and the internal world of the perceiver, which is subjective and partial. Rorty, along with all other twentieth-century philosophers, was aware that familiar physical objects, such as our dog, fence, and children's toys, are actually composed of atoms, which are in turn composed of protons, neutrons, and electrons—things that we will never be able to see directly and whose precise ontological status is far from obvious. That is, even scientific descriptions of nature's elemental building blocks are expressed in terms of symbols (words and images), which are constructs. (We have already pondered, in chapter 8, Werner Heisenberg's comment that the only truly accurate depiction of the atom consists of a partial differential equation.) Therefore, Rorty is saying, it is pointless to debate the extent to which words and symbols may refer to something called "reality." The meaning of words and symbols depends not on ontology but on practical usefulness. Words and symbols express

13. Wikipedia, "Richard Rorty." This article documents citations to Rorty's specific works, including *Consequences of Pragmatism: Essays, 1972–1980* (1982) and *Contingency, Irony, and Solidarity* (1989).

our beliefs, guide our actions, and allow us to communicate with one another, but they have no clear-cut connection to "reality."

> Rorty argues that philosophy has unduly relied on a representational theory of perception and a correspondence theory of truth, hoping our experience or language might mirror the way reality actually is. In doing so, he builds upon the work of philosophers such as Quine, Sellars, and Donald Davidson. Rorty opts out of the traditional objective/subjective dialogue in favor of a communal version of truth. For him, "true" is simply an honorific that knowers bestow upon claims, asserting that they are what "we" want to say about a particular matter.[14]

I believe Rorty is correct in rejecting the representational theory of perception. Insofar as both direct and indirect realism hold that mental representations are an accurate portrayal of physical objects, they are inadequate. The symbols that we use to represent a table, for example, systematically omit any details pertaining to the very large number of atomic events that are incessantly taking place in any table.[15] Rorty is also correct in rejecting the correspondence theory of truth, a theory that centers on an inherent division that is presumed to exist between the mind of the perceiver and the actual state of the world. (I critiqued the correspondence theory of truth in chapter 10 as I discussed the philosophy of William James.) The panexperientialism adopted here is largely compatible with the pragmatic theory of truth, and can be interpreted so as to negate the correspondence theory.

What I do not agree with is Rorty's conclusion that truth must be restricted to the merely communal or the merely verbal. Rorty is correct that, because of the way our sense organs and our brains have evolved, we perceive and describe the world not in ways that are objectively accurate but in ways that tend to be practical in the contexts of human activity. However, Rorty is not correct in reducing this practicality to a mere "internal and linguistic affair." The word *true* is not simply an honorific term whose applicability depends, in the end, on a widespread consensus among human beings. Rather, it is possible, and important, to be able to go beyond or behind language in our conceptions of truth and reality. The main thing to recognize is that the disparity between the alleged subjectivity of symbols and the alleged objectivity of physical conditions

14. Wikipedia, "*Philosophy and the Mirror of Nature*."

15. Whitehead analyzes this issue in *Modes of Thought*, 153–56.

is based on a false dichotomy. The solution to this problem is found when we stop associating facticity only with material conditions and stop defining truth as linguistic consensus, and instead ascribe both facticity and truth to events, viewed as moments of experience. In a nutshell, the answer to Rorty's conundrum is that reality is composed not of external objects to whose nature our ideas must conform but of events or experiences that are by their very nature internally related and perspectival.

The philosophy of perception has chronically suffered from the uncritical acceptance of the assumption—which is actually a metaphysical doctrine—that the world we observe is purely physical, while our conscious awareness of that world is primarily ideational. So conceived, the problem of perception then leads to a question of causality that asks how there can be a causal connection between a physical event and a conceptual symbol. For example, how can a table, which is a physical object whose molecules are held together by physical bonds, cause the image—a symbol—of a single, solid, colored, rectangular object? The usual attempt to answer this question begins by citing the physiological processes in the eye and proceeds to the neurological processes in the visual cortex, which then leads to questions about qualia and the mind–body problem or the "hard problem" of consciousness—at which point most analysts, who still subscribe to a physicalist view of nature, are stymied.

In this book, the answer to the question of perception is based on the following premises: (1) the table is not composed of solid particles but of events; (2) the events are not perceived in one frozen moment but are ongoing in serially and spatially ordered patterns; (3) it is the fundamental nature of these events to occur in fields of influence so that they furnish information about themselves that is *both physical and conceptual* (as noted repeatedly in previous chapters, by *conceptual* I mean information about qualities qua qualities and patterns qua patterns), which is to say that nature itself communicates in ways that are not only physical but symbolic; (4) that for human beings, the information that is conveyed (by light waves, sound waves, and so on) is consolidated and interpreted by our neurological and mental processes in terms of symbols;[16] and (5) though the symbols may be said to be subjective constructs, they are not utterly disconnected from or irrelevant to the physical processes from which they have arisen. What this means is that our acts of perception are not arbitrary attempts to depict physical substances as mental constructs;

16. This statement is supported in chapters 6 through 10, which present an interpretation of quantum mechanics.

rather, they are exemplifications of the relational, symbolical way that nature operates. The fact that there are always multiple ways of interpreting the world around us is not a problem; perspectivism is to be expected in a world that is composed of experiences. It is not a problem that the table we perceive is not like the "real" table which is an enormous swarm of particles, for *"reality" is claimed not for the allegedly material composition of the particles but for our experience of the table and for the multiple (and much simpler) experiences of which the table itself is composed.* As the pragmatic theory of truth suggests, the accuracy of our perception of the table is associated with that perception's usefulness—and if our perception is misleading in this regard, the problem is an empirical one and not a metaphysical one. Whitehead wrote,

> Abstraction expresses nature's mode of interaction and is not merely mental. When it abstracts, thought is merely conforming to nature—or rather, it is exhibiting itself as an element in nature. Synthesis and analysis require each other. Such a conception is paradoxical if you will persist in thinking of the actual world as a collection of passive actual substances with their private characters or qualities. In that case, it must be nonsense to ask, how one such substance can form a component in the make-up of another such substance. So long as this conception is retained, the difficulty is not relieved by calling each actual substance an event, or a pattern, or an occasion. The difficulty, which arises for such a conception, is to explain how the substances can be actually together in a sense derivative from that in which each individual substance is actual. But the concept of the world here adopted is that of functional activity. By this I mean that every actual thing is something by reason of its activity; whereby *its nature consists in its relevance to other things*, and *its individuality consists in its synthesis of other things* so far as they are relevant to it. In enquiring about any one individual we must ask how other individuals enter "objectively" into the unity of its own experience.[17]

So, reality is composed not of substances but of experiences. But this invites an obvious objection. If mental or conceptual information is woven into the fabric of reality, this requires us to include inner imaginings in the category of things that are real. It will next be objected that this attribution of reality to imagination deprives us of two abilities whose importance I have already emphasized—first, the ability to distinguish

17. Whitehead, *Symbolism*, 26–27; emphasis added.

the imaginary from the real, and second, the ability to correct errors. If events of imagination are real, how can we verify, for example, that your dream about a fire-breathing dragon is only an illusion? These objections, however, are not difficult to answer. The dream about the dragon is in fact real in the sense that it is *a real experience*. But it is a simple matter when you awaken to determine that the experience is not associated with any physically existing dragon. The dream is real *as a dream experience*, and it does have physical effects in the cells of your brain and in your bodily responses, which may include an increased heart rate, rapid breathing, perspiration, and the motions of your arms and legs, even while you are still asleep. Indeed, there is probably never a purely mental event or a purely physical event; each event occurs as a dipolar reality. This book's rejection of reductionistic materialism obviates the compulsory attitude that the word *real* can appropriately be used only when an internal experience can be reconciled with an "external" world that, it is claimed, is ultimately derived from nothing more than insensate matter, what Whitehead calls "a collection of passive actual substances with their private characters or qualities."

When Whitehead says that "abstraction expresses nature's mode of interaction and is not merely mental," he means that throughout nature, wholes are prehended as wholes and not merely as an aggregate effect of their individual parts. "Abstraction" means that certain symbols, images, or concepts are used to represent the entirety of an object, but with the omission of all of the detailed properties of the object that are not specified in the symbols being used. If you say, "I let my friend borrow my car," you have uttered a meaningful sentence, but you have not said anything about the identity of your friend, the type of car, or the conditions of the loan. All abstractions are conceptual simplifications. Even at the microscopic level, the object can be responded to (*prehended* is Whitehead's term) as a whole *only because nature involves a level of mentality in the way in which one occasion responds to another*. The verb *to abstract* means "to draw out." We see a red book as a single object partly because the red wavelengths that are emitted by the molecules of the book's cover are all of one color. This process of perception *abstracts* this one color from numerous photonic events and then uses the common characteristic as a way of converting the multitude of physical events into the perception of a single object. This happens not only in human perception but in all situations in which a group of events with one or more common characteristics functions holistically. Whitehead called this process *transmutation*, as discussed in chapter 11.

Obviously microscopic physical processes do not depend on conscious planning or even on the presence of a biological nervous system. Whitehead writes in the quotation above, "When it abstracts, thought is merely conforming to nature—or rather, it is exhibiting itself as an element in nature." The word *thought* is used as a figure of speech; protozoa do not have thoughts, but they do respond with a certain degree of purposiveness to items in their environment. This is what Whitehead is referring to when he writes that "synthesis and analysis require each other." "Synthesis" (responding to wholes as wholes) implies that environmental items are prehended (perceived) as wholes. As Whitehead says, it is not enough simply to describe reality in terms of events, patterns, or occasions. The key is to recognize that this ontology of processive events is what allows one event actually to enter into the composition of its successors. For example, as long as electrons are understood as material particles, one electron cannot enter into another electron. The two electrons simply collide or, more likely, they repel each other by coulombic forces before they can even come close to colliding. In contrast, when electrons are understood as serially ordered events, and when each event is understood to convey information that is partly physical and partly conceptual, it is easy to see how one electronic event can "enter into" other electronic events. This transmission of information is how wholes and patterns emerge in nature and how, for example, the electrons in atoms follow the Pauli exclusion principle and organize themselves into orbitals. This is but to say, as I have noted repeatedly, that nature must possess some ability to transmit and utilize conceptual (or mathematical) information, or else the quantized nature of events at the atomic level is impossible to explain.

We can distinguish between pure imagination and our perception of physical objects because there is a demonstrable ("scientific," if you will) relationship between, say, our perception of a red color and the rose which is its source. The rose really is red (note that red is a symbol), not because the color red is an actual quality (a measurable wavelength) of the molecules that constitute the rose but because our perception of the rose's color is an actual experience,[18] and experiences (not substances)

18. Our ability to perceive colors in a useful way is, of course, dependent upon the assumption that our view of colors closely resembles the visual experience of most other persons and, presumably, of most animals who can perceive colors. Based on

are the basis of reality. This demonstrable, scientific explanation for acts of perception is how we distinguish between imaginary events and externally caused ones. Yes, the rose *really is red*; this statement is not a fiction. Your experience itself is the reality. This obviates Rorty's "mirror of nature" dilemma about the pernicious gap between language and metaphysics. However, even descriptions that sound "scientific" do not guarantee the accuracy of our explanations. For example, John Dalton's "plum pudding" model of the atom gained a level of credence that soon faded in the light of further experimentation. And Rorty has a valid point, that even our current theories pertaining to quarks, quantum chromodynamics, and so forth, are abstractions (or constructs, or symbols), not "literal" depictions of physical facts.

My point is that symbols are ubiquitous in nature. In human life, even simple reflexes, instincts, and our responses to life's basic physical needs are processed in terms of symbols, many of which are unconscious. Feelings such as pleasure, pain, hunger, satiety, and sexual arousal are given content by symbols (which, again, are not necessarily visual). Symbols are reproducible and make it possible for us to sustain various perceptions and types of experience over intervals of time. Without the consistency of experience that symbols supply, we would not be able to respond to our environment coherently. Symbols are often envisaged even unconsciously as a way of entertaining and evaluating various possible outcomes. The idea of "unconscious imagination" may seem counterintuitive, but unconscious imagination is a subject of investigation in many theories of personality and in various schools of psychotherapy. Indeed, we do not have to refer to academic fields of study in order to encounter the topic. For example, unconscious perception is a principle underlying most advertising and marketing.

Obviously our dependence on symbolism can lead to errors in perception, for there is always some subjective aspect when symbols are applied. However these errors are not caused by a categorical gap between a physical environment and a mental mind but by misinterpreted perceptions, insufficient experience, inadequate analysis, the ambiguities of language, and misleading habits of thought. The errors, in other words, are specific to certain circumstances and not due to a neo-Cartesian metaphysical partition between mind and matter. Once we have discarded the fiction of an "objective world" that is "out there," the perspectival or

---

widespread human experience, the color red may be defined as light of a wavelength between 620 and 750 nanometers.

subjective quality of our own mental life may be viewed as an ordinary part of the world. The symbols we use are not inchoate causes of error but a means of relationship with the world around us. Indeed, errors are often valuable as a path to thinking new thoughts and making new plans.[19]

In summary, nature itself—including our own minds—is subjective and experiential, not in the sense of a universal consciousness but in the sense of receiving information that is conceptual as well as physical, responding to that information in a way that contributes to a specific outcome in its own unique circumstances, and thereby endowing nature with a teleological or outcome-oriented characteristic. Critics who find all of this talk about symbols and panexperientialism implausible should immediately be asked to explain why nature tends to produce increasingly complex wholes and how it is possible that they themselves experience such definite opinions and robust feelings about this subject.

Symbolism is involved not only in our moment-to-moment perceptions and decisions. Symbols also provide content for our personal attitudes, the ways in which we approach problems, and our awareness of various hopes and fears. Symbols are vital to our own sense of self, the way we see ourselves and assess our needs and feelings. Thinking, perhaps, of how crucial it is for Native Americans to cultivate a deliberate awareness of their own history, N. Scott Momaday is credited with the saying, "We are what we imagine. Our very existence consists in our imagination of ourselves. Our best destiny is to imagine, at least, completely, who and what, and that we are. The greatest tragedy that can befall us is to go unimagined."[20] Symbols have a vital significance in all cultures. Myths and fairy tales are not merely quaint stories preserved from antiquity but narrative embodiments of personal and cultural feelings, norms, conflicts, and desires, expressed symbolically, as Sigmund Freud, Carl Jung, Edith Hamilton, Joseph Campbell, Paul Ricœur, and numerous others have demonstrated. Freud's doctrine of the Oedipus complex provides an example of the way in which an ancient story can offer insight into psychodynamic patterns that are still encountered in human experience today. Carl Jung posited the existence of a "collective

19. On this point see Hedrick, *Whitehead and the Pittsburg School*, 159–61.

20. Momaday, "We Are What We Imagine." Scott Momaday (1934–2024) was a member of the Kiowa Tribe of Oklahoma. He was a novelist, short story writer, essayist, and poet. His novel *House Made of Dawn* won the Pulitzer Prize for Fiction in 1969. Momaday taught English, creative writing, and Native American oral history at several universities, mostly in the western United States.

unconscious," a pool of symbols and associated feelings held in common by virtually all human beings.[21] Jung called these transcultural symbols "archetypes" because they provide something resembling a pervasive dynamic in human psychology.[22] We noted previously that, since Jung encountered these archetypes in his clinical work with patients of widely varying personal experiences and in widely varying cultures, he regarded the archetypes and the collective unconscious as empirical discoveries.

Words are symbols, and in human experience the symbolism entailed by the use of language is of special prominence. Language presumably originated as a small assortment of primitive sounds but now exists in immense detail in both oral and written form, having been embellished with standardized spellings, syntax, grammar, and style, and with numerous idioms, figures of speech, and other irregularities, and an enormous volume of published literature—and yet all of these aspects of language are made up of what is essentially a system of symbols. Momaday said, "A word has power in and of itself. It comes from nothing into sound and meaning; it gives origin to all things."[23] The fact that "language creates reality" was stressed by twentieth-century feminists, who recognized that established systems of symbols—including not only spoken and written words but workplace practices, financial policies, gender-specific etiquette, and other patterns of social interaction—were being stubbornly maintained by supercilious males in order to preserve a rigid milieu of patriarchy. The so-called glass ceiling, which prevents women from being promoted to upper levels of management, is a symbol, and is undergirded by a system of assumptions and images that also are represented symbolically. Patriarchy itself is expressed as a symbol or, rather, a system of symbols. Women therefore have sought not simply to abolish old symbols but to replace them with new ones. It turns out that symbols are much more than just passively perceived images. Systems of symbols are essential to human life, including both personal and social modes of existence.

As another example, art and music, also, utilize their own sets of symbols—and here we include not only notations printed on a page but sounds, which are audible symbols. Tunes, chords, rhythms, and the timbre produced by specific instruments are all perceived as symbols. Routine customs such as styles of clothing, differing methods of communication, types of home design and furnishings, and various dietary

21. Sharp, "Collective Unconscious," 35–37.

22. Henderson, "Ancient Myths," 95–156.

23. Momaday, *Way to Rainy Mountain*, 33.

practices are present in terms of symbols. Law codes, jurisprudence, medical procedures, and all academic and professional standards are expressed through symbols. The colored lights and lane markers that direct traffic in the street are systems of symbols. Various trades—plumbers, carpenters, electricians, and so on—rely on standardized codes that consist of symbols in the form of words, diagrams, tables, and formulae. There are varying definitions and requirements that each of these codes could adopt, but all are constrained to some degree by the physical conditions to which they must apply. But whether we are considering instincts, tropisms, traditions, languages, laws, regulations, customs, esthetic preferences, professional standards, or academic disciplines, symbols are at the heart of the discussion.

These symbols, being symbols, are always to some extent objective but also always to some extent subjective. The historical idealists were correct that the mind is always involved in a synthesizing and self-interested way even in the most physical types of experience. But the historical realists were right that our experiences and our very lives are circumscribed by physical conditions that we cannot arbitrarily set aside. These physical conditions provide both an objective testing ground for our own actions and beliefs and a common field within which all relationships achieve their mutually adjusted outcomes. But, it is necessary to remember, *our own actions and beliefs—that is, our own experiences—become a part of that larger world, and of its objectivity.* The world is thus based upon both facts and values. It is to the relationship between those two types of reality that we turn our attention in chapter 14.

14

Science, Facts, and Values

THROUGHOUT THIS BOOK I have argued that materialism is not an adequate worldview. In chapter 9 I defended the idea that the pursuit of goals is an inherent aspect of nature, made possible by the ontological status of superposition as a state in which unactualized potentials are held in contrast with one another. Though the mind's use of symbols is undoubtedly coupled with physical activities in the brain, symbols themselves are not merely physical phenomena whose influence can be explained away neurologically or mechanistically. Thus the mind's use of symbols points to the reality of value, or valuation, throughout nature. Teleology (regarding purposes) and axiology (regarding values) are mutually implicated, and both are present in a very simple way even at the atomic level. From this observation we can draw some important conclusions.

The first is an obvious one, that *nature is inherently valuable, or value-producing.* Nature as a whole is valuable because it produces value-seeking entities. Value originates in each individual, and not in some supreme goal or cosmic destiny. This tracing of value to each individual event is an essential aspect of the pluralistic metaphysics that I am advocating. Values, of course, can serve large groups and even entire species, but the interests of a group are derived from the interests of its individual members (and of course the interests of the group as a whole may then influence the interests of the individual members). So, nature produces value-seeking individuals, but nature as a whole has no specific goal. There is no "one far-off divine event, / to which the whole creation moves."[1] Moreover, there is no guarantee that the individual value-seeking processes will be constructive or collaborative.

It is understandable that human beings tend to hope that the vagaries of history will eventually culminate in some enduring meaning. The Western world's monotheistic religions have encouraged the belief that individual values come into being because those values are a part of, or are derived from, a transcendent creative principle, usually identified with the idea of God's will. We remember Augustine's words, "Thou awakest us to delight in Thy praise; for Thou madest us for Thyself, and our heart is restless, until it find repose in Thee."[2] If God is conceived anthropomorphically, as a living Spirit with a mind and a will and characterized by a fundamental goodness, then it seems sensible to imagine that the transient goals and purposes of the world's creatures must have their primordial origin in God's goals and purposes and that God's will or plan is the ultimate measure of their worth. Indeed, by something like an act of philosophical reverse engineering, the simple observation that moral principles do exist among human beings has been cited as evidence of the existence of a moral God.[3]

I consider the idea of God more fully in chapters 15 and 16. For now, however, I should acknowledge that in my own opinion it is not necessary to adopt a specific concept of God in order to accept the existence of value within nature. In this book I have argued that final causes emerge ubiquitously and spontaneously in the natural structure of events, even at the simplest levels of existence. Organisms, even if not conscious, utilize symbols as a way of formulating novel solutions to the demands of existence. In other words, these symbols serve as final causes. Thus we can attribute an axiological or value-laden aspect to nature itself.

One crucial consequence of nature's spontaneous production of values is that ecological and environmental concerns are self-validating. I do not claim that nature is conscious or that nature is a living organism, but there is nevertheless a profound sense in which *nature is alive* in its entirety, as it manifests its primordial tendency to create more complex wholes. Therefore our unfortunate habit of attempting to justify environmentalism by appealing primarily to human interests is not merely

1. Tennyson, "In Memoriam, Epilogue," stanza 36.

2. Augustine, *Confessions* 1.1 (Pusey). Augustine (354–430 CE) wrote his *Confessions* between 397 and 400.

3. Evans, "Moral Arguments." See also Jankowiak, "Immanuel Kant," sec. 5.c ("Postulates of Practical Reason").

shortsighted but self-defeating. The goals and values that support life are intertwined, and have been from the beginning. Activities and conditions that harm one species are liable to harm other species in unforeseen ways. The truth is that teleology and valuation exist throughout nature, regardless of human involvement. Therefore the attitude that we should protect the environment primarily as a way of protecting ourselves is inexcusable. The argument that we are entitled to a preeminent position in nature because we are superior—more complex, more conscious, or have greater moral awareness or greater creative potential—is invalid. An old episode of the television show *The Twilight Zone* depicts Earth under the control of a race of superior aliens who view human beings as a food source.[4] If we suddenly found ourselves in the presence of creatures who were even more intelligent and creative than we are, would we then be willing to surrender our own right to survive and flourish? It is a grave—and perhaps fatal—mistake to view ourselves as privileged above all other species. Our human needs and values cannot simplistically be detached from or elevated above the needs and values of the world around us. Value is not limited to human experience.

This leads to another conclusion, which is that *our habitual separation between facts and values is unwarranted and, in fact, dysfunctional.* The facts–values dichotomy is not merely an academic or intellectual problem. Even in routine conversation, people frequently separate facts and values. It is widely assumed that statements related to subjects such as science, technology, medicine, engineering, agriculture, and finance are based on verifiable facts, while statements related to personal morals, social mores and ethics, esthetics, political beliefs, and religious convictions are based on values, which are presumed to be matters of mere personal preference.

An example of the problematic dichotomization of facts and values is examined in Thomas Haskell's 1987 article, "The Curious Persistence of Rights Talk in the 'Age of Interpretation.'"[5] We live in an era when most people are aware that social mores and ethical norms vary widely from one culture to another and from one historical period to another. Haskell notes that this moral relativism is encountered especially among

4. Bare, "To Serve Man."

5. Haskell, "Curious Persistence," 991. Haskell (1939–2017) was an American intellectual historian. I owe my knowledge of Haskell's work to William Dean, who cites Haskell in *The Religious Critic in American Culture*. I discuss Dean's work more fully in chapter 15.

intellectuals, who deny that there is any universal foundation for laws, morals, or values since all value judgments arise and are applied in specific personal or historical contexts. But it is ironic that the very pluralists and relativists who deny the existence of universal values do affirm one de facto universal value—the value of personal and social human rights, for "the rights of humans exist everywhere and are defensible everywhere."[6] Moreover, people who defend their own rights almost always do so not by referring simply to their own personal wishes but by appealing to larger, more general criteria such as fairness, justice, legal precedent, and established social practice. As a contrasting irony, those who do claim to base their moral and social standards on a universally valid foundation often turn out to be the very persons who do harm by attempting to force their views autocratically upon others. Thus there is a point at which moral relativists turn out to be moral absolutists, and moral absolutists turn out to be morally oblivious.

The view taken in this book is that both views—moral relativism and moral absolutism—are in some ways instructive but need to be reinterpreted. Moral relativism is correct in saying that all value judgments arise in personal and historical contexts, but from this observation it does not follow that moral principles are hopelessly marooned in a kind of foundationless limbo. The personal and historical limitations that inhibit the formation of moral absolutes do still provide conditions that are sufficient for formulating moral precepts that are normative for the situations in which they arise. Morals and mores may never be absolute, but they are nevertheless relevant and amenable to reasonable construction. The insight that supports this assertion is that, though no specific moral values are universal, *valuation, as a phenomenon, is a fact of nature*. Though the scope of what is morally or ethically appropriate is debatable, the criterion of appropriateness is not so idiosyncratic as to be impracticable. The word *normative* does not signify that specific verbal formulations can be deemed applicable for all situations. We have already noted the pervasiveness of perspectivism in epistemology; there is a parallel in ethics.

Moral absolutism, on the other hand, is true to the extent that *all* persons have a duty to pursue moral behavior in their own context and, further, that every individual's moral standards are obligated to take heed of the needs of other creatures and of the environment as a whole. The

6. Attributed to Haskell, as quoted in Dean, *Religious Critic*, 102 (see 102n5).

rights of all creatures, taken together, and even the balance and wholeness of the earth itself, are a universal value.

These comments leave many questions unanswered, but it is clearly beyond the scope of this book to present any comprehensive theory of ethics. My point is simply that because valuation is not a matter of mere, personal interest but, rather, an aspect of nature itself, there is a sense in which moral and ethical norms are always situation-specific and yet another sense in which they are always universal. This affirmation of the compatibility between ethical relativism and ethical universalism is an outgrowth of a worldview that allows the facts–values dichotomy to be deconstructed.

The facts–values dichotomy is perhaps most conspicuous in the perceived divergence between the natural sciences and the humanities. Statements about objects are thought to be more verifiable and to have more lasting meaning than statements about preferences. The sciences, it is said, deal with hard facts such as the molecular weight of carbon dioxide, the names and functions of various bodily organs, the application of Newton's three laws of motion, and so on. The humanities, on the other hand, deal with matters that are said to be debatable: the causes and consequences of various wars, the interpretation of differing styles and periods of art, the appropriateness or inappropriateness of various religious beliefs. In the Western world, at least, and especially among intellectuals of a positivistic temperament,[7] the more factual character of scientific knowledge has imparted an air of superiority to the sciences. Obviously it really is more certain that the formula for sulfuric acid is H_2SO_4 than it is that Henry VIII was a bad (or a good) king.[8] Nevertheless, the sciences have more in common with the arts than is generally acknowledged.

THE FACTUAL ASPECT OF ART

Let us consider art first. Painting may be thought of as completely subjective, for choices about media and colors and the motions of the brush are

7. Logical positivism—the claim that all statements must be either analytic or synthetic in order to be meaningful—obviously contributed to the depreciation of statements about value. But eventually it was realized that the logical positivist creed itself was more nearly a value than a fact.

8. Arguments can be made for either assessment. See Historic Royal Palaces, "Henry VIII."

entirely up to the whim of the artist. However, the idea that painting is totally up to the painter is certainly not accurate—a fact that is obvious if we consider the circumstances in which most paintings are produced. As an example, in a video titled "Turner: Painting *The Fighting Temeraire*," art historian Matthew Morgan offers an engaging analysis of a famous painting by English artist J. M. W. Turner.[9] Morgan describes Turner's place in the history of European painting, his social background, his artistic techniques as compared with those of his contemporaries, and the specific facts about the history of the *Temeraire*, a ship that was launched in 1798 and finally broken up for salvage in 1838. In Turner's time the ship was celebrated especially for her role in the naval Battle of Trafalgar (1805), when the *Temeraire* saved Admiral Nelson's flagship and captured two French ships. The *Temeraire* thus became a prominent symbol of British naval power during the nineteenth century and remains symbolic even today.[10] Turner's painting depicts the ship's final voyage as she sailed into harbor to be reduced to scrap. Morgan describes Turner's use of lighting and color to emphasize certain aspects of the ship's appearance, noting that in some respects Turner's depiction is accurate, but in others Turner deliberately engages in a type of artistic revisionism in order to accentuate the drama and the nostalgia of the event. If, after the video has ended, we were to review Morgan's presentation, I believe we would be impressed by the objective, factual character of almost all of his statements, not merely regarding the ship and its history but pertaining also to Turner's artistic goals and painterly techniques. Very clearly, Turner wished to present certain facts and arouse certain feelings for his viewers. It may be objected that the painting now being discussed portrays a conveniently historical subject and that, among painters, Turner's style itself is more representational than that of the other painters who soon followed. However, the same points I have made about *The Fighting Temeraire* can be made about less literal paintings such as Picasso's *Guernica*, which uses nonliteral depictions of animals and humans to convey an unmistakable message about wanton destruction and overwhelming suffering. It can be argued that even abstract expressionism

9. Morgan, "Turner." Matthew Morgan is museum director of Turner's House in Twickenham, UK, and is an associate lecturer at Birkbeck, University of London. He has worked at the Royal Collection and the National Gallery.

10. A reproduction of *The Fighting Temeraire* appears on the back of the Bank of England £20 note issued in 2020 (Bank of England, "£20 note").

conveys propositions,[11] though those propositions are not as unambiguous as scientific propositions are (or seem to be).[12] My point is that art has a more objective character than we may at first suppose.

THE VALUE-LADEN ASPECT OF SCIENCE

Now let us consider the supposed objectivity of the natural sciences. Perhaps somewhat ironically, there are entrenched differences of opinion about scientific objectivity, and so I will discuss this subject at some length.

In today's world, scientists seem to be thought of as an amorphous but cohesive assemblage of learned specialists who pursue truth dispassionately and can be counted on to supply unbiased, factual information about almost any subject. Headlines typically report that "scientists" have reached vital conclusions about matters ranging from digestive health and weight loss to the age of the universe and the promise (or threat) of artificial intelligence. Our widely held attitude is that science is a vast, interrelated body of knowledge espoused by experts who are somehow aware of and in broad agreement about virtually all the facts of physical existence. During the Middle Ages priests were the primary adjudicators of truth; nowadays we defer to scientists.

In his book *Perfect Symmetry*, scientist Heinz Pagels differentiates between what he calls "first-person science" and "third-person science."[13] First-person science relates to the personal outlook of specific scientists, which cannot help but be influenced by the preferences, opinions, goals, practices, and values of the individuals involved. In contrast, third-person science is what results when scientists "segregate" their privately held beliefs and opinions and report their professional work in the context of their community—material that is "published in professional journals"

11. For example, if interpreted consistently, abstract impressionism includes the rather strict requirement that *no recognizable forms* should be depicted in a painting.

12. On the "Proposition Theory" of artistic meaning, see Beardsley, *Aesthetics*, 369–78 (ch. 8, sec. 20). Beardsley tends to limit the meaning of *proposition* to verbal formulations that can be deemed either true of false (369). Whitehead used *proposition* (or "propositional feeling") much more broadly, as an ontological category applicable to the phases of concrescence, in which some new or unactualized characteristic is envisaged as a possibility. In this book I have argued that this ability of actual occasions to envisage various possibilities is related to the phenomenon of quantum superposition, in which several differing states coexist as potentials.

13. Pagels, *Perfect Symmetry*, 361–71.

and "reported at conferences." Pagels admits that this third-person science is still "culture-bound" and subject to human failings, and yet its intent is to reveal universal, objective truth. Thus, third-person science shows us the world's material order in an unbiased way. Pagels holds that "if science fails in that intent, it fails in any claim, however provisional, it might have to truth."[14]

In general, of course, the emphasis of science on objectivity is appropriate, but this emphasis is not as straightforward as Pagels makes it sound. In the first place, the effort to "segregate" one's privately held beliefs is not so easy to achieve as one might think. It is relatively simple to adopt an *objective style* of writing and speaking, but that objective style does not guarantee an objective content. In fact, an objective style can be observed not only in science but in many diverse disciplines. For example, an objective style typifies most of the papers presented at meetings of the American Academy of Religion. The fact is that the difference between objectivity and subjectivity is, itself, not an entirely objective distinction.

In the second place, even "objective" articles in journals and reports made at meetings have their own subtle ways of being biased. As Michael Polanyi has noted, all knowledge is ultimately personal knowledge and is obtained in the way that it is actually obtained because of the values, goals, and feelings of the persons who obtain it.[15] Peer review and the approval conferred by authorized bodies are significant, but they are not an infallible guarantee of objectivity or accuracy. History provides numerous examples of situations in which prevailing scientific opinion has not furthered the cause of truth but obstructed it. In fact, this book itself may be an example, inasmuch as I am laboring to show that the "weirdness" of quantum physics is not weird and that the existence of teleology and mind are compatible with scientific worldviews—two ideas that are not generally held to be in accord with prevailing scientific opinion.

Third, the notion that science is purely empirical—that is, that it is based ultimately on the results of repeatable experiments—ignores the essential role of theories and philosophical assumptions (assumptions that are too often not analyzed or even acknowledged) in all experimental work. Polanyi writes that

14. Pagels, *Perfect Symmetry*, 362–63.

15. Polanyi, *Personal Knowledge*, 3–17.

> modern man has set up as the ideal of knowledge the conception of natural science as a set of statements which is "objective" in the sense that its substance is entirely determined by observation. . . . That is why scientific theory is represented as a mere economical description of facts; or as embodying a conventional policy for drawing empirical inferences; or as a working hypothesis, suited to man's practical convenience—interpretations that all deliberately overlook the rational core of science.[16]

I believe that Polanyi's phrase "rational core of science" is meant to refer to the necessary role of systematic theoretical-intellectual underpinnings. Another way of making this point is to say, as I have said in previous chapters, that science is founded on the requirement that experiments must be repeatable and that this repeatability results from the fact that nature is mathematical—that is, that "natural laws" have a universal applicability. Polanyi's point, and mine, is that science itself generally makes no attempt to explain this applicability of mathematics or to describe the ontological status of these "laws."

Finally, it is hardly reasonable to claim that science is the definitive path to knowledge or to grant science automatic authority on almost every subject (except the subjects that are less important because they pertain only to "personal values" or "opinions"). The scientific method is regularly lauded as a paradigm and presumed to function as something like an epistemological final court of appeals. Those who view the scientific method this way may be thinking of the ways in which it is science that has been humanity's chief deliverer from all sorts of superstition, delusion, and magical thinking. However, the attitude that beliefs which cannot be proven by science are therefore irrational or meaningless is based on a false dichotomy. There are other paths to truth. Indeed, I submit that the assumption is actually arbitrary and rather odd that science is consummate simply because its statements are more measurable, more precise, or more verifiable. In some situations precision is paramount, but there are other circumstances when it is irrelevant and even intrusive. When Romeo first sees Juliet at the Capulet's party, he exclaims,

> O, she doth teach the torches to burn bright!
> . . . .
> Like a rich jewel in an Ethiope's ear;
> Beauty too rich for use, for earth too dear![17]

16. Polanyi, *Personal Knowledge*, 16.

17. Shakespeare, *Romeo and Juliet*, act 1, scene 5, lines 43, 45–46 (Canterbury

In statements like this one there is a robust type of truth that is not a matter of precise measurement or objective detachment. Romeo's opinion is only Romeo's opinion, but such opinions and feelings are the stuff of life. In fact, despite its understandable pursuit of objectivity, science itself also depends on the personal enthusiasm of its proponents. We will now consider an example from astronomy, seemingly a very objective field of study.

The eminent astronomer Tycho Brahe (1546–1601) was born into an aristocratic family in what is now southern Sweden. In 1562 Tycho (he is generally referred to by his first name) left home for the University of Leipzig. His family expected him to study law, but Tycho was gripped by an innate fascination with the stars. It seemed to him "as something divine that men could know the motions of the stars so accurately that they could long before foretell their places and relative positions."[18] We should note in passing that Tycho felt no hesitancy about associating astronomy with theology. In any case, the fact that Tycho was enthusiastic about astronomy did not make astronomy an easy field to pursue. There were costs of both a financial and a social nature. Telescopes and other equipment had to be constructed and maintained by hand; the stars and planets had to be observed almost every night in order to track their positions accurately, which required access to an observatory, and the calculation of the trajectories of "celestial objects" entailed laborious arithmetic that could only be done with pencil and paper, for there were no mechanical or electronic calculators. (Remember long division?) Moreover, astronomy, though generally respected, was not always viewed as a very practical activity. The relatives and friends of astronomers tended to proffer regular reminders that stargazing did not automatically generate any personal income. Nevertheless, Tycho's social standing gave him access to the financial and political resources that he needed in order to make his way as an astronomer. The result was that, over a period of years, Tycho compiled a body of astronomical data whose thoroughness and accuracy were conspicuously superior to the findings of any of his predecessors.

As Tycho grew older, the demands of his work grew more burdensome. He hired various assistants, but among all of those the one whose accomplishments are still celebrated was Johannes Kepler (1571–1630).

Classics ed.). This play was first published in 1597.

18. In this account I rely heavily on John Gribbin's very engaging *History of Western Science*, 36–89. The statement by Tycho is cited on page 38.

Kepler came from a distinguished family, but his father was a ne'er-do-well, and Kepler's adolescence and early adulthood were plagued by hardship. Like Tycho, Kepler was interested in the stars even as a child. However, the fact that he had learned Latin by the time he was twelve entitled Kepler to study for the Lutheran priesthood, and so, beset by financial worries, it seemed more practical to Kepler to enter the University of Tubingen to prepare for ordained ministry.[19] Even as a theological student, however, Kepler was required to take courses in mathematics, physics, and astronomy—subjects in which he excelled. Then, just as Kepler was poised to graduate and become a Lutheran pastor, the mathematics professor at the Lutheran school in Graz, Austria, died. There is no reason why this should have affected Kepler at all except that, as they searched for a new instructor of mathematics, the academic authorities in Graz contacted the authorities in Tubingen, and the Tubingen authorities recommended their most gifted math student—Kepler. This stroke of fate changed the history of astronomy.

After Kepler came to Graz, it was only a matter of time until he and Tycho became acquainted. In 1600 their relationship became official, for Kepler was granted a position as Tycho's assistant. However, interactions between the two were not always cordial. Kepler could scarcely wait to apply his mathematical skills to the analysis of Tycho's extensive astronomical records, but Tycho was, perhaps understandably, very protective of his data, which had taken thousands of hours of tedious work to compile.[20] But despite their occasional tensions, the two began to work together on *The Rudolphine Tables*, a catalogue of the names and positions of prominent stars and of all of the then-known planets. The catalog was sponsored by, and named for, Rudolf II, the Holy Roman Emperor, and was finally published by Kepler in 1627, some twenty-six years after Tycho's death.

As Kepler gained access to Tycho's data, he made a concerted effort, as many had before him, to understand the orbits of the planets. The paths of the stars were much more predictable than the motions of the planets. We now know that this is because the stars are very far from the earth and because their apparent motion in the sky is caused not only by their own movement but by the earth's daily rotation on its axis and by the earth's (and all the planets') orbit around the sun. Of course the planets

19. In this section I am relying on Gribbin, *History of Western Science*, 56–67.

20. Gribbin, *History of Western Science*, 64–65.

orbit the sun as the earth does, but at differing speeds, depending on their distance from the sun. Kepler took a special interest in Mars, which seems to have a very irregular orbit;[21] this is because Mars is farther from the sun than Earth is and therefore orbits more slowly. A Martian year is equal to about 1.9 Earth years. This means that for part of the year, Earth actually "catches up" with Mars so that Mars appears to move backwards in its orbit. After a sustained period of meticulous measurements and prodigious mathematical analysis, Kepler became convinced that the orbits of the planets were not circular but elliptical. This conviction also involved a certain amount of soul searching and courage, since people in Kepler's time believed that the planets, which were regarded as examples of the perfection of God's creation, must have a circular orbit. Apparently in those days the ellipse was viewed as nothing but a malformed circle.[22]

Odd biases of this type can serve to remind us that Tycho and Kepler lived in a time when there was no clear separation between astronomy and astrology.[23] Newton had not yet formulated his law of universal gravitation, which was published in 1687. People had no concept of the vastness of interstellar space, or that the stars, which appear to be set above the earth in a huge colander-like dome, were actually spread apart at widely varying distances, or that the gravity that makes objects fall to the earth's surface is also what holds the solar system (and entire galaxies) together. Indeed, according to one then-current proposal, the erratic motions of the planets were caused simply by the fact that they were pushed around by angels.[24] Though the Copernican theory was known, most people still believed in the Ptolemaic doctrine that the earth is the stationary center of the universe with the planets moving in orbits within surrounding concentric spheres. The Ptolemaic view did seem to portray an orderly universe that had been created by a wise deity, and the attribution of celestial motions to a foreordained divine plan made it natural to assume that the movement of the stars had some connection with God's control of events on earth. Thus it should not surprise us that both Tycho and Kepler gained a reputation and an income from the practice of casting horoscopes—an activity that they may privately have disdained

21. For a video describing the orbit of Mars and including very helpful graphics, see Welch Labs, "Bizarre Path of Mars."

22. Gribbin, *History of Western Science*, 68.

23. See Wikipedia, "Johannes Kepler."

24. Gribbin, *History of Western Science*, 61.

in its details but whose overall premises they tended to accept.[25] Both men saw astronomy as evidence of divine activity—and certainly not, as many view it today, as the result of a vast, reductionistic, self-regulating mechanism.

This little digression into seventeenth-century astronomy is meant as a reminder that science is not the objective, methodical, sure-footed path to truth that it is usually assumed to be. Like Tycho and Kepler, scientists today may be hindered by their own preconceptions. (Tycho espoused the Ptolemaic view; Kepler became a Copernican.) Like Tycho and Kepler, scientists today are limited by their equipment, their access to financial resources, their personal and family circumstances, and by their own abilities. Back then and still today, luck plays a part. Kepler happened to gain employment when a certain math teacher died. Personal rivalries intervene. Cultural and political situations may help or hinder, as when the interests of the Holy Roman Emperor defined Tycho's and Kepler's work pertaining to the orbits of the planets. Perhaps we could compare this with the "space race" that occurred during the era of John F. Kennedy and Nikita Khrushchev. Scientists' intellectual interests and personal values are of indisputable importance. Tycho and Kepler were both possessed by an intuitive fascination with the stars. It is difficult to imagine that very many first-rank scientific discoveries have been made by persons who were not so personally motivated as Tycho and Kepler were.

Today, scientists are often depicted as impartial intellectuals who pursue truth for its own sake. It is generally assumed that scientists are trustworthy basically because they are brainy—that is, that they are in possession of the truth simply because they are well-informed and highly intelligent. Of course we should know better than to imagine that any one person, no matter how well-versed, can keep abreast of the numerous details pertaining to all the various branches of science. But in any case, we tend to think that the main thing that makes science so successful is neither the preeminence of its intellects (though many are spectacular) nor the superiority of its methods (which are indeed rigorous and ingenious) but the unrelenting, empirical character of its subject matter. Kepler was confronted by the eccentricity[26] of Mars's orbit and took up the

25. Gribbin, *History of Western Science*, 41, 58.

26. This is a pun, since *eccentricity* can mean oddness or incongruity, but in plane geometry the word also applies to the "flattened" shape of an ellipse—the shape of Mars's orbit.

challenge of finding an explanation. Dmitri Mendeleyev (1834–1907), who established the importance of the periodic table, felt compelled to investigate the strange fact that when the elements are listed in order by atomic weight, certain properties tend to repeat themselves. Max Planck was so uncomfortable about his data that he hesitated to publish it, but in the end he could not ignore the quantized aspect of black-body radiation. Aspect, Clauser, and Zeilinger were struck by the weirdness of non-locality and quantum entanglement, but it was a weirdness that could not be explained away and that they themselves therefore felt obligated to examine.

This self-correcting aspect of science's commerce with hard facts has sometimes been taken to mean that scientific opinion should automatically be trusted. However, this trust is not always warranted. Though science does, over time, support the advance of truth, the science of various historic periods has advocated many famous mistakes, as we are reminded by Tycho's and Kepler's belief that there is a direct connection between celestial motion and the outcome of earthly events. Of course Tycho and Kepler were merely maintaining the beliefs of their predecessors; but science's native conservatism or trust in established doctrines is not a problem that has gone away in more recent years. Rather, it remains characteristic. In the early 1800s, Thomas Young (1773–1829) experimented with interference phenomena and published experimental evidence that light is a wave rather than a particle—but other scientists refused to believe it. They were "angered by the suggestion that anything Newton said could be wrong" and therefore they "ridiculed the idea that you could make darkness by *adding* two beams of light together."[27] In 1864, when John Newlands (1837–1898) first published evidence of the periodicity of chemical properties, he suffered the fate of "being savagely ridiculed by his peers, who said that the idea of arranging the chemical elements in order of their atomic weight was no more sensible than arranging them in alphabetical order of their names."[28] In a commencement address given at Caltech in 1974, physicist Richard Feynman acknowledged the tendency of scientific beliefs and practices to be skewed in a conservative direction:

> We have learned a lot from experience about how to handle some of the ways we fool ourselves. One example: Millikan

27. Gribbin, *History of Western Science*, 441.
28. Gribbin, *History of Western Science*, 409.

> measured the charge on an electron by an experiment with falling oil drops,[29] and got an answer which we now know not to be quite right. It's a little bit off because he had the incorrect value for the viscosity of air. It's interesting to look at the history of measurements of the charge of an electron, after Millikan. If you plot them as a function of time, you find that one is a little bit bigger than Millikan's, and the next one's a little bit bigger than that, and the next one's a little bit bigger than that, until finally they settle down to a number which is higher.
>
> Why didn't they discover the new number was higher right away? It's a thing that scientists are ashamed of—this history—because it's apparent that people did things like this: When they got a number that was too high above Millikan's, they thought something must be wrong—and they would look for and find a reason why something might be wrong. When they got a number close to Millikan's value they didn't look so hard. And so they eliminated the numbers that were too far off.[30]

Millikan himself exemplified scientific conservatism intensely. In chapter 6 we mentioned Millikan as the physicist who tested Einstein's theory about the photoelectric effect. We noted that Millikan disagreed strongly with Einstein's proposal that light can be viewed as a particle (as did all other physicists at that time) and that Millikan was actually attempting to discredit Einstein's explanation. It was certainly ironic when, instead, Millikan's test results verified Einstein's proposal, which led to Einstein's reception of the Nobel Prize in 1921.[31]

Sometimes this conservative bias among scientists involves the social prejudices of their period. Amalie "Emmy" Noether (1882–1935) was an extraordinary mathematical genius[32] whose work is still studied by mathematicians and physicists, but the advancement of her career was constantly impeded by the severe male chauvinism of her era.[33] Male scientists have felt that they can safely belittle or ignore women in

29. Physicist Robert A. Millikan determined the net charge on the electron in an experiment involving falling drops of oil whose velocity was influenced by varying static charges.

30. Feynman, as quoted in Wikipedia, "Oil Drop Experiment."

31. See Hofstadter and Sander, *Surfaces and Essences*, 461.

32. See Wikipedia, "Emmy Noether," and Angier, "Mighty Mathematician." Noether conceived a hypothesis, now known as "Noether's Theorem," that relates two fundamentals of theoretical physics: cosmological symmetry and the universal laws of conservation.

33. See *Encyclopædia Britannica*, 15th ed. (1974), s.v. "Noether, (Amalie) Emmy."

order to maximize their own professional standing. Rosalind Franklin (1920–1958) was the British chemist and X-ray crystallographer whose work was essential to understanding the molecular structures of DNA and RNA,[34] but she was not given credit by Francis Crick and James Watson, the two scientists who won the 1962 Nobel Prize. It is now known that Crick and Watson gained their basic insight about DNA structure from Franklin's research.[35] Also in 1962, Rachel Carson (1907–1964) published *Silent Spring*, which exposed the dangers of DDT, once hailed as an insecticide that would rid the world of insect pests.[36] When I was a child in the Deep South in the 1950s, the "DDT truck" would regularly drive slowly through our neighborhood on summer nights, spraying a cloud that was intended to eliminate mosquitoes. No one stopped to wonder what effect the chemical would have on those species of birds and fish that depended on the insects as a food source. Not surprisingly, Carson—not only her ideas but also Carson herself—was roundly criticized and ridiculed by representatives of the chemical industry.

> Those who stood to be professionally undermined by the policies her ideas would soon help to shape criticized her vehemently. Dr. William J. Darby authored a review of *Silent Spring*—titled "Silence, Miss Carson"—in the Oct. 1, 1962, issue of *Chemical & Engineering News*, published by the American Chemical Society. Darby criticized Carson for not adopting the views of "responsible, broadly knowledgeable scientists" and recommended that "in view of her scientific qualifications in contrast to those of our distinguished scientific leaders and statesmen, this book should be ignored."[37]

In cases such as Noether's, Franklin's, and Carson's, it is evident that the supposed objectivity of scientific truth and the alleged superiority of the scientific method contribute to a clubby atmosphere in which it

34. Nature Education, "Rosalind Franklin." See also Brangham et al., "DNA's Double Helix."

35. "It has been argued that Watson and his colleagues did not properly acknowledge Franklin for her contributions to the discovery of the double helix structure. Robert P. Crease notes that 'Such stingy behaviour may not be unknown, or even uncommon, among scientists.' Franklin's high-quality X-ray diffraction patterns of DNA were unpublished results, which Watson and Crick used without her knowledge or consent in their construction of the double helix model of DNA" (Wikipedia, "James Watson").

36. DDT is dichloro-diphenyl-trichloro-ethane (Wikipedia, "DDT").

37. American Chemical Society, "Rachel Carson's *Silent Spring*." Note that this favorable recognition of Carson is by the same organization that originally rejected her work.

is easy to belittle or ignore the contributions of women—or anyone else who deviates from scientific orthodoxy or discredits scientific ideas that have become the basis for lucrative industries.

The widespread attitude that science is simply a matter of reporting "facts" that have no inherent association with values has repeatedly encouraged scientists to approach their work with the idea that conscience can be suspended or ignored, as is evident in actual cases. The idea is apparently that the objectivity of science frees science from the obligation to assess human and natural values, which are, after all, not considered a matter for scientific inquiry. This assumption is based on a non sequitur, but it is nevertheless an evident premise in science's track record. Pesticides and herbicides, various explosives,[38] nuclear fission and fusion, the extraction and burning of fossil fuels, certain coolants or refrigerants, heavy metals, plastic particles, dioxin contaminants, various pharmaceuticals, chemical fertilizers that wash from fields into rivers, lakes, and oceans, the factory conditions entailed by the manufacture of these items, and, finally, the complex social and economic conditions that have recently resulted from electronic methods of information processing and social media—all of these science-dependent creations were produced with a chief interest in imminent financial gains and frequently with only a token assessment of the possible harms that might ensue. It is sometimes almost as if science were automatically granted immunity from unintended consequences.

One now encounters the attitude that past errors have been corrected and that better safeguards are currently in place so that today's science is purely objective and factual—but this is simply implausible. No human activity is truly value-free. In fact it is now of vital importance to ask whether any scientific discovery can ever be regarded as truly "value neutral." Even something so objective and factual as a table of trigonometric values would never have been compiled unless it could be put to some use—surveying the land, designing a skyscraper, or hitting a target with an artillery shell. The phrase "put to some use" suggests that even a trigonometry table can eventually be implicated in ethical questions. Consider that surveying, which seems innocuous, is usually done so that land can be divided into parcels and sold, which is deemed inappropriate in some cultures. Among some Native Americans, for example, the

38. Alfred Nobel invented dynamite mainly for use in construction projects and was unhappy when one of its major applications turned out to be the manufacture of bombs and artillery (Kravitz and the ACS Committee on Ethics, "Dynamite").

land is regarded not as a thing that can be owned but as a gift for all people from the Great Spirit. The construction of skyscrapers also is questionable because of the amount of material, energy, and capital that it requires. And the moral difficulty of the artillery shell is, we assume, self-evident.

All of these observations may sound like a wholesale indictment of science, but that certainly is not my intent. The earlier chapters of this book make it clear that I have a heartfelt fondness for science. My point now is not that science is exceptionally problematical either ethically or epistemologically but simply that science is just as rooted in questions of value, meaning, and personal commitment as any other field of human activity. Scientists themselves are often just as motivated (and sometimes more motivated) by personal values, commercial interests, and the praise of their peers as are any artists, writers, historians, or philosophers. Actually, science has often progressed—or failed to progress—because of the same kinds of personal and social factors that one might expect to encounter in politics or commerce. Science does tend to cling to its traditional doctrines, as in many cases it should. But my point is that scientific knowledge is generally limited by the same kinds of restraints that pertain to other fields of study—the intrinsic ambiguities of language, concepts, and symbols; the inescapable narrowness of personal and social perspectives; and the fact that life itself requires all persons to make choices that, on a de facto level, are axiological (about value), ethical (about benevolence and justice), and metaphysical (about the perceived nature of reality), whether or not these choices are recognized and appraised.

In fact, if science does have a problem of its own, it is that its focus on measurable facts creates a false sense of objectivity that makes science unwilling to face the typical arbitrariness of its metaphysical presuppositions. In support of this point I have already quoted Michael Polanyi. Whitehead also can be quoted:

> Science has never shaken off the impress of its origin in the historical revolt of the later Renaissance. It has remained predominantly an anti-rationalistic movement, based upon a naïve faith. What reasoning it has wanted, has been borrowed from mathematics which is a surviving relic of Greek rationalism, following the deductive method. Science repudiates philosophy. In other words, it has never cared to justify its faith or

> to explain its meanings; and has remained blandly indifferent to its refutation by Hume.[39]

Whitehead is not saying that science is irrational. His use of the term *anti-rationalistic* refers to the fact that scientists have usually not wanted to admit that their work is, unavoidably, grounded in assumptions, such as physicalism or reductionism, that are philosophical in character. As an example of this repudiation of philosophy, in the first chapter of this book I described an encounter with my college advisor, a chemist who adopted an attitude of mock pity when I informed him that I planned to take a course in philosophy. In chapter 2 I cited physicist Stephen Hawking's claim that "philosophy is dead" because "it has not kept up with modern developments in science," adding that "scientists have become the bearers of the torch of discovery in our quest for knowledge."[40] In chapter 8 I quoted Niels Bohr's well-known words, "There is no quantum world. There is only an abstract quantum mechanical description. It is wrong to think that the task of physics is to find out how Nature *is*."[41] Also in chapter 8, I referred to a relatively recent poll among physicists indicating that "nearly 90 years after [quantum] theory's development, there is still no consensus in the scientific community regarding the interpretation of the theory's foundational building blocks."[42] All of these examples illustrate the attitude that science can be and perhaps even should be neatly detached from philosophy, though, in fact, scientific work always involves assumptions that are at least implicitly axiological, ethical, and metaphysical.

We have already cited some situations when scientific dogma has actually been the enemy of truth. Especially today, because scientific knowledge is so reliable, so scholarly, and so prestigious, it can be counterproductive and even foolhardy to question scientific orthodoxy, even when that orthodoxy needs very much to be questioned. In many cases it is typical to encounter the claim that *religion*—and not science—is the more frequent source of error. Sometimes that has been true but not as often as we are led to believe. For example, much has been made of the image of poor Galileo, persecuted by church leaders who refused to

39. Whitehead, *Science and the Modern World*, 16 (ch. 1). The phrase "refutation by Hume" refers to philosopher David Hume's claim that causation itself—which is at the heart of the scientific method—is not directly observable.

40. Hawking, *Grand Design*, 5.

41. Petersen, "Philosophy of Niels Bohr," 12; emphasis added.

42. Moskowitz, "Physicists Disagree."

accept the scientific implications of his observation of Jupiter's moons; but in fact the erroneous belief originated not with theologians but with earlier scientists, Aristotle and Ptolemy.[43] The church's mistake was not that it defended bad science based on religion, or particularly on the Bible, as is sometimes alleged, but simply that the religious authorities tried—responsibly, they thought—to incorporate an established scientific idea (Ptolemaic astronomy) into the Christian doctrine of creation. When the scientific orthodoxy was eventually doubted, the theologians and the pope then felt a natural obligation to defend an outdated scientific theory. Yes, the theologians were conservative—but so were the scientists.[44] When we remember the *scientific* opposition encountered by Copernicus, Young, Newlands, Carson, and even Einstein and Planck, is it not clear that scientists very often have felt that same need, usually attributed to religion, to defend the established ideas of the past?

All of this leads to the crucial, culminating question of what it is that is still widely believed today that will later be recognized as erroneous. The answer proposed in this book is that the long-standing assumptions of materialism, reductionism, and antiteleology that are still generally embraced by many scientists are just as incorrect as the old Ptolemaic astronomy and the plum pudding model of the atom. The main purpose of this book has been to present an outline of a philosophy of nature that is based instead on process, relationality, and the transmission of information that is not only physical but mathematical and conceptual so that meaning and value are built into nature itself. There are three basic principles: (1) Reality is composed not of matter but of events or occasions. (2) These events manifest not only a physical aspect but an informational (mathematical or conceptual) aspect, culminating in an integrative, dipolar unity. (3) The nature of this integrative process manifests an outcome-oriented tendency, which is the primordial source of value or teleology in nature. What in physics is called superposition is the functional equivalent of concrescence in Whitehead's metaphysics.

We can conclude this chapter with the observation that there are no value-free activities or lifestyles. Life itself calls us to conscience. Neutral

43. See Riebeek, "Planetary Motion." Aristotle is usually thought of today simply as a philosopher, but he produced seminal ideas in fields that today would be regarded as scientific. See Schwarcz, "Aristotle."

44. For Whitehead's entertaining account of the conflict between Galileo and Pope Urban VIII, see Whitehead, *Essays in Science*, 227–42 ("The First Physical Synthesis"). For a more recent account correcting several errors, see Wolf, "Truth About Galileo."

objectivity has its place, which is chiefly to avoid biased reasoning and to negate self-interest disguised as expertise. But it is not possible—nor would it be desirable—to become completely objective or self-transcendent. Ontologically, we are selves, and there is an extent to which nature herself has assigned us the duty of self-protection and, even more, of self-improvement, and concomitantly, the duty of protecting the environment upon which we all depend. The fact is that we live in a world where ontology and axiology, being and value, are intertwined. In other words, we live in a world where we are bound to accomplish *something*—even if it is only that we succumb to the deceptive appeal of safe repetition. Taken day by day and hour by hour, we receive virtually unlimited opportunities for novelty, for meaning, for renewal, for redemption. On the other hand, taken as a whole, each of us has only one life to live. In this one life, life's actual purpose and meaning are partly determined by us individually but also by the facts and values embedded in our personal and social history. Our response to these questions of fact and value cannot be postponed indefinitely. There is a saying that "life is not a dress rehearsal."

Michael Raposa reminds us of William James's 1895 meditation on the question, "Is life worth living?"[45] James concluded that "life is either a 'real fight in which something is eternally gained for the universe by success,' or it is a trivial game from which 'one may withdraw at will.' As evidence for the first possibility, [James] observed that 'it *feels* like a real fight—as if there were something really wild in the universe which we . . . are needed to redeem.' The very first task, then, is 'to redeem our own hearts' by overcoming both our fears and our lack of conviction."[46]

In this book I have argued that metaphysically real forms of purpose and value emerge spontaneously throughout nature. As human beings, we inherit these purposes, meanings, and values in a complex personal and social journey that we can scarcely anticipate, ranging in content from striking beauty to appalling tragedy. What are we to make of this? In our final two chapters I discuss some perspectives regarding religion and offer some options for belief.

45. Referring to James's essay by the same name.

46. Raposa, "Divided Self of William James."

15

The Problem of God

THE PURPOSE OF THIS chapter and chapter 16 is to show that science and religion are not necessarily incompatible and, even more, that in many ways they may actually be based on compatible kinds of thinking. Both rely on what Whitehead called "symbolic reference"—the use of mental symbols to experience and interpret the world. This is not to say that science and religion really boil down to the same thing but, rather, that the two are not separated by an essential and impenetrable divide, as has often been supposed. Instead, there is continuity between the two. The fundamental insight here is that, as I have said repeatedly, reality consists not of an objective, material world "out there" but of experience. Experience involves physical facts, of course, but static physical objects are not the ultimate basis of experience. Nature is regarded as dynamic, relational, and informational. Religions are a way of interpreting the physical facts of the world; they allege that somehow, the world is grounded in the sacred. But is there any sense in which this sacred interpretation can be objective? Or do religions merely create ideas in the same way that a novelist creates a fictional scenario? It is with these questions that we begin this chapter.

When I began my theological education in the early 1970s it was still a common practice for seminaries to offer an introductory course in "philosophy of religion." Philosophy of religion and theology were two different fields, distinguished principally on the basis that philosophy of religion operated within an academic discipline that was grounded in reason and experience, which were thought to be unbiased underpinnings, while theology was an expression of some faith stance, whether broadly monotheistic or supportive of some specific religious tradition. In other words, for theology certain religious beliefs were granted a provisional acceptance, but philosophy of religion expected its criteria and conclusions to avoid parochialism. Instead, philosophy of religion attempted to seem reasonable for persons of any faith or even of no faith at all. Philosophy of religion included a consideration of topics such as the nature of religious experience and religious truth claims, the function of religions (what purpose do they serve?), theodicy (if God is good, why does God allow evil?), the nature and existence of miracles (if accepted), questions about life after death, and, most of all, an analysis of various historical attempts to prove the existence of God. For centuries, philosophy of religion's evident assumption was that if a proof were logically valid, then any fair-minded person who understood the proof would be compelled by the sheer force of the argument to become a believer. That this assumption was still in place in the mid-twentieth century is attested by the occurrence of a memorable debate about the existence of God between Frederick Copleston and Bertrand Russell, which was aired on the BBC in 1948 and again in 1959.[1]

Looking back, it is now interesting to see how quickly these attitudes about philosophy of religion have eroded and how extensively seminaries and departments of religion have altered their curriculum. To be sure, courses that apply philosophical resources to religion are still offered, but the interest in proofs and in alleged objectivity has dwindled away. There are several reasons for this. In the first place, philosophers of religion have increasingly had to consider the manifest plurality of the world's religions as evidence that it is simply not possible to speak in one neutral, objective way about the core content of religious truth. The syncretistic claim that all religions are simply differing paths to the same goal is inaccurate. A landmark in the examination of this issue was the publication of *The Absoluteness of Christianity and the History of Religions* (1911) by the German liberal theologian Ernst Troeltsch (1865–1923). In this book, Troeltsch argued that all religions must be studied in their own context and evaluated, to some extent, by their own standards of truth. However, as the title implies, Troeltsch also argued that Christianity remains the most truthful and most adequate expression of faith. That it is difficult or impossible to reconcile these two claims became increasingly evident,

1. Copleston was a Catholic priest; Russell was a professed nonbeliever. See Wikipedia, "Copleston–Russell Debate."

especially as the twentieth century brought electronic communications, faster and cheaper modes of travel, and more frequent, accessible forms of cross-cultural contact.

In the second place, the twentieth century witnessed an increasing awareness of the innate ambiguities of language itself. Logical positivism held that no statements are meaningful unless they are either analytic or synthetic and then claimed that most *religious* statements are neither analytic nor synthetic.[2] Though logical positivism has long since waned in influence, its spirit lives on in more recent philosophies of linguistic analysis. In chapter 13 I mentioned the ideas of Richard Rorty and said that numerous scholars have demonstrated that the relationship between language and reality is far more complicated than it might at first seem. Presumably these linguistic complications would pertain especially to language about God and religious beliefs. People now wonder how *talking* and *thinking* about God or religion can really *prove* anything about God or *validate* any religious claims to truth.

In the third place, feminists, liberationists, postmodernists, and deconstructionists have shown that all language entails subtle ways of embodying and maintaining personal and social power structures.[3] It has been noted that religion, especially, has served to maintain class structures based on inequality and oppression. Karl Marx's well-known aphorism that religion "is the *opium* of the people" illustrates this idea.[4] Among feminists it has been recognized that allegedly neutral language about reason and experience conceals harmful patriarchal assumptions. Rosemary Radford Ruether writes that in order for progress to be made,

> hierarchicalism, as the established and only possible human order, had to be thrown into question. . . . Consciousness is much more of a collective social product than modern individualism

2. In chapter 2 I noted that "analytic" statements are true by definition because of the very meaning of their words. "A rectangle has four sides and four angles" is an analytic statement. "Synthetic" statements are statements that can validated by observation or measurement. "Lead has a density of 11.34 grams per cubic centimeter" is a synthetic statement. The logical positivists maintained that the statement "God exists" is neither analytic nor synthetic and is therefore meaningless. However, it was soon noted that the positivist statement that all meaningful statements must be either analytic or synthetic is itself neither analytic nor synthetic.

3. Feminism and postmodernism combine forces in the writing of Luce Irigaray. See Whitford, *Irigaray Reader*, 5–7, and esp. sec. 2. Note that the feminist "hermeneutic of suspicion" sometimes continues to appeal to reason and experience, when suitably rehabilitated, as useful resources in the tasks of criticism and construction.

4. Marx, introduction to O'Malley's *Critique of Hegel's Philosophy*; emphasis original.

> realizes. No one can affirm an idea against the dominant culture unless there is a least a subcultural group that gives people both the ideas and the social support for an alternative position. The dominant ideology and social order have to have become weakened and discredited enough that such countercultural groups can build up their position and survive.[5]

However, not all feminists and liberationists reject philosophical language categorically. Some continue to use the language and methods of philosophy to further the cause of liberation. In contrast, postmodernism is more prone to reject unconditionally the claim that language can be objective. French philosopher Michel Foucault (1926–1984) has argued that in any society the ideas that become accepted truths tend to support the interests of the privileged. "The theme that underlies all Foucault's work is the relationship between power and knowledge, and how the former is used to control and define the latter. What authorities claim as 'scientific knowledge' are really just means of social control."[6] Similarly, deconstructionist Jacques Derrida (1930–2004) has asserted that "the whole philosophical tradition rests on arbitrary dichotomous categories (such as sacred/profane, signifier/signified, mind/body), and that any text contains implicit hierarchies, 'by which an order is imposed on reality and by which a subtle repression is exercised, as these hierarchies exclude, subordinate, and hide the various potential meanings.'"[7]

Any discussion about religion or about the reality of God would seem particularly vulnerable to these criticisms.[8] With this postmodernist outlook in mind, then, it is difficult to justify any sympathetic interest in philosophy of religion's traditional attempt to formulate neutral or unbiased claims about God or religion.

Finally, recent decades have been characterized by an increasing awareness that some kinds of truth are based not solely on rationality or objective evidence but, in important ways, on personal feelings and commitments. Pertaining to religion, this awareness can be traced to the existentialism of Søren Kierkegaard (1813–1855) and, especially in America, to William James's seminal 1897 essay "The Will to Believe,"[9]

5. Ruether, *Sexism and God Talk*, 184.

6. Stokes, *Philosophy*, 187.

7. Wikipedia, "Jacques Derrida," citing Lamont, "Dominant French Philosopher," 590.

8. On this point, see McFague, *Metaphorical Theology*, esp. the preface to the second printing and ch. 1.

9. On the similarity between Kierkegaard and James, see Diamond, *Contemporary*

in which James famously argued that there are some ideas that become true precisely because we believe them to be true.[10] However, this notion of "personal truth" as a deliberate alternative to the usual correspondence theory did not achieve popular expression in North America until the counterculture movement of the 1960s, and it did not exercise a major influence in academia until after that, in the 1970s and later. In the United States recently, among some groups the "personal truth" pendulum has swung too far, as many people arbitrarily dismiss any ideas they dislike by labeling them "fake news" and insist on the truth of whatever they wish to believe, even if their beliefs contradict direct evidence.[11] But beyond this cavalier mistreatment of truth at a widespread level, there is a well-reasoned recognition among scholars that all religious beliefs are to some extent perspectival and context-dependent.[12] This philosophical-theological "historicism" holds that religions are always at least partly dependent upon the historical context in which they arise and continue to exist, and that all religious beliefs and practices are based not on impartial, universal truths but are more nearly manifestations of the need of various individuals and groups to embrace some sort of ultimate commitment, which may be formulated and expressed in widely differing ways.

In this book, these four criticisms—the criticism based on religious pluralism, the linguistic criticism, the feminist-postmodernist criticism, and the pragmatic-historicist criticism—are taken seriously. However, I do not take these criticisms to mean that philosophy of religion is simply no longer feasible. Truth claims (especially, in our present discussion, *religious* truth claims) may admittedly not be absolute or universal, but the validity of these claims may, and indeed *must*, still be evaluated in their own context. Philosophy of religion remains important because, in the first place, it is essential to be able to identify and correct errors—and errors there are. Religious precepts have served not only as the basis for all sorts of silly ideas but as the source of pernicious injustice and even appalling brutality. As a Christian, I must acknowledge that even today, some forms of Christianity bar women from ordination and positions

Philosophy, 163–65.

10. James, "Will to Believe."

11. At the time of this writing, large numbers of Americans continue to believe that the 2020 US presidential election was "rigged" or "stolen," that there is no connection between global warming the burning of fossil fuels, and that certain long-established vaccines cause more health problems than they prevent.

12. See Wheeler, *Religion*, esp. ch. 6.

of leadership, blame women for the "sin of Eve," and counsel women to submit to the authority of their husband even when that husband is abusive.[13] Even today, some forms of Christianity reject LGBTQIA+ persons and condemn non-Christians to hell. Even today, there are versions of Christianity that doubt the effectiveness of vaccines, oppose birth control, and prohibit abortion even when the life of the mother is in danger. There is no need to prolong this list. It is clear that religious truth claims, like the claims of any other field of human activity, must be subjected to fair-minded criticism on the basis of reason and evidence. Indeed, religions themselves proclaim the norms of love and justice—and then too often ignore these norms in their de facto practices and beliefs. Our assessments of religious truth claims may always to some extent be biased and inconclusive, but those assessments are not necessarily so flawed or uncertain as to be effectually meaningless or inappropriate.

Beyond a critical function, philosophy of religion can also still have a function that is constructive. The results of our constructive efforts will never constitute the timeless, universal truths that were the goal of our philosophical and theological forebears. Our constructive philosophy will always be limited by personal and social contexts and by the necessarily subjective aspect of our thoughts, our language, and our symbols. Obviously these limitations are not simply incidental obstacles; they are woven into the very fabric of human experience. However, this is not necessarily a drawback. In this book I have argued that it is our experience itself that is real, not some error-free "God's-eye view" of a universe that is ultimately nothing more than a vast collection of bare facts. Nature itself creates values and perspectives. Therefore our conversations with other persons who hold other viewpoints is not something to be regretted but a potential source of enriched meaning and deeper truth.

With these thoughts in mind, I will now describe the traditional arguments for God's existence, not with an anachronistic expectation that their logic is so powerful as to compel assent but with the idea that *each argument points to some aspect of nature that may be regarded today as religiously meaningful*. It is, of course, up to you, the reader, to form your own opinion about the outcome of this procedure. As a way of focusing

13. See 1 Cor 14:34–35, 1 Tim 2:8–15, Eph 5:22–33, and the contemporary advice of Focus on the Family, "Submission of Wives to Husbands." I personally have known women who were counseled by their pastor to be patient with and pray for an abusive husband. As an example of the unequal status of women in the Hebrew Scriptures (the Old Testament), see Lev 12:1–5.

the discussion, we will restrict our attention to four historic arguments: the cosmological, the teleological, the ontological, and the moral. Many books have been written about these four arguments; the treatment that they receive here can consist of only a brief summary.

THE COSMOLOGICAL ARGUMENT

In Western thought the cosmological argument for the existence of God is suggested by various Greek philosophers but may be traced especially to Aristotle. Aristotle discusses his idea of God in several works, but his *Metaphysics* is a good place to start. His argument is based on physical motion, not only on earth but of the stars. The argument is "cosmological" because it is based on the general appearance and behavior of the cosmos. Seeming to anticipate Newton's first law of motion, Aristotle asks, "For how will motion come about if there be no actual mover? . . . For movement certainly does not originate in chance, but something must always be responsible for it. . . . And since a moved mover is intermediate, there is, therefore, also an unmoved mover, being eternal, primary, and in act. . . . The first mover, accordingly, is a necessary being, that is, it is well that it is necessary, and it is thus a first principle."[14] Basically Aristotle is saying that the motion of all things in nature must originate in some starting point, for he views the idea of an infinite regress of causes as simply incoherent. His idea of God is thus encapsulated in his concept of an "unmoved mover" who is the originator of all motion.

Aristotle's works were preserved during the Middle Ages by Islamic scholars, notably Avicenna (d. 1037) and Averroes (1126–1198). During the twelfth century these works became available to Christian theologians, and at the hands of Albertus Magnus (ca. 1200–1280) and Thomas Aquinas (1225–1274), Aristotelianism became a transformative intellectual resource for Christian theology. The theology of Aquinas, now known as Thomism, continues to provide a foundation for much Roman Catholic thinking.[15]

14. Aristotle, *Metaphysics* 12.6–7, 1072 (257–59).

15. Aristotle's idea of "substance" (*ousia*) was mentioned briefly in chapter 2. "Substance" has been especially important both in Western philosophy to denote the idea of an inner reality or essence (Descartes's distinction between "mental substance" and "physical substance") and, especially since Aquinas, in Catholicism, as a crucial aspect of sacramental theology ("transubstantiation"). In the latter case, it is held that though the "accidents" (nonessential, outer properties) of the bread and wine (their appearance and taste) remain the same, the inner reality or substance becomes the actual body and

Thomas's primary work, the *Summa Theologica*, is arranged in a format of questions and answers. In question 2, article 3, Thomas asks whether God exists and states that God's existence can be proved in five ways. The first three ways are usually understood as differing forms of the cosmological argument. The first is clearly based on Aristotle's discussion of the unmoved mover. Thomas writes, "Therefore, whatever is moved must be moved by another. . . . But this cannot go on to infinity. . . . Therefore it is necessary to arrive at a first mover, moved by no other; and this everyone understands to be God."[16] The second way is based on the existence of efficient causes. As an Aristotelian, Thomas was familiar with Aristotle's fourfold system of efficient, material, formal, and final causes, which I described in chapter 2. Thomas notes that every event has an efficient cause, and since there cannot be an infinite regress, there must be an ultimate source of this system of efficient causation. "Therefore it is necessary to admit a first efficient cause, to which everyone gives the name God." The third proof is based on the contrast between existence and nonexistence, or on the idea of contingency. The things that we observe in the world, and even we ourselves, do not exist *necessarily*; they are contingent. But if everything were contingent, Thomas says, "then at one time there was nothing in existence," since contingencies cannot be projected backward infinitely. "Therefore, we cannot but admit the existence of some being having of itself its own necessity, and not receiving it from another, but rather causing in others their necessity. This all men speak of as God." In the first two ways, Thomas employs an obvious line of reasoning, referring to a chain of cause and effect that stretches backward in time, depending on a first cause chronologically. But in the third way, Thomas argues that the entire system of contingent cause and effect must be based upon some noncontingent source.[17] There are, naturally, modern versions of the cosmological argument. For example, the question of "contingency" was prominent especially in Copleston's

blood of Christ. It is interesting that this substance-oriented thinking is also the basis of the Catholic opposition to abortion. It is reasoned that the human soul is a substance that, like a chemical, is either present or not present and that, for the lack of a more obvious alternative, the soul must enter the body at the moment of conception. Substance theories tend to oppose philosophies of ongoing process or gradual emergence.

16. Quotations pertaining to the five ways are from Thomas Aquinas, *Summa Theologica* 1.12.3, as quoted in Pegis, *Introduction to St. Thomas Aquinas*, 25–27, unless otherwise noted.

17. Concerning the two forms of the cosmological argument, see Diamond, *Contemporary Philosophy*, 282–84.

dispute with Russell.[18] However, the ideas of Aristotle and Aquinas give us an adequate overview.[19]

Thomas's fourth way is "taken from the gradation to be found in things." This argument, which is perhaps the most obscure of the five, is based on the observation that all things in nature can be described in terms of gradations related to various descriptive qualities: hot and hottest, and so on, "so that there is something which is truest, something best, something noblest, and consequently, something which is most being, for those things that are greatest in truth are greatest in being, as it is written in *Metaphysics* ii."[20] The argument seems to be somewhat Platonic in its recognition that natural objects, and nature as a whole, can be described in terms of categories or classes that are applicable both logically and empirically. "Therefore there must also be something which is to all beings the cause of their being, goodness, and every other perfection; and this we call God." I provide additional comments about this argument below.

THE TELEOLOGICAL ARGUMENT

Thomas's fifth way is a version of the teleological argument, often referred to as the "argument from design." Thomas says that this proof "is taken from the governance of the world. We see that things which lack knowledge, such as natural bodies, act for an end, and this is evident from their acting always, or nearly always, in the same way, so as to obtain the best result." Thomas here is thinking of final causes or result-oriented behaviors including instincts, and noticing that these are effective even in objects that "lack knowledge," such as animals, plants, and physical systems. (I discussed final causes and instincts in chapter 12.) Thomas reasons that God must be the source or the designer behind these natural "actions toward an end."

A more well-known statement of the teleological argument comes from the Anglican clergyman William Paley (1743–1805), who proposed

18. See Hick, *Philosophy of Religion*, 283–91. The entire text of the Copleston–Russell debate is contained in this chapter.

19. For additional material on the cosmological argument, see Diamond, *Contemporary Philosophy*, 272–99; Ferré, *Modern Philosophy of Religion*, ch. 5, esp. 136–38; and Neville, "Cosmological and Ontological Contingency."

20. Thomas Aquinas, *Summa Theologica* 1.12.3 (Pegis, 27). Thomas's reference is to Aristotle, *Metaphysics* 1.1, 993b30. Aristotle expands on this point at 1072b17–20.

the example of a watch found lying on the ground. Paley compared the watch to a stone and reasoned that whereas the stone may be the result of random processes, no one would think that the watch was an accidental occurrence, for "when we come to inspect the watch, we perceive—what we could not discover in the stone—that its several parts are framed and put together for a purpose, e.g., that they are so formed and adjusted as to produce motion, and that motion so regulated as to point out the hour of the day."[21] After discussing several instances of design in nature, Paley concluded with the example of the eye: "Were there no example in the world of the contrivance except that of the *eye*, it would be alone sufficient to support the conclusion which we draw from it, as to the necessity of an intelligent Creator."[22] A more recent affirmation of the teleological argument is provided by F. R. Tennant in his two-volume *Philosophical Theology*.[23]

THE ONTOLOGICAL ARGUMENT

Unlike the cosmological and teleological arguments, the ontological argument is said to be an a priori argument because it does not depend on empirical evidence. It is, rather, purely logical. The classic expression of the argument was stated by Anselm of Canterbury (ca. 1033–1109). Anselm began by noting that we can imagine both things that do exist and things that do not exist and that, regarding good things, it is better to exist than not to exist. Then Anselm proposed the concept of some being that is absolutely perfect—without flaw. This being, he then said, *must necessarily exist*, since if it did not exist, the failure to exist would itself be a flaw, a defect, and our original stipulation was that the thing in question must be flawless. Therefore, unless we wish to deny, in principle (and what would that principle be?), the existence of a perfect being, we must accept that a perfect being does exist. And that's the whole argument.

This logic seems simple, but persons who hear it for the first time usually want to stop and consider the matter. Some find the reasoning unconvincing or even frivolous, but others have found it compelling. In his own time Anselm was opposed by a French monk, Gaunilo of

21. Paley, *Natural Theology*, 46.

22. Paley, *Natural Theology*, 54.

23. See esp. vol. 2, ch. 4, available in Hick, *Philosophy of Religion*, ch. 18 ("Cosmic Teleology").

Marmoutiers (fl. eleventh century), who, as a counterargument, posed the example of a "perfect" island. We may imagine an island that has no flaws, said Gaunilo, but this does not prove that such an island actually exists. Anselm's rebuttal was simply that an island is a finite, physical entity and that no finite, physical entity can be absolutely flawless; the example of an island, therefore, is irrelevant. In ensuing years the ontological argument was defended in various forms by Descartes, Spinoza, Leibniz, and Hegel, and it has been revamped in our own time by scholars including Charles Hartshorne (1897–2000), Kurt Gödel (1906–1978), Norman Malcolm (1911–1990), and Alvin Plantinga (b. 1932).[24] The argument invites us to try to understand what it means to be "perfect" and what it means to "exist." For example, there is a sense in which numbers exist and a sense in which they are perfect—but it is doubtful that their ontological status and their mode of perfection exemplify the type of existence and perfection that most believers would want to attribute to God.

THE MORAL ARGUMENT

The moral argument for God's existence begins with the observation that there is a moral aspect built in to human life. In other words, instead of appealing to evidence observed in nature generally, the moral argument appeals to evidence found in human thoughts and feelings, which are, after all, also products of natural creation. The point of the argument is not that most people are basically moral; certainly there are many who are not. The point is that some forms of behavior are better (healthier, kinder, more helpful, and so on) than others and that human beings generally live in a context in which they are aware of choices related to actions and goals that are better or worse. Though specific social mores vary widely among human societies, virtues like bravery, loyalty, and kindness, and vices like greed and cruelty, are recognized throughout the world. Every time we ponder issues related to right and wrong, we appeal to a higher standard of behavior that we assume everyone is aware of, at least in principle. This points to a preexisting source of valuation. This source, it is argued, is God.

The most well-known advocate of the moral argument was Immanuel Kant (1724–1804). As a consequence of his comprehensive rejection

24. Wikipedia, "Ontological Argument." See also Hick, *Philosophy of Religion*, chs. 3, 7, 25, and 31; Diamond, *Contemporary Philosophy*, ch. 11; and Hartshorne, *Anselm's Discovery*.

of metaphysical philosophy, Kant rejected the cosmological, teleological, and ontological arguments, but on a practical level he contended that moral decisions are an unavoidable aspect of life. Kant held that the existence of human conscience—or at least the common *need* to have a conscience—points to the existence of a Creator who is benevolent and just. Kant lived before psychological theorists like Freud attributed the emergence of conscience to social forces and inner psychodynamics. However, we can guess that Kant would argue that conscience and morality are still necessary elements of human life, however we may explain their origins.[25]

As something like an extension of the moral argument, we can also imagine an *esthetic* or *axiological* argument for God's existence. All persons have some inborn, intuitive appreciation for beauty—things that are experienced as pleasing or desirable—and a natural aversion for the opposite of beauty—things that are experienced as displeasing or repulsive. For example, human babies are born with a desire to be fed, held, and kept warm, and with a fear of loud noises and jolting motions. As babies develop, these likes and dislikes are quickly expanded in number and in complexity. Where does this native sense of value (both positive and negative) originate? The reductionist claim that these feelings can be explained away as nothing more than products of physical biology has already been rejected (see especially chapters 2 and 9). Rather, the existence of appreciative valuation within nature seems to point to a definite source of all life. As Aquinas would put it, "this we call God." In fact, it is possible to interpret Thomas's fourth argument, the argument from gradation, as an esthetic or axiological argument. In this book I have maintained that there is a tendency in nature to produce what Whitehead called "worthwhile things."[26] The fact that, as historicists would say, "worthwhileness" is determined not by some eternal cosmic principle but by the creatures themselves (whether individuals or groups) does not diminish the force of the argument. What Thomas called "gradation" is carried out not only by human beings but throughout nature—for example, among termites as they build their mounds, among mockingbirds as they copy the songs of other birds, and even at atomic and molecular levels of nature, as I concluded in chapter 9.

25. See Neiman, "Immanuel Kant."

26. Price, *Dialogues*, 370.

Now, what are we to make of the cosmological, teleological, ontological, moral, and axiological-esthetic arguments? Considering the criticisms that have already been noted—linguistic, feminist, postmodernist, and pragmatic-historicist—it is hardly feasible in our own era to claim that any of the historic arguments can prove God's existence as a simple fact, in the same way that we would prove that aluminum is lighter than iron or that the square root of sixteen is four. And besides, there is another difficulty that I have not mentioned up until now. It is that, though it may not be stated explicitly, people typically assume that the word *God* refers to a being who is characterized by certain traits. In cultures that have a historical tradition of monotheism, God is typically thought of as a personal spiritual being who has a mind and a will, who "hears" prayers and "sees" what happens on earth, who is usually said to be omniscient and omnipotent, and who in fact controls the ultimate outcome of history. The problem is that these personal divine attributes are not actually established by any of the proofs. At most, the proofs suggest only that there are factors at work within nature that are related to the origination of causation, teleology, and the emergence of value and goodness, and that these factors might reasonably be associated with the consistent exemplification of some sort of creative principle. The idea of a divine personal Being with the traits listed above stems not from the logic of the "proofs" themselves but from the religious assumptions about God with which the proofs have historically been associated. Ironically, therefore, it turns out that Thomas's pivotal assertion in the proofs comes not in his stepwise reasoning but in the way he concludes each proof by simply stating that "this everyone understands to be God."

In chapter 1 I indicated that this book embraces a naturalistic metaphysical perspective. This perspective consists of a one-order theory of reality in which there is no supernatural or spiritual realm from which a deity might "look down" (Ps 53) or intervene. Interventionist theology, with its implied concept of a God who interrupts the natural order rather than functioning as an integral part of that order, is denied categorically. The notion that God is characterized by qualities that are personal or anthropomorphic—including psychological anthropomorphisms such as having a mind or a will—is also denied. In fact, in this book, religious anthropomorphism and theistic "personalism"[27] are deemed to be un-

27. Most historical types of monotheism are personalistic in the sense that they view God as a single being with a conscious mind and a will. The term *personalism* may also refer to a specific type of theology, influenced by Hegelian idealism, that

sustainable for the following reasons (which can be stated here only in abbreviated form).

(1) The Purpose of Creation

If there is a personal God who created the world deliberately, then we might expect that an overall purpose for this creation would be suggested by natural history, by the content of human experience, or both. However, such a purpose is, at best, ambiguous. The truth is that natural history, human history, and our own lives are characterized by the recurrent emergence of cross-purposes, conflicts, and other random events that are simply pointless. Some persons have argued that such cross-purposes and randomness occur because God has intentionally designed the world to be what the theologian Irenaeus of Lyon (ca. 130—ca. 202) called "a veil of soul making"—a kind of workshop or proving ground for moral and spiritual development.[28] The problem is that such a proving ground turns out to be empirically indistinguishable from a world in which there is no overarching purpose at all.

(2) The Problem of Evil

As a corollary to questions about the purpose of the universe, "the problem of evil" asks specifically why a benign deity would allow innocent persons to suffer. One would expect that a personal God who is loving and just and who is aware of conditions on earth would act to protect innocent people from harm. The existence of a personal God who *could* prevent undeserved anguish but fails to do so constitutes the problem of evil in its most intractable form. In philosophical theology there are several generally recognized types of theodicy (defenses against the problem of evil), but none has ever attained a consensus endorsement. As theologian William Dean has said, "The long history of theodicy suggests that its task has never been quite accomplished."[29]

---

flourished at Boston University in association with the work of Borden Parker Bowne (1847–1910), Edgar S. Brightman (1884–1954), Georgia Harkness (1891–1974), and Peter Bertocci (1910–1989).

28. See Hick, *Philosophy of Religion*, 506–21 ("The Irenaean Theodicy").

29. Dean, *Religious Critic*, 144.

(3) The Question of Prayer

If God is a conscious being with a will, then prayer as a means of direct communication with this God should bring about observable results, but such results are not consistently forthcoming. Understandably, cases in which definite answers are said to have materialized tend to be reported more frequently, but there are surely a great many other cases when no results at all are in evidence.

(4) A Psychologically Motivated Fiction

The claim that God is personal seems to many to be an obvious psychological projection of human traits onto the universe. Since prehistoric times human beings have believed in gods who have personal qualities, but during the past three hundred years philosophers and social critics such as Feuerbach (1804–1872) and Freud[30] (1856–1939) have propagated the notion that an omniscient, forgiving deity is merely a projection or a construct, the result of a neurotic attempt to bestow authority upon various social mores and alleviate certain psychodynamic tensions.

(5) An All-Experiencing Consciousness

It is problematical to attribute consciousness to God. Consciousness is a temporal, situation-dependent function, not an eternal, transcendent one. If we consider it carefully, it is evident that consciousness refers to the way we finite creatures focus our attention on aspects of our experience that are felt to be important in our present circumstances. In chapter 12 I quoted Marjorie Suchocki's assertion that "consciousness is itself the narrowing of focus, the creation of a foreground and a background by lifting 'this' rather than 'that' into significance."[31] But in contrast to our personal need to focus our attention on one thing or another, God is said to be everywhere, transcending all circumstances and having unlimited knowledge—in other words, to have a constant, uninterrupted awareness of everything. A conscious God would see all perspectives at once,

30. For Freud's analysis of religion, see his *Moses and Monotheism* (1939); *The Future of an Illusion* (1927); and *Civilization and Its Discontents* (1930), 28, 31–32 (final pages of sec. 2).

31. Suchocki, *Fall to Violence*, 38.

hear all sounds at once, and feel all feelings at once. God would thus somehow constantly experience even things such as odors and flavors, physical feelings like satiety, hunger, desire, revulsion, pain, and pleasure, and emotions such as fear, joy, ennui, shame, lethargy, and so on. God knows the appetite of the big fish who eats the little fish and the misery of the little fish who is eaten. In other words, God's attention would not need to be focused anywhere because God's consciousness is focused everywhere—which is the equivalent of not being focused on anything. I submit that using the word *conscious* to describe this very puzzling state of affairs is little more than a way of modeling God in the image of humanity, albeit in an ostensibly perfected state.

(6) A Perfect Divine Will

Our argument against consciousness can be adapted in order to assess the proposition that God has a perfect will. What can it mean, in a pluralistic universe, to claim that the Creator of that universe does only good things? One of my teachers, Harvey Potthoff, used to ask first-year theology students, "What is God supposed to do when both the pitcher and the batter are praying?" One student cleverly responded, "Foul ball!" Droll—but of course the world cannot proceed for very long on a strict diet of foul balls. The fact is that no choice or action—not even God's—can be said to be "perfectly good" from every standpoint. Certain theologians have objected that this challenge to God's goodness is based on our own self-oriented understanding of what is good—that, in other words, it is a manifestation of our own hubris. For example, it might seem good to us to exterminate all mosquitoes and cockroaches—but would that be good for nature as a whole? Certainly it would not be good for the mosquitoes and the cockroaches. The implied question is, Can God be expected to place human needs ahead of all others? This question emerges when God answers Job out of the whirlwind:

> Where were you when I laid the foundation of the earth?
> Tell me, if you have understanding.
> Who determined its measurements—surely you know!
> Or who stretched the line upon it?
> On what were its bases sunk,
> or who laid its cornerstone
> when the morning stars sang together
> and all the heavenly beings shouted for joy?

. . . .

Is it by your wisdom that the hawk soars,
and spreads its wings towards the south?
Is it at your command that the eagle mounts up
and makes its nest on high?
(Job 38:4–7, 39:26–27)

Reflecting on passages like this one, many have suggested that mature faith calls us simply to accept the fact that whatever God wills is, by definition, good, even if the results are not to our liking. This seems to be the argument of the apostle Paul in Romans 9, and it is a prominent belief in the theology of Calvinism.[32] Paul writes, "Whether we live or whether we die, we are the Lord's" (Rom 14:8). These are words of valiant commitment to a good higher than our own security. But on the other hand, if we attribute every event to God's will, isn't it possible to become too acquiescent, too passive? We cannot help but ask whether this type of brave, unquestioning resignation is really what faith requires. When we are forced to confront all of the pain, failure, tragedy, and death that life entails, we are flung back again upon the problem of evil.

There is also another problem related to an allegedly perfect divine will. The conviction that God does have a definite will leads all too easily to the attendant conviction that one's own group knows exactly what that will is and what it requires of human beings. History shows that this outlook is a sure recipe for trouble, leading to the formation of religious movements that are self-righteous, moralistic, coercive, and predisposed to violent forms of enforcement and punishment. The whole idea of a "perfect" divine will is therefore not only illogical but dangerous.

(7) An Ontological Peculiarity

A God who is a ubiquitous, invisible, eternal, all-knowing personal spirit would clearly be in a special ontological category—a category not limited by physical or temporal conditions. Augustine recognized this and held that God is wholly beyond space and time, seeing all of history from the

32. John Calvin (1509–1564) and his followers interpreted the sovereignty of God to mean that God controls everything absolutely, and that whatever God wills is good simply because God wills it. I suggest that this Calvinist axiology is actually more defensible than our typical procedure of speaking casually of good and evil as if these terms have some objective, universal meaning, when what we really have in mind is our own cultural and personal values.

perspective (if that's a suitable term) of one eternal moment.[33] Religious dogma has historically claimed that God's spiritual realm is also inhabited by other beings such as angels and, in some versions of the story, by a predestined collection of our deceased friends and relatives. But whether God is said to be alone or in company, God's eternal mode of existence is by definition radically transcendent and therefore not a type of being that we have any impartial means of verifying. This type of nonverifiability was illustrated in the mid-twentieth century by philosopher John Wisdom (1904–1993), who invented a parable, which was subsequently elaborated by Anthony Flew (1923–2010), about two jungle explorers who encounter a clearing where flowers are blooming. One explorer believes that there must be a gardener who tends the flowers, but the other views the flowers as a chance occurrence. Various attempts to detect a gardener turn out to be inconclusive, and the first explorer continues to believe that there is a gardener, while the second explorer does not. The parable is usually applied to the question of God's existence, suggesting that there is no evidence that can convince either believers or nonbelievers to change their minds.[34] The point, which is still valid today, is that there is no generally satisfactory way to determine whether God's exceptional ontological status is anything more than a figment of some people's religious imagination.

In view of these seven objections, it is now tempting to state categorically that it would be best to eliminate all personalistic imagery from any claims about God's nature. Personalistic concepts of God lend themselves to patriarchal attitudes and authoritarian abuse. They are derived from outmoded traditions and forced reasoning, they exacerbate the problem of evil, their internal coherence is fragile at best, and they are often sustained by the unconscious influence of neurotic motives. Finally, they strain the credulity of a great many thoughtful persons.

On the other hand, astute readers may notice a critical problem with my seven objections to theological personalism. It is that they have been formulated by relying on the very same methods that have been employed by traditional philosophy of religion—that is, by appealing to reason and experience as unbiased grounds for argumentation. The seven objections proceed brashly without taking account of the pluralistic, linguistic, postmodernist, liberationist, and historicist objections to

33. Augustine, *Confessions* 11.5–6, 11.11–28.

34. See Hick, *Philosophy of Religion*, ch. 32 ("Theology and Falsification"); and Wikipedia, "Parable of the Invisible Gardener."

all statements that purport to reveal unbiased philosophical truth. The seven critiques do have a slight advantage that may seem strange to non-theologians—namely, that they do not attempt to state what God is but only what God is *not*. Stretching back to antiquity there is a tradition called "apophatic theology," from the Greek *apophemi*, "to deny," which holds that due to the limits of human knowledge, the true nature of God's perfectness must simply be deemed ineffable. Confronted by God's unavoidable mysteriousness, the method associated with apophatic theology is the *via negativa*, the "way of negation." For example, it can be said confidently that God is *not* a golden calf (Exod 32) or a stone statue (Acts 17:22–29), but not precisely what God *is*.[35] Of course, most apophatic theologians of antiquity would still have affirmed that God is a personal spirit with a preordained plan for human history, and they would not have agreed with the seven arguments against personalism that I have just presented.

In any case, the bottom line is that, because of the pluralistic, linguistic, postmodernist, and pragmatic-historicist critiques, the theistic proofs are not convincing verifications of the existence of a God who is universally worthy of worship. Even if the proofs *were* valid, there are formidable objections to the assumption that what they reveal has any clear linkage with the personal God of traditional monotheism. What the proofs do more modestly suggest is that there are aspects of nature that are potential sources of religious meaning. The presence within nature of orderly causation, design or purposiveness, and gradation or valuation does often inspire a religious response, sometimes even among persons who do not embrace traditional theism.[36] My proposal now is that though the cosmological, teleological, ontological, moral, and axiological-esthetic arguments do not establish the existence of a personal divine being, they do serve a heuristic function, drawing our attention to aspects of nature that merit religious interpretation.[37] Philosopher Frederick Ferré proposes that

35. See Wikipedia, "Apophatic Theology."

36. Philosopher Donald A. Crosby rejects traditional notions of God but is eloquent in his argument that nature itself is a source of religious meaning. Some of Crosby's books are *More Than Discourse: Symbolic Expressions of Naturalistic Faith* (2014); *The Thou of Nature: Religious Naturalism and Reverence for Sentient Life* (2013); *Faith and Reason: Their Roles in Religious and Secular Life* (2011); *Living with Ambiguity: Religious Naturalism and the Menace of Evil* (2008); *Novelty* (2005); and *A Religion of Nature* (2002).

37. On the heuristic value of the proofs, see Ferré, *Modern Philosophy of Religion*,

> religious language may contribute to cognitive inquiry without itself being subject to the same rules of inquiry. We recall Kant's granting to theistic discourse, even within the domain of pure theoretical reason, certain "heuristic" roles. Imagining God as a cosmic designer . . . can give shape and direction to our observations, pointing our attention toward the intimate interconnections between things and spurring us on to search for intelligible principles in every art of nature. Again, picturing God as the one Supreme Being can give a needed sense of unity to our inquiries and place an important demand against all our explanations that they [must] strive toward . . . the "ideal of coherence."[38]

In other words, the proofs are valuable not because of the force of their logic but because they draw our attention to certain aspects of nature—and it is nature, not special revelation, that is here understood as the ultimate basis of our knowledge and of any faith that we espouse.

Nevertheless, concerning any concept of God, at present we unfortunately seem to be stuck between the proverbial rock and a hard place. Traditional concepts of God as a divine Person seem to be all metaphor with no underlying basis, and heuristic suggestions are a poor substitute for deeply held convictions. If we do attempt to go beyond metaphors and rehabilitate a belief in a personal deity, we face difficulties. If we attempt to describe God as an objective reality in the manner of the correspondence theory of truth, the fact of religious pluralism will remind us that there is no single, universal religion to which alleged truth may conveniently be attributed. Linguistic analysts will accuse us of overestimating the trustworthiness of language and argue that our claims are unverifiable. The postmodernists will point out that any statements about an omnipotent being whom we ought to worship cannot help but establish the type of narrow-minded orthodoxy for which religions are notorious and create a power imbalance that favors believers over nonbelievers. And the pragmatic historicists will say that we have mistakenly ignored the influence of our own contexts and traditions and have failed to appreciate the role of personal choice, which is, after all, always an aspect of religious faith.

Despite all of this, however, we are not without options. There are still theological ideas and methods that are open to us. In this book's final

193–94, 206–7, 366–70.

38. Ferré, *Modern Philosophy of Religion*, 366–67.

chapter I will describe several possibilities for theological construction. Such theological construction is crucial, for despite all rational, logical, and empirical criticisms, human beings still find in their own experience that it is possible to have what has traditionally been called a personal relationship with God or the Sacred. My own conviction is that in today's world that relationship needs very much to be renewed.

16

The Glory of God

IN THIS, OUR CONCLUDING chapter, I present some ideas about God and about the overall meaning and function of religion. These ideas are intended to be in harmony with the organismic, naturalistic philosophy that has been developed in the previous chapters. The overall goal is to contribute to a philosophy of nature, a worldview, in which the natural sciences and the humanities, including even religion, can both be affirmed. Here, then, are some resources, some building blocks, for a naturalistic religious perspective.

(1) God as the Ground of All Being

The first building block is Paul Tillich's idea of *God as the Ground of all Being*. Tillich (1886–1965) was a Lutheran minister and scholar who left his teaching position in Germany when Hitler came to power, subsequently teaching at Union Theological Seminary in New York, at Harvard University, and at the University of Chicago. Tillich was a rare theologian who achieved a level of public recognition; in March 1959, his portrait appeared on the cover of *Time* magazine. Tillich's theology is complex, subtle, and detailed; his three-volume *Systematic Theology* runs to 892 pages.[1] Some have accused Tillich of being a philosopher rather than a theologian, but this charge is unwarranted. Tillich relied on philosophy—especially existentialism—to frame the questions that

1. Tillich, *Systematic Theology*.

theology needs to address, but he relied on the Christian tradition as the primary source of answers.

One of Tillich's main ideas, and the idea that we will now consider, is his distinction between the conventional idea of God as *a* being—albeit a unique, divine one—and the idea of God as *being-itself* or as the *ground of all being*.

> The statement that God is being-itself is a nonsymbolic statement. It does not point beyond itself. It means what is says directly and properly; if we speak of the actuality of God, we first assert that he is not God if he is not being-itself. . . . Theologians must make explicit what is implicit in religious thought and expression; and, in order to do this, they must begin with the most abstract and completely unsymbolic statement which is possible, namely, that God is being-itself or the absolute. However, after this has been said, nothing else can be said about God as God which is not symbolic.[2]

What does Tillich mean by being-itself, or the ground of all being? (Both phrases are used.) If I may propose an analogy, consider what is signified by the word *theater*. Theater includes a stage with sets and various props; there are actors, a director, a playwright, a producer, and other persons, including an audience. Then, there are the standards, practices, and traditions associated with theatrical activity. Consider *all* of this, not limited to a particular time or place but all human enactments of drama, from the present time and stretching back to antiquity. We are now speaking of theater in its most generalized sense. In our analogy, what Tillich is asserting is that God is not a specific character in the play—not even a very special character. Neither is God the playwright, the director, nor the producer. God is theater itself, or the basis of theater, in its entirety. To be clear, God is not simply the large, loose assemblage of all the actors, writers, stage settings, and so on—not a piecemeal collection. Translated from the theater analogy to religion, that would be a type of pantheism. Leaving the theater and returning to Tillich's theology, we are asked to contemplate being-itself—all of existence—as a single category. There is a sense in which this is a high-level abstraction, but there is also a sense in which being is utterly concrete. Thus Tillich speaks of God as the ground of all being, the absolute.

2. Tillich, *Systematic Theology*, 1:238–39. Tillich did not believe that "Being-Itself" ought routinely to be substituted for "God" in ordinary speech, and especially not in religious settings, in which traditional symbolic language is preferable.

Tillich's concept of God has been criticized for being too impersonal, too intellectual, and too inaccessible for religious purposes. These criticisms may to some extent be a matter of taste, but in any case, despite these objections there are distinct advantages to Tillich's concept of God. In the first place, the usual debates about whether God exists are obviated.[3] Even though one may question the details of various ontologies, it is hardly reasonable to say that being-itself does not exist. A second advantage is that our seven objections (in chapter 15) to the idea of a personal God become largely irrelevant, for Tillich's God is not an individual being who has attributes such as consciousness or a specific plan for the world. A third and, in my view, most significant advantage is that, when taken seriously, the idea of God-as-being-itself inspires a sense of consummate majesty. Meditating on being-itself, one may be reminded of the words of the *Gloria Patri*: "As it was in the beginning, is now, and ever shall be: world without end. Amen." About Tillich's God, Malcolm Diamond writes,

> The truly ultimate cannot be the personal God of theism. . . . God cannot be an individuated and characterizable [example] of consciousness and power, that is, an individual person, because any such entity is, according to Tillich, necessarily limited. It [a personal God] cannot be the truly ultimate. Therefore, God cannot be a being alongside others to which the traditional theological attributes can be applied.[4]

If a Scripture passage is wanted, we may cite one of the holiest moments in the Bible, the narrative of Moses's experience before the burning bush. "But Moses said to God, 'If I come to the Israelites and say to them, "The God of your ancestors has sent me to you," and they ask me, "What is his name?" what shall I say to them?' God said to Moses, 'I AM WHO I AM.'[5] He said further, 'Thus you shall say to the Israelites, "I AM" has sent me to you.'" (Exod 3:13–14 NRSV) The divine name "I AM" is said to be difficult to translate, but it is a form of the verb *to be*. This sounds remarkably Tillichian.

3. Tillich analyzes and rejects the ontological, cosmological, and teleological arguments in *Systematic Theology*, 1:204–10.

4. Diamond, *Contemporary Philosophy*, 336. Diamond provides a helpful summary of Tillich's ideas in part 5 of his work (chs. 13–15).

5. By long-standing convention, YHWH, the transliterated letters of God's name ("I AM") in Hebrew are written in small capitals.

(2) Empirical Theology

A second building block is designated by the phrase *empirical theology*. The word *empirical* in this context does not mean simply that certain kinds of experience are emphasized. Rather, empirical theology refers to a school of thought that rejects the customary theological method of beginning with a conventional concept of God and then attempting to prove that that God does exist. Instead, empirical theology proceeds by surveying human experience and then asking whether there is any aspect of experience that might reasonably be identified with the sacred or the divine, or as an ultimate source of goodness or creativity. Another way to phrase this question is to ask what it is that *ultimately* gives life its purpose, its meaning, and its value—if life is indeed meaningful or valuable. Any proposed answer to this question suggests a way of thinking about God. Since this method seeks to identify God with some aspect of experiential reality, it should be obvious that this way of defining the idea of God obviates the difficulties that we encountered pertaining to the traditional theistic proofs and to the idea that God is a personal being. As I often used to hear Charles Milligan say, "I believe that God is the universe; and nobody seems to doubt that the universe exists."

In modern Christian thought, empirical theology can be traced especially to the German theologian Friedrich Schleiermacher (1768–1834). Beginning in 1810 Schleiermacher served as the pastor of *Driefaltigkeitskirche* (Trinity Church) in Berlin, where his sermons gained wide recognition. Schleiermacher simultaneously was a professor of theology at the University of Berlin. He and Hegel were contemporaries and, to some extent, rivals. Schleiermacher published several books, including his systematic theology, *Der christliche Glaube* (*The Christian Faith*), which is still in print (750 pages in English).[6] He also made groundbreaking contributions to the fields of hermeneutics and historiography. For example, in 1806 he produced a new collection and interpretation of the fragments of Heraclitus, whose philosophy we considered in chapter 10.

Schleiermacher is known as an empirical theologian because he believed that we experience God directly. In his analysis of experience, Schleiermacher noted that there is a constant distinction or tension between things that we do cause and things that we do not cause. "Thus

6. A new translation of *The Christian Faith* by Terrence Tice, Catherine Kelsey, and Edwina Lawler was published by Westminster John Knox Press in 2016.

in every self-consciousness there are two elements, which we might call respectively a self-caused element (*ein Sichselbstsetzen*) and a non-self-caused element (*ein Sichselbstnichtsogesetzthaben*)."[7] Schleiermacher observed that we also have a direct feeling of the boundary between these two types of causation—a boundary that is dynamic, moving to and fro between relative freedom and relative compulsion within the overall context of experience. This larger context makes possible not only our dependencies but also our freedom, and since we have an awareness of this larger context itself, we experience a *feeling of absolute dependence*—a feeling of the total causal matrix upon which both our self-caused and our non-self-caused experiences depend.[8] This feeling of absolute dependence is, Schleiermacher says, the same thing as our feeling of God. "As regards the identification of absolute dependence with 'relation to God' in our proposition: this is to be understood in the sense that the *Whence* of our receptive and active existence, as implied in this self-consciousness, is to be designated by the word 'God.' . . . [To] feel oneself absolutely dependent and to be conscious of being in relation with God are one and the same thing."[9]

It may seem that Schleiermacher's "Whence of our existence" resembles Tillich's notion of God as "being-itself," but Tillich distances himself from Schleiermacher at this point, saying that being-itself is different from *our experience* of being-itself.[10] It is possible to view Schleiermacher's feeling of absolute dependence as an experiential parallel to the cosmological argument. The feeling of absolute dependence amounts to something like an awareness that, beyond the contingent nature of our own freedom and the contingent nature of various factors in our environment, there is always an inclusive, noncontingent causal matrix within which all contingent events occur. This reasoning is basically the gist of Thomas Aquinas's third proof, which I described in chapter 15. Schleiermacher's genius is his recognition that this noncontingent causal matrix is not just a conclusion reached by deductive thought à la Aquinas but a feeling that is experienced directly. At this point I will share a personal example.

I have a nephew who, with his wife, won a lottery that allowed them to join a group on a Grand Canyon raft trip down Arizona's Colorado

7. Schleiermacher, *Christian Faith*, 13.

8. Schleiermacher, *Christian Faith*, ch. 1, sec. 4.

9. Schleiermacher, *Christian Faith*, 16–17.

10. Tillich, *Systematic Theology*, 1:42.

River. Boating through the Grand Canyon is not a casual endeavor but a two-to-three-week expedition that requires stamina, resolve, and prior experience in white water rafting. The Grand Canyon is, on average, about a mile deep and ten miles wide. During the trip there are no contacts with the world outside, and it is not possible to quit and go home if you get tired or discouraged. All food and supplies are hauled along in the rafts. The river currents are swift and powerful, and there are sections where rafts may easily be capsized or wrecked upon the rocks. Days are spent in constant navigation and exertion, and nights are spent camped on sandbars next to the river.

After several days, one night my nephew had an unexpected experience. As he sat beside the river at night, he heard only the constant sound of the rushing water, reverberating off of thousand-foot cliffs whose rock layers represent millions of years of geologic history. In the darkness high above the canyon walls he could see not only individual stars but the delicate, dim whiteness of the Milky Way Galaxy stretching across the sky. It is not hard to imagine that a person in this setting could be deeply moved. Anyone who rafts in white water is constantly reminded of the distinction that Schleiermacher cited between self-caused factors (rafting skills, for example) and non-self-caused factors (the force of the water and the potential danger of the rocks). For my nephew, however, from beyond those contrasting factors there came an experience of wholeness or transcendence. No single element in his surroundings caused the experience; it was the unity and the timing of the whole setting. In the darkness and majesty of that moment, he found himself overtaken by an awareness of his own small but significant place in an unimaginably large universe. He writes,

> For me, that moment I had described where I felt transcendence was some kind of acceptance of these forces. What I had control over and what I did not; and that I had to find a way to accept not being fully in control of my destiny, but also continuing to move through the trip. Sitting quietly in the bottom of the Grand Canyon, next to the rushing river, looking at the rock that has been exposed from millions of years of work, and looking up at a sea of stars that are millions of light-years away had a profound effect on me. Accepting how small you are in the universe, how little time you have, and reflecting on the challenges and fear you have overcome (physical and mental) has a way of re-grounding your basic humanity.[11]

11. Email correspondence between Kyle Bartsch and the author, Nov. 12, 2024; used

Schleiermacher called this the feeling of absolute dependence and identified it as an experience with religious significance. I hasten to add that Schleiermacher did not regard the feeling of absolute dependence as a special phenomenon restricted to what has been called "religious experience" or "mysticism."[12] The feeling of absolute dependence is accessible even in ordinary moments because it is an experience of the total context in which our freedom is exercised.[13] I have cited my nephew's experience because it illustrates absolute dependence so vividly.

Schleiermacher's emphasis on experience found fertile ground in the United States. The pietism of much American religion, exemplified in the first and second Great Awakenings (approximately 1730–1755, and 1790–1840), is primarily an expression of religious feeling. Another natural fit for Schleiermacher was the typical American interest in the practical as opposed to the theoretical. As a manifestation of this practical bent in America, when William James delivered his Gifford Lectures on religious experience (1900–1902) he defined religion not in terms of doctrines, biblical content, church history, or beliefs about God and the supernatural, but as "the *feelings*, acts, and *experiences* of individual men in their solitude, so far as they apprehend themselves to stand in relation to whatever they may consider the divine."[14] I have italicized James's words *feelings* and *experiences* to draw attention to his continuity with Schleiermacher.

During the first decades of the twentieth century the empiricism of Schleiermacher, James, and of American culture as a whole spawned a school of thought known as "empirical theology" at the University of Chicago Divinity School. Empirical theologians in Chicago included George Burman Foster (1858–1918), Shailer Mathews (1863–1941), Gerald Birney Smith (1868–1929), Edward Scribner Ames (1870–1958), Shirley Jackson Case (1872–1947), and others. Most of these scholars defined empiricism not in reference to the emphasis on experience found in American pietism but as a method of scholarly inquiry that they sometimes described as "scientific," modeled to some degree after the

---

with permission. Though I think of Kyle as a nephew, he is actually my first cousin's son.

12. For the classic definition of religious experience and mysticism, see William James, *Varieties of Religious Experience*, esp. lectures 9 and 10 on "conversion" and lectures 16 and 17 on "mysticism."

13. Schleiermacher, *Christian Faith*, 16.

14. James, *Varieties of Religious Experience*, 42 (lecture 2).

empiricism of psychology and sociology.[15] To them, empiricism meant more or less the same thing it had meant to James: the investigation of what religion signified—that is, how religion functioned—in the lives of people who were actually religious. Ideas of God in empirical theology were formulated by giving a naturalistic (nonsupernatural) interpretation to the beliefs and feelings that the Chicago theologians observed in religious persons. In other words, virtually all of these theologians rejected the customary supernaturalism of traditional monotheism and replaced it with a worldview of religious or philosophical naturalism.

The second generation of empirical theologians included three seminal figures, Henry Nelson Wieman (1884–1975), Bernard Meland (1899–1993), and Bernard Loomer (1912–1985).[16] Wieman defined God in experiential terms in his book *The Source of Human Good*, in which he developed the concept of God as *creative interchange* or *the creative event*. He wrote, "We shall now try to demonstrate that the creative event is the actual reality doing the work in history which has been mythically attributed in the Christian tradition to a transcendental person, even though the real source of human good is neither transcendental nor a person."[17] Wieman emphasized a distinction between *created good* (specific instantiations of creativity) and *creativity itself*, noting that religion has too often focused on the former when it should have been attentive to the latter.[18]

Like Wieman, Bernard Meland found religious significance in nature's tendency towards creativity, complexity, and wholeness, but he was less willing than Wieman and the Whiteheadian "process theologians" to try to condense that significance into any system of clearly defined concepts. If Meland's theology can be associated with one representative aphorism, it is his statement that "we live more deeply than we can think."[19] This depth dimension is experienced through an attitude that Meland referred to as "appreciative awareness." This dimension of depth warrants religious devotion, but that devotion is based not only on

15. For the definitive essay on the Chicago school of empirical theology, see Bernard Meland, "Empirical Tradition," 1–62. For another very helpful description of the Chicago school see Cobb, *Process Theology*, ch. 2.

16. The God concepts of these three theologians are described and incisively critiqued in Dean, "Empiricism and God," 120–28, and Frankenberry, *Religion and Radical Empiricism*, ch. 4.

17. Wieman, *Source of Human Good*, 268.

18. See Frankenberry, *Religion and Radical Empiricism*, 122–24.

19. Meland, *Fallible Forms*, 24.

rational assessment but on intuition, on feelings of reverence, and on ethical commitment. Meland observed that the human race did not invent itself. It is nature that has produced us and endowed us with the ability to live lives of kindness, generosity, growth, and creativity. Thus there is an element of grace within nature that "does for us what we cannot do for ourselves."[20] By affirming our need for a greater-than-human source of grace, Meland stoutly repudiates humanism,[21] but he also repudiates traditional supernaturalism since grace (or God) is found not beyond nature but within it.

Bernard Loomer has been the scholarly source of several original and very suggestive theological ideas. One is a new understanding of divine power. In an address called "Two Conceptions of Power,"[22] Loomer stated that most understandings of power center upon unilateral control, whether exercised upon objects or upon people.

Unilateral power is exemplified by individual or corporate efforts to dominate, manipulate, and exercise control. Power is typically seen as a zero-sum endeavor in which gains by the powerful are balanced by the losses of others. In contrast, relational power regards power as a collaborative process. It is focused on constructive interaction and mutuality. Relational power may seek to influence others, but it is also present in the ability to allow oneself to be influenced by others without feeling that one's own integrity has been diminished. Loomer observed that people of mature religious faith have typically idealized relational power rather than unilateral power.[23]

This differentiation between unilateral and relational power has also been emphasized by feminists, who sometimes refer to a contrast in human relationships between "power over" and "power with." However, Loomer carried this distinction beyond human-to-human relations by wondering how the idea of relational power could pertain to God. Relational power balances the *power to act* with the *power to endure*—that is, the power to be influenced by others without sacrificing one's own

20. In fact Meland wrote an article titled "Grace: A Dimension Within Nature?"

21. *Humanism* in this sense refers to a twentieth-century movement that disallowed religion and supernaturalism (the self-avowed humanists tended to conflate the two) and suggests that any solutions to human problems must be achieved chiefly or even solely through human efforts.

22. This paper was subsequently published in the journal *Process Studies*, vol. 6, no. 1 (Spring 1976), and is now available online: see Loomer, "Two Conceptions."

23. Loomer originally used the term *linear* instead of *unilateral.*

integrity.[24] Thinking not about how much God could control but about how much God could endure, Loomer defined the word *S-I-Z-E* (with hyphens) as a technical term to designate "the volume of life you can take into your being and still maintain your integrity."[25] Loomer then concluded that God has ultimate SIZE because nothing can compromise God's integrity—or else God would not be God. Loomer's reasoning here is not that God is a kind of superman who is personally invulnerable but that God can withstand anything because God *is* everything and therefore does in fact *include* everything. "God should be identified with the totality of the world, with whatever unity the totality possesses."[26] This, as Loomer acknowledged, is a type of pantheism.[27] He wrote,

> The world is God because it is the source and preserver of meaning; because the creative advance of the world in its adventure is the supreme cause to be served; because even in our desecration of our space and time within it, the world is holy ground; and because it contains and yet enshrouds the ultimate mystery inherent within existence itself. "God" symbolizes this incredible mystery. The existent world embodies it. The world in all the dimensions of its being is the basis for all our wonder, awe, and inquiry.[28]

Loomer's theology has been called "brave and instructive" and been praised for its candor in admitting that God is "morally ambiguous." However, this praise should be tempered because, in the first place, there is nothing very profound or instructive about the simple recognition that the universe contains both good and evil, and, in the second place, moving from that recognition to statements about the moral quality of *the universe as a whole* commits a category mistake, or the fallacy of

24. In Christian theology, this type of power—the power to endure—is more often associated with Christ than with "God the Father," especially in kenotic theories of the atonement. *Kenosis* refers to Christ's act of "self-emptying," based on Phil 2:4–8. The idea is that Christ willingly emptied himself of divine privilege in order to procure our salvation. Applied to Christian soteriology, the reasoning here is subtle, if not in fact forced. In any case, Loomer's interests were not christological but philosophical.

25. Loomer, "S-I-Z-E," 6, quoted in Dean and Axel, *Size of God*, 1.

26. Loomer, "Size of God," 20.

27. The term *pantheism* is not in itself a very precise concept. As Charles Milligan noted, there are various types of pantheism—either static or dynamic, and either monistic or pluralistic.

28. Loomer, "Size of God," 42.

composition.[29] Words such as *good* and *evil* are meaningful only when applied to a concrete situation (whether actual or hypothetical) in which comparative values or competing interests are involved. The universe itself does not exist within a context; it *is* the context. To refer again to Tillich's theology, being-itself does not manifest characteristics like "good" or "evil." It is true that nature as a whole may be said to be good in the general sense that it makes specific instantiations of goodness possible. That is, it is reasonable to claim that a complex, evolved universe is better than a universe filled with nothing but cosmic dust. But in a complex universe in which life and success are possible, death and failure are necessarily also possible. And perhaps that is actually Loomer's point.

One other empirical theologian who deserves to be mentioned is one of my teachers, Harvey H. Potthoff. Though not widely known in academic circles, Potthoff influenced a large number of students during his years on the faculty of the Iliff School of Theology in Denver. Potthoff had been a student of William Henry Bernhardt, whose PhD work at Chicago was supervised by Shailer Mathews. In addition to the Chicago influence, Potthoff also studied with Alfred North Whitehead himself during Whitehead's final year of teaching at Harvard.

As an analytical typology, Potthoff defined four commonly encountered models of God: the Planner and the Plan is the obvious personalistic model; the Ocean and the Wave is found in the Neoplatonism of Plotinus and in some forms of mysticism; the Speaker and the Word model describes the theology of Karl Barth and neo-orthodoxy; and finally there is the Organismic Model, which Potthoff himself preferred.[30] In his book *God and the Celebration of Life*, Potthoff identified God as "the Wholeness-Reality."

> The hypothesis being suggested, then, is that we think of God not as an isolated being or process or function, but as the Wholeness-Reality, experienced in trust, devotion, and hope. . . . God may be spoken of variously as *the ground of wholeness, the character of reality in its wholeness, the dynamic reality making for wholeness*. . . . In this holistic approach, power and value,

29. A "category mistake" occurs when some quality or behavior is attributed to an object to which the quality or behavior in question cannot be meaningfully applied, such as, for example, "an immoral oak tree." The "fallacy of composition" errs by mistakenly attributing some characteristic of the parts to the whole: "Since every member of the team has two eyes, the whole team must have two eyes."

30. I learned of these four models during several years of study with Potthoff. I am not aware of any published resource in which he presented these models.

> process and structure, limitation and potentiality, integration and disintegration, part and whole, life and death are held together in *patterned process* by virtue of the reality of God. . . . It is consistent with this approach to conceive of God as the integrity of reality, the power of being and becoming, implicated in all events and experiences as "the intangible fact at the base of finite existence."[31]

Potthoff wished to avoid a definition of God that was merely a concept or an abstraction. He suspected that Wieman's "Creative Event," though helpful in many ways, was such an abstraction, for it defined God primarily in reference to human needs and human personality and focused on one particular type of process. In contrast, Potthoff regarded the Wholeness-Reality as being rooted in physical nature. Atoms themselves are organismic or holistic, and this holism continues to occur naturally in the evolution of life and in the human experience of wholeness.[32]

Potthoff defended the religious applicability of the Wholeness-Reality by showing how our experience of it can function as ground, grace, and goal. We experience the Wholeness-Reality as Ground when we experience "a dimension of reality which transcends or is more than the tangible and measurable—a dimension of order, dependability, integrity, directionality, creativity, wholeness, and whole-making."[33] We experience grace not "as an infused potency or favor arbitrarily bestowed by an external God," but as "the power for being and becoming more whole, appropriated in the midst of life," and especially as we are renewed by "redemptive processes by which situations counted hopeless or evil issue in new structures of hope and worth."[34] God is known as goal when we experience "the Real-Other who lays claim" on our lives, summoning us into service, and leading us toward a destiny that is real and yet undetermined at present. We cannot predict the future, but we can "discern

31. Potthoff, *God and the Celebration*, 192–93. Potthoff was not alone in making the idea of wholeness a central concept. Jan Smuts (1870–1950), a South African military and political leader, is noteworthy for his use of the concept. See Smuts, *Holism and Evolution*.

32. For example, it is interesting that in the Bible the phrase "made whole" is a synonym for physical and emotional healing. See Jer 19:11, Matt 9:22 and 15:28, Luke 8:50, John 5:4–9.

33. Potthoff, *God and the Celebration*, 205.

34. Potthoff, *God and the Celebration*, 208–9.

a purposefulness inherent in reality—to bring forth the possibilities in persons and events, moving on toward wholeness of linkages."[35]

(3) Feminist Theology and Liberation Theology

There is one more type of religious empiricism to be examined, if only briefly, but which absolutely must not be overlooked. I am referring to feminist theology and liberation theology. These theologies may or may not use the traditional language and conceptualities of established religions, but whether or not they do, they reconstruct traditional religious themes in the light of the experiences of those who have been oppressed, exploited, or marginalized on the basis of gender, race, or economic class. It is this reference to *their own experience* as an essential criterion for theological construction that enables feminists and liberationists to identify and shake off the oppressive aspects of existing religious beliefs and practices. This reference to experience is what makes feminist and liberation theologies empirical. But their empiricism is different from the empiricism of the Chicago school and other instances of religious liberalism.

German feminist theologian Dorothee Sölle speaks of "the hermeneutical privilege of the oppressed" as "a central idea for all theologies of liberation."[36] This hermeneutical privilege is not some sort of personal superiority of insight but rather a viewpoint from which prejudicial attitudes and systemic injustices can more easily be discerned, a viewpoint that, paradoxically, is *advantaged* precisely because it has been *disadvantaged*. It is based on the repeated observation, by people who are in a position to recognize the fact, that those who are not disadvantaged have a natural tendency to drift into a state in which they minimize or fail to understand the plight of others, especially if those others are not like them. Sölle tells of an experience in 1972 when she was in Chicago at a large church where Jesse Jackson and James Cone were speaking. The leader called out, "I am," and the congregation—perhaps three thousand

35. Potthoff, *God and the Celebration*, 210–11.

36. Sölle, *Thinking About God*, 69. Some authors use the phrase "epistemological privilege of the oppressed" instead, but the word *hermeneutical* is more appropriate than *epistemological*. The issue is not a theoretical one about how knowledge is gained but a practical and ethical one about how social bias (racism, sexism, classism) and oppression are recognized and interpreted.

people—repeated, "I am!" This was followed by "I am Black," "I am beautiful," "I am respected," and "I am honored." Sölle writes,

> I was sitting next to an old Black woman, with wrinkled, work-worn hands. And now this old woman next to me said, "I am beautiful." I think that for the first time I then understood what beauty is: this feeling of worth, of strength, of knowledge, of being grounded in human worth. . . . And this old cleaning woman who, as I then saw, also had no teeth and certainly no money to buy false teeth, said that she was beautiful. She knew it. "I am a child of God" was the foundation of this liturgy of liberation.[37]

Then James Cone began to speak. "To know God means to be on the side of the oppressed, to be one with them and to have a share in the goal of liberation. We must become Black with God." Sölle continued in her own mind:

> If God is on the side of the oppressed, then God cannot be "white," just as he cannot be "man." . . . In that case we must become Black with God, become woman with God. . . . And each of these statements can be transferred to feminist theology or other models of liberating theology: we must become woman with God, become poor, become Black, become old, and recognize God in the many faces of the oppressed. . . . The soil in which such a theology comes into being is the suffering community.[38]

A good many white people—and especially white men—dislike this kind of language, if not publicly, at least privately. Conventional Christianity says, "God is not only on the side of the oppressed; God loves everybody." "God is not Black. God is a spiritual being and has no skin color." "God—or, my goodness, Goddess?—has no gender. How can God be a woman?" Thoughts such as these rest on the presupposition, which I have repeatedly argued is unsustainable, that God is an objective being whose traits can be observed or deduced and stated impartially in the manner of the correspondence theory of truth. In this view, people who claim that God is Black or God is woman are simply incorrect. Traditional religious outlooks also miss the point by being too literal-minded. Blackness and femaleness are not descriptions of God's skin color or gender; as Sölle observes, they are "ontological symbols of oppression." The

37. Sölle, *Thinking About God*, 96.

38. Sölle, *Thinking About God*, 97.

real point is that a "theology which does not articulate the suffering community, does not speak from it, think from it, feel from it, is de facto a theology of oppression. Theology cannot be neutral or avoid taking sides with either the oppressor or the oppressed."[39] This reminds us of Martin Luther King's warning that injustice stems not only from "the vitriolic words and the violent actions of the bad people, but [from] the appalling silence and indifference of the good people."[40] Regarding matters of oppression and social justice, the attitudes and beliefs of conventional religion have too often made us not more aware but more oblivious.

Naturally there are theologies invented by white males that do tend to support the goals of feminism and liberation. For example, Wieman's "Source of human good" attempts to describe the source of good for all humans, and it rejects oppression. Meland's "depth dimension" affirms ethical commitments and social justice. We have already noticed that Loomer's distinction between unilateral power and relational power has a parallel in the feminist distinction between "power over" and "power with." And Potthoff's concept of God as the Wholeness-Reality would surely lead us to recognize and reject the fragmenting, marginalizing quality of racism and sexism. However, it must be admitted that in the historical sources of empirical theology, as in liberal theology generally, opposition to racism and sexism is too often implicit rather than explicit and is sometimes presented as an afterthought rather than as a principal concern—though there are exceptions.[41]

There are reasons for the differences between the empiricism found in empirical theology and the empiricism of feminists and liberationists. These differences have to do with overall purposes and goals. In its spirit and general outlook, modern empirical theology may be viewed as a descendant of what has historically been called "Christian apologetics."[42] Apologetic theology has nothing to do with expressing regret for wrongdoing but is rather the attempt to argue on intellectual grounds in support of religious beliefs, generally appealing to reason and evidence. In

39. Sölle, *Thinking About God*, 97.

40. King, *Testament of Hope*, 270.

41. One thinks especially of Walter Rauschenbusch (1861–1918), whose *A Theology for the Social Gospel* (1917) influenced many pastors and helped to bring about positive changes. It is interesting that Rauschenbusch was the maternal grandfather of philosopher Richard Rorty. Though the ideals of the social gospel usually constituted a minority report in American denominations, social gospelers should nevertheless be recognized as a redemptive influence.

42. Editors of the Encyclopædia Britannica et al., "Apologetics."

Christianity, an early example of apologetics can be found in Paul's speech to the Athenians (Acts 17). Another early example is Justin Martyr (ca. 100—ca. 165), who reasoned that all truth is one and that therefore even philosophers such as Socrates and Plato can be regarded as Christians. Schleiermacher, too, may be viewed as an apologist; one of his most influential books, *On Religion: Speeches to Its Cultured Despisers*,[43] attempts to defend Christian beliefs against the skeptical criticisms of the Enlightenment. Like Schleiermacher, the twentieth-century American empirical theologians did not create a creedal theology specifically for orthodox church folk; rather, they undertook to rehabilitate outdated ideas and articulate new expressions of faith that could appeal to "the modern mind"—that is, to anyone, or at least to anyone who was moderately well-read.

But oddly, their attempt to be reasonable and inclusive constitutes not only a commendable asset but a surprising problem. There is an extent to which, by attempting to produce a set of ideas that any intelligent person could believe—a "plain vanilla" theology, if you will—the empirical theologians left out the concerns and the people who were not plain vanilla, who were not "the average man on the street," but who, it can be argued, constitute the very people and the very concerns that do most urgently need to participate in any benefits or liberation that religion can offer. Luce Irigaray exposes this problem, which is hidden in the assumption that philosophy (or theology) can be objective, or written for everybody.

> [It] is indeed precisely philosophical discourse that we have to challenge, and *disrupt*, inasmuch as this discourse sets forth the law for all others, inasmuch as it constitutes the discourse on discourse.
>
> Thus we have had to go back to it in order to try to find out what accounts for the power of its systematicity, the force of its cohesion, the resourcefulness of its strategies, the general applicability of its law and its value. That is, its *position of mastery*, and of potential reappropriation of the various productions of history.
>
> Now, this domination of the philosophic logos stems in large part from its power to *reduce all others to the economy of the Same*. The teleologically constructive project it takes on is always also a project of diversion, deflection, reduction of the other in the Same. And, in its greatest generality perhaps, from

43. *Über die Religion: Reden an die Gebildeten unter ihren Verächtern* (1831).

> its power to *eradicate the difference between the sexes* in systems that are self-representative of a "masculine subject."[44]

Irigaray's allusion to a self-referential masculine subject can remind us that even Black theology and liberation theology have tended to view the male experience as normative and thus have not always understood or embraced the cause of feminism. Therefore, as Mary Daly puts it, if there is to be a creative eschatology—that is, a new and more adequate basis for hope—it "must come by way of the disenfranchised sex."[45]

Sometimes the disagreement between Black liberation and women's liberation has simply been a result of the divide-and-conquer strategies of white males who deliberately present women and Blacks with conflicting options. This was the situation in 1869 when Frederick Douglass, who consistently supported women's rights, nevertheless declined to support Elizabeth Cady Stanton and Susan B. Anthony's position that women's suffrage should be included with the proposed enactment of voting rights for Black men. Douglass believed that violence against Black men made their need for the franchise more imperative.[46]

But as Irigaray indicates, the tendency of males to normalize their own thoughts and feelings typically makes it more difficult for all men, regardless of race, to recognize sexism. My own father, James S. Conner (1918–1988), was a Southern, white minister who opposed racism. In January 1963, he was one of twenty-eight (out of over seven hundred!) Methodist ministers in Mississippi who signed a statement favoring improved race relations. The statement was published in the *Mississippi Methodist Advocate* and then reprinted on the very next day by every daily newspaper in the state.[47] All twenty-eight suffered negative career repercussions, and some had to relocate to other states. Many, including my father, had to deal with credible threats of violence.[48] In contrast, a few years later when women began to graduate from seminary and apply for ordination, my father and his ministerial colleagues did not see this

44. Whitford, *Irigaray Reader*, 122–23; emphasis original.

45. Daly, "Why Speak About God?," 214.

46. BlackPast, "Frederick Douglass."

47. Reiff, *Born of Conviction*, 93.

48. This whole series of events is described in Reiff, *Born of Conviction*. This story and others like it should remind us that the stereotype of religious liberals as an assemblage of naïve optimists who blithely spout cheerful platitudes while having little genuine effect on social problems is disturbingly inaccurate. I was thirteen at the time and remember the events vividly.

as a situation in which equality and justice figured prominently. Instead, they seemed to see the ordination of women mostly as an unfamiliar change in the workplace that had to be tolerated and, sometimes, as something to joke about. The bishop and the district superintendents undoubtedly felt inconvenienced by the "problem" of whether there would be any churches who would willingly accept a female minister. Throughout society, in those days most men felt that the concerns of the women's movement were overstated. To most men, women seemed fortunate to be able to stay at home and do as they pleased while men were expected to "go out and get a job." In this attitude we men managed to be both self-congratulatory and ethically obtuse, all at once. This relative obliviousness to the experience of women was and still is found not only among white men but among Black and Hispanic men.

Today we say that "things are better." To the extent that this is true, it is mostly because during the past fifty years women have continued to show us the ways in which we men still institutionalize male patterns of thinking and acting. But despite these lessons, sexism persists. In fact in some ways things have worsened. In the present chapter I have been referring to theology and religion, but beyond religion and throughout the world women continue to encounter gender-based stereotypes, condescending attitudes, unequal pay, unequal career opportunities, the opinion that homemaking and childcare are not real work, intrusion by male ideologues pertaining to birth control and women's health, sexual harassment, sexual exploitation, and sexual violence. And, if these symptoms of sexism are still pernicious in the United States, evidence suggests that they are often even worse in other parts of the world.

I have said that the motive underlying the empirical theology of Schleiermacher and the Chicago school was intellectual clarity or apologetics—that is, the attempt to make religious beliefs coherent and persuasive for an educated, contemporary audience. It is now clear that in contrast, the empirical aspect of feminist theology and Black theology is prompted not by an interest in intellectual clarity but by religious conscience and the pursuit of social justice.

Imagine that you are a person whose life includes daily exposure to bias and discrimination based on your gender, your race, or your sexual orientation. Then, imagine also that you are part of a religious community that claims, as most religions do, to offer some form of "salvation"—that is, a constructive approach or maybe even a comprehensive answer to life's most pervasive sources of sin, pain, distress, and failure.

But next, imagine that your present religious community interprets sexism, racism, and similar forms of injustice as only secondary problems, as matters of merely occasional significance. The foremost goal, according to your religious community, is something other than social justice—possibly eternal salvation or "going to heaven," correct biblical interpretation, the promise that your sins are forgiven, denominational priorities and programs, or maybe simply beliefs that are intellectually plausible or up-to-date! Given this scenario, you can scarcely be blamed if you soon conclude that your present religious community is existentially irrelevant, that it is not meeting your needs or, as far as you can see, the needs of society, and that in fact its version of salvation is mostly just a sentimental reference to an anachronistic tradition whose goals focus not on the modes of personal and social liberation that are truly needed, but on reassuring formulas and collective complacency.

This assessment may sound harsh, but it explains why feminist theology and Black theology must necessarily be included as the third major building block in our current project of theological construction. It is also evident that environmental justice must be included in any proposed concept of the Sacred. Though I have not mentioned it in this chapter, in previous chapters I have argued that values arise not only in human life but in the most rudimentary elements of nature. Nature itself must therefore be protected as hallowed ground.

(4) God as a Sacred Convention

We come now to a fourth and final theological building block. It is the notion of *God as a sacred convention*, formulated by theologian William Dean in his book *The Religious Critic in American Culture*.[49] The meaning of "sacred convention" gives us a clearer understanding of the relation between our concepts of God and the reality to which those concepts refer.

In chapter 13 I discussed the indispensable functioning of symbols in our mental life. I said that all conscious experiences or qualia—not only visual images but sounds, smells, tactile sensations, visceral feelings, imagined scenes and emotions, and so on—are constituted of symbols. I acknowledged that this definition of symbols is broader and more inclusive than other possible definitions, and referred readers to Whitehead's

49. Dean, *Religious Critic*. In what follows I rely especially on chs. 7 and 8.

Symbolism: Its Meaning and Effect for further detail. In our acts of perception, symbols are used by our minds to simplify or abstract from a massive welter of sense data in a way that enables us to comprehend our environment and respond to it. Thus, a pine tree is initially seen when a great many photons trigger nerve impulses in our retinas. In chapter 13 I described the way that our minds employ symbolism to consolidate these impulses into the experience of a single object—in this case, a tree. Symbols are also involved when we dream, when we remember the past, and when we imagine future events and actions.

It is obvious that symbols do not simply reproduce physical objects. The role of symbols in all of our conscious life is constructive or inventive. The pine tree itself is not really a solid object with green needles and a brown trunk. It is a very large number of atoms—and, for that matter, there is no simple way for us to conceptualize the atoms, either. Our idea of the pine tree and of the atoms, too, is merely that—an idea. And yet, the subjective aspect of this idea does not mean that our experiences are arbitrary, imaginary, or unreal. In fact, in this book I have rejected materialistic theories of reality and argued that experience—even imaginary experience—is what is real. We awaken from a dream and say, "Oh, it was only a dream"—but we *really did have* the dream. Usually, it is a simple matter to determine whether what we dream or imagine is based on actual events in our environment.

Moreover, because our own sense organs and brains closely resemble those of other human beings, symbols are experienced in a mutual way, and thus symbolism enables us to communicate meaningfully with one another. (Symbolism is also used for communication in nonhuman species and even for intercommunication between species.) Our capacity for imagination enables us to share ideas and to add symbols to symbols in creative synthesis. New patterns of thought and action evolve. We ponder, interpret, and create—and to a considerable extent we do these things not only individually but collectively. These novel syntheses are not unreal simply because they have originated in our minds. They still have a bearing on actual events. This is where conventions come in. William Dean writes,

> Conventions are social traditions developed through several generations, as opposed both to universal structures and to arbitrary acts of the subjective will enforced by the arbitrary exercise of power. Conventions offer less than the

> authority of the absolute and more than the authority of the isolated individual. . . .
>
> A convention, in short, lives through time as a tradition, but as a tradition that is continually augmented by new criticisms and precedents. A convention is a reality apart from its causes, even though its causes contribute to it. . . . Pragmatically stated (where meaning is determined by effects), the independence of a convention from its causes is evident in the fact that a convention can have effects different from the effects that would be predicted if one knew everything that contributed to the convention. Equally, a convention is a reality apart from its effects, so that it is not simply what it is interpreted to be, even though what it is interpreted to be can alter its identity. . . .
>
> It is this independent effectiveness that makes a convention public. The convention is public because it transcends, works beyond, those private persons who invent it.[50]

From this description it is obvious that we human beings are constantly immersed in a multitude of conventions, and yet we interpret them, use them, develop them, and acquiesce to them so naturally that we are scarcely even aware of their role in our lives. Language, for example, with all of its word meanings, syntax, and rules of grammar, is a convention (or perhaps a complex of conventions). Red, green, and yellow traffic signals, lane markers, crosswalks, and rules of the road are based on laws, but we follow them intuitively as conventions. Manners (shaking hands, nodding, saying hello), customs (wearing clothing, and of certain types in certain settings), and facial expressions (smiling, frowning, rolling one's eyes, looking disinterested) are conventions. There are countless conventions pertaining to professional activities such as accounting, engineering, medicine, dentistry, law, music, literature, and teaching. Games such as basketball and football are almost entirely convention-based. The meaningful quality of our holidays is generated by conventions: fireworks on Independence Day, cards and gifts at Christmas, turkey and dressing on Thanksgiving.

These examples make it clear that conventions are not merely optional embellishments in life but, more nearly, essential parameters. Though conventions vary in importance, in many cases they are actually necessary for public safety and social interconnection. Dean notes that conventionalism is similar to the ideas advanced by Peter Berger

50. Dean, *Religious Critic*, 104–5.

and Thomas Luckmann in *The Social Construction of Reality*.[51] However, Berger and Luckmann's social constructions are just that—constructs or projections, which have no reality of their own. For Dean, conventions function in history (that is, in actual human circumstances) in ways that are not reducible to psychological projections. It is true that there is an aspect of arbitrariness or peculiarity pertaining to most conventions. Consider the idiosyncrasies and idioms that characterize all languages, or the specific colors and design of each nation's flag, or the particulars of each nation's currency, money supply, and banking regulations. But despite the element of arbitrariness in these examples, there is also an aspect of necessity. Though not preordained, the rules, once adopted, must be followed carefully, especially in certain cases. For example, in some countries motorists drive on the right side of the road, and in others, on the left, but it is of extreme importance for all drivers in a given region to follow the same pattern. We see that the specifics that develop could have been defined in many different ways but that, once a certain way has been established, it can be very difficult or even impossible to change it.

In the view that I am developing, conventions may be understood as complexes of shared symbols that appear within groups of people because of some common interest or concern, so that the resulting conventions then influence both individuals and groups in ways that are more or less appropriate but not always predictable. The point is that conventions make a real and sometimes spontaneous difference in the world. Thus, ideas and feelings acquire an efficacy of their own, especially when they are maintained throughout a group. This is true not only among human beings. In chapters 8 and 9 I developed the idea that the information shared throughout nature is not only physical but conceptual and that each actual occasion integrates the two types. This means that biological and even physical systems can create and transmit patterns and trends, amounting to something like a protoconventionalism.[52] The information

51. Berger and Luckmann, *Social Construction of Reality*.

52. Dean elaborates on this issue by distinguishing between epistemological conventionalism and ontological conventionalism (see Dean, *Religious Critic*, 107–25). Epistemological conventionalism depends on processes (presumably *human* processes) of knowing or recognition, while ontological conventionalism is present in nature, as I have just affirmed. In Whitehead's philosophy, which replaces static being with dynamic becoming, the epistemological-ontological distinction is blurred because the ontology of concrescence always involves prehension, which is the basis of Whitehead's epistemology. I think Dean is mistaken in calling Whitehead an epistemological conventionalist (see Dean, *Religious Critic*, 112–13). Whitehead's use of conventions is ontological, based on the reality-creating, prehensive activity of all actual occasions.

transmitted by the DNA molecule is an example. Another way of talking about the use of symbols within conventions is to say that Aristotelian formal causes are experienced and shared *as forms* and not merely as predetermined repetitions of efficient (mechanistic) causes. My point is that conventions are not peculiar to human beings; they have their origins in the real functioning of formal causes throughout nature.

As I have said, conventionalism can be viewed as a way of understanding the relationship between actual events (history) and ideas or goals. By now you may already have guessed the way in which conventionalism is relevant to religion. The trappings of religion—stories, moral precepts, rituals, physical objects, hymns, scriptures, orders of worship, types of prayer—are all understandable as sacred conventions. Indeed, as Dean proposes, the very idea of God may be understood as a sacred convention:

> Conventionalism, I will argue, offers a post-structuralist[53] way to account for a public sense that something sacred works through the public's spiritual culture, that it expresses their scriptures and traditions, that it fosters and judges their spiritual culture, and, most importantly, that the sacred is real, that it cannot be reduced to mere projections of people's hopes and fears.[54]

Dean observes that religions arise naturally and spontaneously in human cultures. This occurs not because a supernatural Being inspires certain people in a privileged way but because human beings have a natural need to find significance, purpose, and hope in a larger context, though daily existence often seems haphazard, improvised, and lacking in cumulative meaning.

> People have questions about the meaning of parts of their history, and seem unable to answer those questions unless they acquire a sense of the whole of their history. Religion refers to a sense of the whole; religion refers to that appreciation for the whole that gives an adequate context to parts of people's personal or public history. The sacred is whatever charges that

53. Post-structuralism is an interpretive outlook that identifies power structures that are implicit or taken for granted in literature, social mores, and other modes of cultural expression; post-structuralism then seeks to discredit or dismantle those structures. Post-structuralism holds that modes of speaking, writing, and decision-making have a tendency to evolve in ways that protect the interests of privilege and class. Dean here is arguing for ways of perceiving and talking about the divine or the sacred in ways that avoid the pitfalls of structuralism.

54. Dean, *Religious Critic*, 102.

> sense of the whole with evaluation; the sacred is understood to be ultimately important and, as such, is the source for religion's normative claims. The sacred provides the narrative direction, the thrust against which, and in collaboration with which, a person's deepest emotions gain and expend their energy. None of this requires that the sacred is a being or is supernatural or that the word "God" is anything other than a term of reference pointing to whatever it is that accomplishes those functions.[55]

Conventionalism avoids the customary belief that some doctrines, mores, scriptures, and rituals (but not others) are valid as "historical mirrors of a God beyond history."[56] In conventionalism, doctrines, mores, scriptures, and rituals are indeed allowed—in fact, they are necessary—but they are viewed simply as typical components within the religious convention that one has adopted or in which one has found oneself. "A convention is a public construction, always formed by an objective public past interacting with a current subjective creativity, but it is not reducible to either of those causes or to their combined effects. A convention is as real as the American Constitution is real."[57]

> The sacred . . . is no less real and independent of people than the Constitution is real and independent of people. The life and mystery of the sacred become evident when the sacred acts in unpredictable ways, something the ancient Hebrews well understood. . . . But the sacred, thus understood, is also relative in a way William James recognized when he said "mental interests . . . help to *make* the truth which they declare." Recognizing both relativism and realism, James, said, first, that religion, if it is not to be a game of private theatricals, must affirm that our efforts count, so that "God himself, in short, may draw vital strength and increase of very being from our fidelity" and, then, that "god is real since he produces real effects."[58]

Thus, when civil rights marchers sing "We Shall Overcome," the liberating power that they sing of is not based merely on wishful thinking. Liberation has its roots in ontological freedom and in ontological goal-seeking,

55. Dean, *Religious Critic*, 132.

56. Dean, *Religious Critic*, 135.

57. Dean, *Religious Critic*, 136.

58. Dean, *Religious Critic*, 137. The quotations of William James are from his essay "Is Life Worth Living" (*"The Will to Believe" and Other Essays*, 55; *Varieties of Religious Experience*, 407 [lecture 20]).

and the liberating force that they claim and that strengthens them is not merely a psychological projection but an ontological reality.

There is one point at which Dean and I seem to diverge.[59] This point has to do with the ontological status—the reality—of the God to which the sacred convention refers. Dean believed that the convention itself is real as a social phenomenon and that this level of reality is sufficient—in fact it is the basis—for adherence to any religion. Dean notes that the needs that prompt the formation of sacred conventions—needs for meaning, hope, inspiration, and wholeness—are not a conscious human creation but arise within us spontaneously as aspects of the way our species evolved—that is, of the way that nature produced us. Thus Dean and I agree that human beings did not invent our thirst for wholeness and cumulative meaning any more than we invented our thirst for water. The problem is, I suspect, that God's alleged origin in personal and social needs will not be enough for most people, who want to embrace faith in something greater than merely an expression of human wishes and feelings. Regarding this "something greater," what are our options? As a way of addressing this question, it will be helpful to distinguish between different types of conventions.

Some conventions depend almost entirely on human choices. At a formal dinner, the place setting may include a salad fork and a spoon. In the United States at least, the fork is normally placed on the left side of the plate, and spoons are placed on the right. Within some groups these are meaningful conventions to observe, but hardly anyone would be distressed if they were not followed strictly.

There are other conventions that have important connections with certain physical conditions. A map shows highways with intersections and bridges in specific locations. The appearance of any map is based on conventions. Roads of a certain type are depicted in a certain color, distances are provided in either miles or kilometers, rivers and lakes are blue, and so on. But imagine that a map depicts a reliable route in a remote area where, in fact, the road has become impassable or a bridge has washed out. In this case, major delays or serious accidents may be the result. The convention has incorporated mistaken information. The definitive reality behind conventional mapmaking is the physical land itself, with its layout of roads and intersections.

59. It is an honor to say that William Dean and I were friends for over thirty years until his death in 2023. During those years we discussed many things, but this is one topic that we never resolved.

Another illustration of this type of convention is the electrical code. This code, which is in force in most counties in the United States, pertains to matters such as what types of wiring, switches, and connections may be used in various circumstances, what circuit breakers are necessary, and the methods of connection to the power grid. In the interests of safety, few if any exceptions to this code are allowable. On the other hand, however, in Europe the entire code is different. This is partly because the standard voltage supplied to customers is 220 instead of 110, as in the United States. In Europe, even the plugs and the wall outlets are designed differently, and a small transformer is necessary to convert the voltage for American appliances. In this example the definitive reality behind various electrical conventions is the raw power of electricity itself. Electrical power can be as enormous and uncontrolled as a lightning bolt, but this great power has practical use for human beings because we have adopted the systematized methods of electrical engineering and manufacturing, which may be viewed as a type of convention.

The obvious question now to be asked is, Is there a larger reality or power behind sacred conventions? Or are sacred conventions more like the conventions related to the placement of knives, forks, and spoons on the dinner table? I believe that the answer to this question is that there really is a definitive reality behind sacred conventions—but this definitive reality must be described with nuance, and without the least suggestion of dogmatic finality of statement.

We have concluded that the usual assumption that God is a conscious, male, personal spirit is untenable—but that there are other possibilities. Tillich and the empirical theologians, including William Dean, associate the sacred with our experience of the whole, with being-itself, described in various ways. British mystic Evelyn Underhill's definition of worship illustrates this reverence for the universe itself, with little immediate interest in doctrine or other cognitive content:

> Worship, in all its grades and kinds, is the response of the creature to the Eternal: nor need we limit this definition to the human sphere. There is a sense in which we may think of the whole life of the Universe, seen and unseen, conscious and unconscious, as an act of worship, glorifying its Origin, Sustainer, and End. . . . Where conscious, [worship's] emotional colour can range from fear through reverence to self-oblivious love. . . . For worship is an acknowledgment of Transcendence; that is to say, of a Reality independent of the worshipper, which is

> always more or less deeply coloured by mystery, and which is there first. As Von Hügel would say, it is "rooted in ontology": or . . . it at least points to man's profound sense of dependence upon "the spiritual side of the unknown."[60]

There are various ways of identifying what we need, religiously, from this whole. For example, in the midst of the transience of our own experience, we seek that which endures. As we cope with life's unforeseeable interplay between what we can control and what we cannot control, we seek cumulative meaning. Even in the certain knowledge of our own death, we seek a basis for hope. In a world where oppression, injustice, war, and the desecration of the earth itself are all too real, we seek a transcendent source of liberation, justice, restoration, and peace. Martin Luther King said, "The arc of the moral universe is long, but it bends towards justice."[61] King's words lift us up, but at the same time we want to know how long that arc is and whether we can really count on the justice that is promised.

In chapter 15 I said that the arguments for God's existence may not be valid as proofs but that they do have a heuristic value as aids to discovery. All of the arguments refer to the universe as a whole (even the ontological argument, which refers implicitly to the whole in its search for perfection). The cosmological argument reminds us that the universe is radically interconnected. We now know that this interconnection is not based merely on mechanistic collisions but on universal fields of influence. Harvey Potthoff said that in the first lecture that he heard Whitehead give at Harvard, Whitehead began by striking a match and talking about the flame. By the end of the lecture, Whitehead had described that flame in such a way that the whole universe was involved. There is a kind of ultimate dependability and majesty in such an interrelated universe, even if it does not and can not safeguard us in the ways we might prefer. Such a universe does create a web of activity in such a way that our own accomplishments will always be woven into the world's history, even long after other human beings have forgotten us.

The teleological argument and the axiological or esthetic arguments point to the reality of goals, outcomes, purposes, and values in nature. This book has defended the view that reality is composed of events, that these events traffic in information, and that some of the information is

60. Underhill, *Worship*, 13. She is quoting Baron Friedrich von Hügel (1852–1925), an Austrian Catholic religious writer and Christian apologist.

61. King, "Where Do We Go."

mathematical-conceptual in nature. The influence of this type of information is not deterministic but depends on an element of self-causation. This allows complexity to increase, eventuating in deliberate purposiveness. Valuation, aspiration, and self-transcendence are therefore not mere epiphenomena but are real factors in history, and urge us forward even now.

The moral argument reminds us that our own capacity for conscience and self-sacrifice, our own passion for justice, and our own halting steps towards liberation and peace are not separate from nature but are instilled in us by nature. We did not originally create ourselves, nor did we suddenly appear in Eden ready-made, as the presumptive culmination of an anthropomorphic God's plan of creation. Nature—the universe—produced us within the same creative matrix that produced all other creatures, and to whom we are therefore kin. Indeed we are kin even to the dust, the atoms, from which we are made. Human yearnings for compassion, freedom, justice, and peace are not epiphenomenal, subjective dreams. They are products of nature and history, just as are all of our other characteristics. The moral argument reminds us that the arc of the universe does bend toward justice because behavior that is heedless, dictatorial, or cruel tends to defeat itself—first, because it poisons the social order upon which it itself depends, and second, because it persistently tends to arouse the conscience and the resistance of all those persons who have not been seduced by its delusions. God does help us by making it possible for us to help ourselves. Sacred conventions have made an essential contribution to liberation by fostering collective unity, sustaining hope, and inspiring courage and vision among the oppressed.

Critics of religion have contended that a hope in other-worldly salvation has had a pacifying effect on those for whom the only real hope lay in actual social transformation. To some degree this criticism may be valid, but it will not hold up as a blanket denunciation of religion, which has also served to encourage and guide liberating action. The liberating God becomes real in these yearnings, struggles, and victories, and the judgment of God becomes real in the self-defeating consequences of oppression. But of course injustice still persists, and the sources of justice that I am describing may not seem very comforting in the face of history's recurrent tragedies and atrocities. However, though it is only a rhetorical question, it is instructive to ask ourselves what our existence *would* be like if a supernatural deity did intervene every time an innocent person suffered. And in that scenario, of course, a just God would be required to protect all persons equally. I suggest that the result would be a form of existence that would be too safe, too guaranteed, and too regimented to be meaningful. In a life that is truly free and self-creative, there must also be an element of risk.

This brings us to the ontological argument, which reminds us that even if the fully perfect is never possible in this world of transience and finitude, perfection continually haunts our imagination. And this state of being haunted by dreams—that is, by unrealized possibilities—is as it should be and must be. The chronic persistence of yearning is in itself a form of perfection.

But how do all of these yearnings and imaginings relate to the whole? Our attempts to describe the cosmos suggest that it has a number of physical regularities. Though current views vary, it is generally believed that a primordial multitude of photons or radiant energy expanded and cooled to the point at which particles formed. The particles are of distinct types with uniform qualities. Certain forces—gravity, electromagnetism, the strong force, and the weak force—exist in such a way that atoms and molecules are able to form. The Pauli exclusion principle, which is purely mathematical, mysteriously applies to the arrangement of electrons in each atom. Without all of these details, there would be no suns, no planets, no chemical compounds, and certainly no life. But all of these factors are indeed real and depend on what cosmologists call initial conditions. However, there is no way to investigate the cause of the initial conditions because they are, well, *initial*. The point is that all of these conditions work together to create a natural system that is organismic—that is, that produces wholes. As conventions pertaining to electricity are derived from electrical power and conventions pertaining to maps are derived from the land and its features (such as roads and bridges), in this book, sacred conventions are based on the underlying holism of nature. But that, of course, allows for a tremendous diversity in the ways that sacred conventions actually develop in history.

Meditating on ideas such as these, some years ago the phrase *total causal matrix* came into my mind. This was not a product of scholarly reflection but merely a casual thought. However, as it turns out, that phrase has stuck with me, and when I want an intellectual-sounding substitute for the word *God*, this phrase is usually what I use. It has become my equivalent of Tillich's *being-itself*, of Wieman's *creative interchange*, of Loomer's *totality*, and of Potthoff's *Wholeness-Reality*. "Total Causal Matrix" does have some advantages. The word *total* suggests the whole, the wholeness of the universe itself, a wholeness that serves as a central touchstone regarding religious feelings and needs. The word *total* also indicates that we humans—our values, our decisions, our actions; our failures and our successes, our careful plans and our unwitting coincidences—are included in the course of world history. As individuals we are small, but we are ontologically as significant as any other individual. We do make a difference, and God regularly works through us. Despite all of our intellectual explanations, there is a sense of mystery—religious mystery, if you will—pertaining to the ways that Nature produces meaning.[62] Bernard Loomer calls our attention to this mystery with his assertion, noted above ([X-REF]), that God is "the totality . . . [and] *whatever unity the totality possesses*."[63] I submit that there is no clear-cut way to demystify either the source or the extent of the Totality's Unity. The word *causal* suggests radical interconnectedness or relatedness, and the way that all events arise in the context of influences from their past. Even in our solitariness, there is a sense in which we remain connected. *Matrix* is a particularly fortunate word. At first I associated *matrix* with the importance of the physical fields that fill the universe. I also associate it with the matrix mechanics of Heisenberg, Born, and Jordan, and the mathematically connected tables (matrices) that they used to solve quantum riddles. But more recently I have remembered that the word *matrix* is of female gender in Latin and that it has a history in English of referring to the uterus—the uterine lining in which the fertilized ovum embeds itself and is nourished. This, if I may say so, seems to be a particularly fertile religious metaphor. I mention this idea of God as the Total Causal Matrix not in the hope of popularizing a phrase but because the phrase itself is an example of the novel terminologies and insights that are generated by conventions—in this case, a blending of sacred conventions with scientific-philosophical conventions.

The way that religious language originates in conventions (which always have a constructed, social aspect) reminds us that no statements about God can appropriately be interpreted as literal descriptions of a supernatural being whose objective characteristics are known only to certain persons who represent superior religious traditions. Truth comes in forms other than the statement of bare facts. In my library there is an

62. Concerning religious mystery, helpful texts are Kaufman, *In the Beginning*; Kaufman, *God, Mystery, Diversity*; Keller, *On the Mystery*; and Wildman, *Effing the Ineffable*.

63. Loomer, "Size of God," 20; emphasis added.

old copy of the autobiography of Benjamin Franklin that includes several miscellaneous articles. One is titled "Remarks Concerning the Savages of North America"—but the title is satirical, for Franklin's purpose is actually to reveal several ways in which the Indians are more civilized than many "civilized" white people:

> A Swedish minister having assembled the chiefs of the Susquehanna Indians made a sermon to them, acquainting them with the principal historical facts on which our religion is founded—such as the fall of our first parents by eating an apple, the coming of Christ to repair the mischief, his miracles and suffering, etc. When he had finished an Indian orator stood up to thank him. "What you have told us," says he, "is all very good. It is indeed bad to eat apples. It is better to make them all into cider. We are much obliged by your kindness in coming so far to tell us those things which you have heard from your mothers.[64] In return, I will tell you some of those we have heard from ours. 'In the beginning, our fathers had only the flesh of animals to subsist on, and if their hunting was unsuccessful they were starving. Two of our young hunters having killed a deer made a fire in the woods to boil some parts of it.'"

As the story continues, the two hunters are about to eat when a young woman descends from the clouds. The two men politely offer her some meat, for which she thanks them and then promises a reward. If they will return to this place "after thirteen moons," they will find a new form of nourishment. They do return at the appointed time and find maize and kidney beans growing, which then become an essential food source for their people. Franklin continues the story:

> The good missionary, disgusted with this idle tale, said: "What I delivered to you were sacred truths; but what you tell me is mere fable, fiction, and falsehood." The Indian, offended, replied: "My brother, it seems your friends have not done you justice in your education; they have not well instructed you in the rules of common civility. You saw that we, who understand and practice those rules, believe all your stories; why do you refuse to believe ours?"[65]

64. In many Native American groups it has been customary for women to maintain the group's history, which is repeated orally and memorized.

65. Franklin, *Autobiography*, 198–99.

The Native American apparently understood what the missionary did not—that religious stories become true not by conforming to some set of physical facts but because of their gradual inclusion as vital ingredients in a sacred convention. This type of truth is not subject to alteration by individuals; it emerges slowly as a kind of social force. Conventions are not immune to criticism. Aspects of a religious convention that seem unhealthy, unrealistic, or uninteresting may undergo alteration or simply be allowed to fade away—or else the criticisms and alterations become parts of the convention. On the whole, a religious convention endures not because it is rational or provable but because of the true meaning, the true hope, and the true personal and social transformation that it tends to inspire. The nature of sacred conventions reminds us that religion is not based merely on verbal propositions. Religious faith is not like the intellectual devotion of Lucretius to material Nature or the stoic devotion of Marcus Aurelius to the Logos. Sacred conventions go beyond what is basically an intellectual-esthetic appreciation of cosmology.

Dean cites the ideas of William James in order to remind us that our religious conventions actually contribute to the reality of God, just as humanly devised electrical conventions contribute to the reality of what electricity can do, and just as accurate maps enable travelers to boldly go where they have never gone before. It is true that some religious beliefs and behaviors can be deemed true or false, functional or dysfunctional, on relatively objective grounds, but religious beliefs are often more like marriage vows, vocational goals, and personal standards of conduct. It is the essence of faith always to involve personal choices and loyalties. This does not make faith a mere opinion. It makes it a truth that we ourselves help to establish and uphold. The absolute dichotomy between God's actions and our own actions is replaced by the insight that the activity of God can include our own activity. The Total Causal Matrix that created the universe may not be personal, but, through history, it produced the personal. When we rely on personal metaphors in our worship services God becomes personal in about the same way that our use of electrical conventions allows nonpersonal electrical power to transmit personal messages. This comparison may not be a precise equivalence, but it is more than just an analogy.

Another way of looking at this is to remember that in the epistemology developed in this book, all conscious experiences involve symbols. On a sunny day, you look at the sky and see that it is blue. But is it really blue? A strict physicalist will point out that the apparent color is really caused by radiant energy of a certain range of frequencies and that the color blue is a mental invention. But since, in our epistemology, symbols are a part of the way that nature operates, our experience of blue is itself real. The sky, in that case, is truly blue. Within limits, the same kind of reasoning can be applied to our experience of the sacred. The sacred convention of which we are a part is full of symbols and practices that enable us to experience God. The personal relationship that we can experience with God is as real as the blue color that we can see in the sky or the love that we can feel for our family.

In this perspective, God helps us not only in the ways that nature tends to heal itself (including our own bodily, psychological, and spiritual forms of healing) but by means of our own ability to do a better job of protecting nature and protecting one another. Indeed, it ought to haunt our consciences that we have the knowledge and the resources needed to feed the world's hungry, to provide universal healthcare, to end war, to redirect our vast expenditures on arms to more beneficial ends, to improve education, to provide better working conditions for laborers, and to end global climate change. We also have the ability to make realistic, positive choices about sources of personal meaning and hope.

The stories of Jesus are true for me because they really do bring me into the presence of God. This is the character of sacred conventions. They really do create situations of forgiveness and renewal and really do encourage oppressed people to resist. The language of faith takes on a life of its own. As a Christian, every winter I look forward to the Advent season and Christmas. As the nights lengthen during December I listen expectantly during Sunday services for the assurance that "the people who walked in darkness have seen a great light" (Isa 9:2). I find myself hoping that one of the hymns will be

> O little town of Bethlehem,
> how still we see thee lie!
> Above thy deep and dreamless sleep,
> the silent stars go by.
> Yet in thy dark streets shineth
> the everlasting light.
> The hopes and fears of all the years
> are met in thee tonight.[66]

66. Written by Phillips Brooks, 1868.

Are these words mere sentimentality? Or is it possible that an everlasting light somehow *can* shine even in a dark street? The stories that most Christians hear at Christmas time are not based on historical fact. Only two of the four Gospels mention Christ's birth at all, and those two differ substantially. Matthew 2 tells the story of the wise men (there were three gifts but not necessarily three men) and of King Herod's slaughter of the innocents. Luke, in contrast, includes Mary's wonderful *Magnificat* (Luke 1:46–55) and reports that a choir of angels sang on a dark night to a bucolic group of shepherds. It seems quite unlikely that any of these reported events actually took place, but as a preacher I can say with assurance that they do make excellent sermon material. Indeed, because they do grip our imaginations, these stories have been repeated for two thousand years with the same sincerity of the Native American who tried to enlighten the Swedish minister regarding the origins of maize and kidney beans.

Regarding the impartation of Christian beliefs, it has been noted that Jesus never verbalized a system of doctrinal principles or ethical norms. Instead, he told parables (Matt 13). Because they engage us existentially, the parables are more memorable, more instructive, and more inspiring than any number of didactic, high-level abstractions or any pedantic recitation of history just because it is "really true." The parables remind us that our Western, practical-minded preoccupation with verifiable facts and scientific objectivity is often an obstruction to deeper meaning. For example, the type of truth contained in works of fiction—such as Alice Walker's *The Color Purple*, Harper Lee's *To Kill a Mockingbird*, and Herman Melville's *Moby Dick*—is more relevant and profound than the dry data found in many textbooks. And Whitehead agreed with Plato that "deeper truths must be adumbrated by myths."[67]

The fictional truths of religion sometimes produce factual history that is stranger than fiction. For example, what seems most astonishing is not the improbability of the Christmas myths but the even greater improbability of what are, most likely, the real facts. A baby is born to an obscure family in an obscure town; of course no shepherds or wise men come. Now, imagine that you are visiting with Mary and Joseph at this point in the story, trying to convince them regarding what their baby is destined to accomplish. They will never believe any of what you tell them. Amazingly, without formal education, the child memorizes large sections of his people's scriptures and embraces a messianic calling; he successfully

67. Whitehead, *Modes of Thought*, 10.

recruits disciples and attracts large crowds; but then, just when he is off to a wonderful start, he dies, reportedly by a brutal, unjust execution. However, instead of fading from memory, he inspires a faith that spreads rapidly. Two thousand years later, billions of people remember his parables and know at least the general outline of his life's story. This improbable history and the impracticality of the high ideals that are associated with it have generated a worldwide religious movement. The existence of this movement illustrates the power of sacred conventions, in which factual reporting is often not a relevant concern. As Whitehead says,

> The essence of Christianity is the appeal to the life of Christ as a revelation of the nature of God and of [God's] agency in the world. The record is fragmentary, inconsistent, and uncertain. It is not necessary for me to express any opinion as to the proper reconstruction of the most likely tale of historic fact. Such a procedure would be useless, without value, and entirely out of place in this book. But there can be no doubt as to what elements in the record have evoked a response from all that is best in human nature. The Mother, the Child, and the bare manger; the lowly man, homeless and self-forgetful, with his message of peace, love, and sympathy: the suffering, the agony, the tender words as life ebbed, the final despair: and the whole with the authority of supreme victory.[68]

All of this demonstrates that there can be another type of truth—a truth about what there is within human existence that is sacred and therefore about what must be the basis of our hope and the focus of our aspirations. As William James observed, those hopes and aspirations themselves help to establish the truth of what we believe.

Even if only implicitly, everyone adopts some hermeneutic, some approach to the interpretation of life itself. This interpretation is not a vague assumption in the background but a matter of familiar attitudes and daily practice. My wife also is a pastor. If one of us says to the other, "God in her great wisdom has brought us to this day" or "to these circumstances," we do not mean literally that a personal, supernatural deity has deliberately manipulated certain events; rather, it is our way of reminding ourselves of our commitment to view life not as a random mixture of decisions, luck, and fate but as a whole that manifests cumulative value, and therefore, as a meaningful gift. The stories, symbols, and convictions of the Christian convention reinforce our commitment. Life's meaning is

68. Whitehead, *Adventures of Ideas*, 214 (ch. 10).

not merely a product of subjective religious preferences. It arises within and from the Total Causal Matrix, of which we are a part.

To repeat, all people adopt some existential hermeneutical stance, whether or not they formulate that stance deliberately. Such stances make an enormous difference in a person's life. I will conclude this book with three summary recommendations regarding the interpretation of life. All three make comments pertaining to religion, and all three couch those comments in the context of an awareness of scientific cosmology. The first statement is by philosopher-mathematician Bertrand Russell, a noted antagonist of religion:

> That man is the product of causes which had no prevision of the end they were achieving; that his origin, his growth, his hopes and fears, his loves and his beliefs, are but the outcome of accidental collocations of atoms; that no fire, no heroism, no intensity of thought or feeling, can preserve a life beyond the grave; that all the labors of the ages, all the devotion, all the inspiration, all the noonday brightness of human genius, are destined to extinction in the vast death of the solar system; and the whole temple of Man's achievement must inevitably be buried beneath the debris of a universe in ruins—all these things, if not quite beyond dispute, are yet so nearly certain, that no philosophy which rejects them can hope to stand. Only within the scaffolding of these truths, only on the firm foundation of unyielding despair, can the soul's habitation be safely built.[69]

I will make no comment about the meaning of this statement or regarding the assumptions upon which it is based but will simply offer, as a contrast, the following statement by Alfred North Whitehead, Russell's teacher and coauthor in the writing of *Principia Mathematica*:

> It was a mistake, as the Hebrews tried, to conceive of God as creating the world from the outside, at one go. An all-foreseeing Creator, who could have made the world as we find it now—what could we think of such a being? Foreseeing everything and yet putting into it all sorts of imperfections to redeem, [for] which it was necessary to send his only son into the world to suffer torture and hideous death; outrageous ideas. The Hellenic religion was a better approach; the Greeks conceived of creation as going on everywhere all the time within the universe. . . . There is a general tendency in the universe to produce worthwhile things,

69. From Russell's *A Free Man's Worship* (1923), as quoted in Pagels, *Perfect Symmetry*, 363.

> and moments come when we can work with it and it can work through us. But that tendency . . . is by no means omnipotent. Other forces work against it.
>
> God is in the world, or nowhere, creating continually in us and around us. This creative principle is everywhere, in animate and so-called inanimate matter, in the ether, water, earth, human hearts. But this creation is a continuing process, and "the process itself is the actuality," since no sooner do you arrive than you start on a fresh journey. Insofar as man partakes of this creative process does he partake of the divine, of God, and that participation is his immortality, reducing the question of whether his individuality survives death of the body to the estate of an irrelevancy. His true destiny as cocreator in the universe is his dignity and his grandeur.[70]

Finally, here is a statement by theologian Rosemary Radford Ruether:

> We should not pretend to know what we do not know, or to have had "revealed" to us what is the projection of our wishes. . . . What we [do] know is that death is the cessation of the life process that holds our organism together. Consciousness ceases and the organism itself gradually disintegrates. . . .
>
> What then has happened to "me"? In effect, our existence ceases as individuated ego/organism and dissolves back into the cosmic matrix of matter/energy, from which new centers of the individuation arise. It is this matrix, rather than our individuated centers of being, that is "everlasting," that subsists underneath the coming to be and passing away of individuated beings and even planetary worlds. Acceptance of death, then, is acceptance of the finitude of our individuated centers of being, but also our identification with the larger matrix as our total self that contains us all. . . . To the extent to which we have transcended egoism for relation to community, we can also accept death as the final relinquishment of individuated ego in the great matrix of being.
>
> All of the component parts of matter/energy that coalesced to make up our individuated self are not lost. Rather, they change their form and become food for new beings to arise from our bones. . . . Is this merely the disintegration of centers of personality into an "impersonal" matrix of the all? If the interiority of our organism is a personal center, how much more so is the great organism of the universe itself? That great matrix that supports the energy-matter of our individuated beings is

70. Price, *Dialogues*, 370–71.

> itself the ground of all personhood as well. That great collective personhood is the Holy Being in which our achievements and failures are gathered up, assimilated into the fabric of being, and carried forward into new possibilities.
>
> We do not know what this means. . . . It is not our calling to be concerned about the eternal meaning of our lives, and religion should not make this the focus of its message. Our responsibility is to use our temporal life span to create a just and good community for our generation and for our children. It is in the hands of Holy Wisdom to forge out of our finite struggle truth and being for everlasting life. Our agnosticism about what this means is then the expression of our faith, our trust that Holy Wisdom will give transcendent meaning to our work, which is bounded by space and time.[71]

Ruether refers to the universal matrix as *everlasting* (though she does place quotation marks around "everlasting"). Russell, on the other hand, is certain that the universe will end in ruins. If I may say so, Russell's idea is something of a downer. Friends and colleagues have asked me whether I think the whole universe will run down. I always answer that if all of this could happen once, it can happen again. This seems reasonable enough and tends to make us feel better, though I suspect that Ruether is right: we do not know what this means. However, as she says, our agnosticism itself is an expression of our faith. I have been asked whether I believe in heaven. I have my doubts. It has been said that some people who yearn for endless life don't know what to do with a rainy afternoon. Ludwig Wittgenstein is said to have commented that the real question about life after death is not whether it exists, but even if it does, what problems that really solves. However, I am not prepared to make any rash pronouncements. It seems to me that the real point of faith in God is not to figure out heaven but to trust the Process.

I sometimes tell people that I am a quantum mystic. I look at objects that seem dense and solid and then experience a sense of wonder, based on my modest understanding that these objects are really a constantly moving, exponentially large swirl of high-speed particles—particles that are themselves not solid matter but are self-replicating processes occurring in incomprehensibly rapid succession. I feel especially amazed to think that all of the electrons make it possible for their atoms to bond as they do because they spontaneously form orbitals based on the Pauli

71. Ruether, *Sexism and God Talk*, 257–58.

exclusion principle.[72] Then I marvel that all of life's inevitabilities and novelties, all of its challenges and disappointments, its successes and failures, all of its experiences of redemption, and all of its moments of transcendent beauty and deep meaning have arisen from these tiny, fleeting processes. I realize that in life's welter of randomness, our existence in its wholeness, our accumulation of days, has still been meaningful. This is a gift. Then I feel thankful to God—the Total Causal Matrix.

Sometimes people ponder the imponderable vastness of interstellar space and conclude that our little planet and our little lives can hardly amount to much. But this, I think, is a non sequitur. Is the vastness of space really a valid criterion for the assessment of life's value? An awareness of space has naturally been very important to us because of the ways in which our survival has depended on space, or at least, on land: as hunter-gatherers, as farmers and ranchers, and always as members of tribes whose territorial boundaries seemed to need safeguarding. But a major emphasis on space is superficial. Perhaps there is a deeper perspective to be gained from the discovery of nonlocality and from the reality of quantum entanglement: space is, after all, not so ultimate. Information is shared between entangled entities instantaneously, regardless of distance. In that sense, at least, information can partake of transcendence. Maybe, as Ruether says, Holy Wisdom really can reveal a transcendent meaning to our lives, spatiotemporal creatures though we are. Life is not merely a way of producing a long, detailed list of facts. It is a hermeneutical project, requiring an act of commitment. In your project, I wish you the best.

72. One of the clearest chemical explanations of why atoms form bonds is provided in Ash, "Why Do Atoms Form Molecules?"

Bibliography

Alexander, H. G. *The Leibniz–Clarke Correspondence*. New York: Manchester University Press, 1956.

American Chemical Society. "Legacy of Rachel Carson's *Silent Spring*." https://www.acs.org/education/whatischemistry/landmarks/rachel-carson-silent-spring.html.

American Institute of Physics. "Digital Content on Track to Equal Half 'Earth's Mass' by 2245." ScienceDaily, Aug. 11, 2020. www.sciencedaily.com/releases/2020/08/200811120205.htm.

Angier, Natalie. "The Mighty Mathematician You've Never Heard Of." *New York Times*, Mar. 26, 2012. https://www.nytimes.com/2012/03/27/science/emmy-noether-the-most-significant-mathematician-youve-never-heard-of.html?smid=url-share.

Antic, Ivan. *The Physics of Consciousness: In the Quantum Field, Minerals, Plants, Animals and Human Souls*. Self-published, Samkhya, 2021.

Aristotle. *Metaphysics*. Translated by Richard Hope. Ann Arbor: University of Michigan Press, 1960.

Arts of Thought. "Carl Jung on Synchronicity." May 30, 2020. https://artsofthought.com/2020/05/30/carl-jung-synchronicity/.

Ash, Arvin. "Atoms Touch. Then Quantum Physics Does Something Unexpected!" Uploaded Jan. 31, 2026. YouTube video, 14:03. https://www.youtube.com/watch?v=jXVrCne1738.

———. "Why Do Atoms Form Molecules? The Quantum Physics of Chemical Bonds Explained." Uploaded Aug. 29, 2020. YouTube video, 13:24. https://youtu.be/azI-_S6g8C8?si=GxEWCA8Q_qcDYTzB.

Atmanspacher, Harald. "The Pauli-Jung Conjecture and Its Relatives: A Formally Augmented Outline." *Open Philosophy* 3 (2020) 527–49. https://doi.org/10.1515/opphil-2020-0138.

Atmanspacher, Harald, and Christopher A. Fuchs, eds. *The Pauli-Jung Conjecture: And Its Impact Today*. Exeter, UK: Imprint Academic, 2014.

Augustine. *The Confessions of St. Augustine*. Translated by Edward B. Pusey. New York: Dutton, 1907.

Auxier, Randall E., and Gary L. Herstein. *The Quantum of Explanation: Whitehead's Radical Empiricism*. New York: Routledge, 2017.

Azarian, Bobby. "The Romance of Reality: How the Universe Organizes Itself to Create Life, Consciousness, and Cosmic Complexity." Next Big Idea Club, Aug. 24, 2022. https://nextbigideaclub.com/magazine/romance-reality-universe-organizes-create-life-consciousness-cosmic-complexity-bookbite/35458/.

Bainton, Roland. *Here I Stand: A Life of Martin Luther*. Nashville: Abingdon, 1950.

Ball, Philip. *Beyond Weird: Why Everything You Thought You Knew About Quantum Physics Is Different*. Chicago: University of Chicago Press, 2018.

———. "Experiments Spell Doom for Decades-Old Explanation of Quantum Weirdness." *Quanta*, Oct. 20, 2022. https://www.quantamagazine.org/physics-experiments-spell-doom-for-quantum-collapse-theory-20221020/.

———. "Quantum Leaps, Long Assumed to Be Instantaneous, Take Time." *Quanta*, June 5, 2019. https://www.quantamagazine.org/quantum-leaps-long-assumed-to-be-instantaneous-take-time-20190605/.

———. "The Universe Is Always Looking." *The Atlantic*, Oct. 20, 2018. https://www.theatlantic.com/science/archive/2018/10/beyond-weird-decoherence-quantum-weirdness-schrodingers-cat/573448/.

Bank of England. "£20 Note." Last modified Aug. 4, 2025. https://www.bankofengland.co.uk/banknotes/polymer-20-pound-note.

Barad, Karen. *Meeting the Universe Halfway: Quantum Physics and the Entanglement of Matter and Meaning*. Durham, NC: Duke University Press, 2007.

Barbour, Ian G. *Issues in Science and Religion*. Englewood Cliffs, NJ: Prentice-Hall, 1966.

———. *Myths, Models, and Paradigms: A Comparative Study in Science and Religion*. New York: Harper & Row, 1974.

Bare, Richard L., dir. "To Serve Man." Season 3, episode 24 of *The Twilight Zone*. Aired Mar. 2, 1962.

Bargh, John A. "How Unconscious Thought and Perception Affect Our Every Waking Moment." *Scientific American*, Jan. 1, 2014. https://www.scientificamerican.com/article/how-unconscious-thought-and-perception-affect-our-every-waking-moment/.

Barrett, Nathaniel. "Pragmatist Inquiry and the Problem of Value." In *Religion in Multidisciplinary Perspective: Philosophical, Theological, and Scientific Approaches to Wesley J. Wildman*, edited by Le Ron Shults and Robert C. Neville. Albany: State University of New York Press, 2022.

Bartlett, John. *Bartlett's Familiar Quotations*. 14th. ed. Boston: Little, Brown, 1968.

BBC Earth Science. "Can Slime Mould Solve Mazes?" Uploaded Dec. 5, 2018. YouTube video, 2:32. https://www.youtube.com/watch?v=HyzT5botNtk.

Beardsley, Monroe. *Aesthetics: Problems in the Philosophy of Criticism*. New York: Harcourt, Brace & World, 1958.

Becker, Adam. "Quantum Particles Aren't Spinning. So Where Does Their Spin Come From?" *Scientific American*, Nov. 22, 2022. https://www.scientificamerican.com/article/quantum-particles-arent-spinning-so-where-does-their-spin-come-from/.

Bell, John S. *Speakable and Unspeakable in Quantum Mechanics*. 2nd ed. Cambridge, UK: Cambridge University Press, 1987.

Berardelli, Phil. "Quantum Physics Gets 'Spooky': Experiment Shows an Effect That Appears to Smash the Speed of Light." Science, Aug. 13, 2008. https://www.science.org/content/article/quantum-physics-gets-spooky.

Berger, Peter L., and Thomas Luckmann. *The Social Construction of Reality: A Treatise in the Sociology of Knowledge*. Garden City, NY: Doubleday, 1966.

Bergson, Henri. *Creative Evolution*. Translated by Arthur Mitchell. Westport, CT: Greenwood, 1975.

———. *Introduction to Metaphysics*. Translated by T. E. Hulme. Indianapolis: Bobbs-Merrill, 1949.

Berkeley, George. "Three Dialogues Between Hylas and Philonous, in Opposition to Sceptics and Atheists." Abridged by Richard Taylor. In *The Empiricists: Locke, Berkeley, and Hume*. Garden City, New York: Dolphin, n.d.

Berry, Drew. "Animations of Unseeable Biology." Filmed May 2011. TED Talk video, 8:51. https://www.ted.com/talks/drew_berry_animations_of_unseeable_biology #t-248355.

Berry, Drew, and Etsuko Uno. "DNA Animation (2002–2014)." WEHImovies, uploaded July 16, 2018. YouTube video, 7:19. https://www.youtube.com/watch?v=7Hk9jct2ozY.

BlackPast. "(1888) Frederick Douglass on Woman Suffrage." Jan. 28, 2007. https://www.blackpast.org/african-american-history/speeches-african-american-history/1888-frederick-douglass-woman-suffrage/.

Bobroff, J. "Quantum Superposition of States and Decoherence." Wikipedia, uploaded Dec. 1, 2014. Video, 2:57. https://en.wikipedia.org/wiki/File:Quantum_superposition_of_states_and_decoherence.ogv.

Bohr, Niels. "Can Quantum-Mechanical Description of Physical Reality Be Considered Complete?" In *Causality and Complementarity*. Vol. 4 of *Philosophical Writings of Niels Bohr*. Woodbridge, CT: Ox Bow, 1998.

Borenstein, Seth, et al. "Three Physicists Share Nobel Prize for Work on Quantum Science." Associated Press, Oct. 4, 2022. https://apnews.com/article/science-health-stockholm-nobel-prizes-e5dbe8322b2cfe33d2dc4ad1af393518.

Brangham, William, et al. "Why Discovery of DNA's Double Helix Was Based on 'Rip-Off' of Female Scientist's Data." Featuring Howard Markel. *PBS NewsHour*, Sept. 30, 2021. https://www.pbs.org/newshour/show/watson-cricks-breakthrough-dna-discovery-was-based-on-rosalind-franklins-work.

Brewer, Grace. "Sneaky Orchids and Their Pollination Tricks: Orchids Are Top Tricksters When It Comes to Pollination." Kew: Royal Botanic Gardens, Feb. 4, 2024. https://www.kew.org/read-and-watch/orchid-pollination-tricks.

Bush, Evan. "Scientists Push New Paradigm of Animal Consciousness, Saying Even Insects May Be Sentient." NBC News, Apr. 19, 2024. https://www.nbcnews.com/news/amp/rcna148213.

Calvo, Paco, et al. "Plants Are Intelligent, Here's How." *Annals of Botany* 125 (2020) 11–28. https://doi.org/10.1093/aob/mcz155.

Chalmers, David. "Facing Up to the Problem of Consciousness." *Journal of Consciousness Studies* 2 (1995) 200–219. https://philpapers.org/rec/CHAFUT.

ChemEurope. "Wolfgang Pauli." https://www.chemeurope.com/en/encyclopedia/Wolfgang_Pauli.html.

Chodos, Alan. "June, ca. 240 B.C. Eratosthenes Measures the Earth." APS News, June 1, 2006. https://www.aps.org/apsnews/2006/06/eratosthenes-measures-earth.

Clayton, Philip. *Mind and Emergence: From Quantum to Consciousness*. New York: Oxford University Press, 2004.

Cobb, John B., Jr. *A Christian Natural Theology: Based on the Thought of Alfred North Whitehead*. Philadelphia: Westminster, 1965.

———. *Process Theology as Political Theology*. Philadelphia: Westminster, 1982.

Collingwood, R. G. *The Idea of Nature*. Oxford: Clarendon, 1945.

Conner, David. "Beyond Emergentism." In Wheeler and Conner, *Conceiving an Alternative*.

———. "A Functional-Empirical Approach to the 'Whitehead Without God' Debate." In *New Essays in Religious Naturalism*, edited by W. Creighton Peden and Larry E. Axel. Macon, GA: Mercer University Press, 1993.

———. "The Plight of a Theoretical Deity." *Process Studies* 41 (2012) 111–32. https://doi.org/10.2307/44798998.

———. "Quantum Non-Locality as an Indication of Theological Transcendence." *American Journal of Theology and Philosophy* 27 (2006) 259–84. https://www.jstor.org/stable/27944383.

———. "Whitehead the Naturalist." *American Journal of Theology and Philosophy* 30, (2009) 168–86. https://www.jstor.org/stable/27944471.

Copleston, Frederick. *Descartes to Leibniz*. Vol. 4 of *A History of Philosophy*. Paramus, NJ: Newman, 1958.

———. *Greece and Rome*. Vol. 1 of *A History of Philosophy*. Paramus, NJ: Newman, 1946.

CosmicWatch. "Cosmic Muons." http://www.cosmicwatch.lns.mit.edu/about.

Crosby, Donald A. *The Specter of the Absurd: Sources and Criticisms of Modern Nihilism*. Albany: State University of New York Press, 1988.

Daly, Mary. "Why Speak About God?" In *Woman Spirit Rising: A Feminist Reader in Religion*, edited by Carol P. Christ and Judith Plaskow. New York: Harper Collins, 1992.

Davies, Paul. *The Mind of God: The Scientific Basis for a Rational World*. New York: Simon & Schuster, 1992.

Davies, Paul, and John Gribbin. *The Matter Myth: Dramatic Discoveries That Challenge Our Understanding of Physical Reality*. New York: Simon & Schuster, 1992.

Day, R. A., Jr., and A. L. Underwood. *Quantitative Analysis*. 2nd ed. Englewood Cliffs, NJ: Prentice-Hall, 1967.

Deacon, Terrence. *Incomplete Nature: How Mind Emerged from Matter*. New York: Norton, 2013.

Dean, William. "Empiricism and God." In *Empirical Theology: A Handbook*, edited by Randolph Crump Miller. Birmingham, AL: Religious Education, 1992.

———. *The Religious Critic in American Culture*. Albany, State University of New York Press, 1994.

Dean, William, and Larry Axel. *The Size of God: The Theology of Bernard Loomer in Context*. Macon, GA: Mercer University Press, 1987.

De Broglie, Louis. "The Concept of Contemporary Physics and Bergson's Ideas on Time and Motion." In *Bergson and the Evolution of Physics*, edited and translated by Pete A. Y. Gunter. Knoxville: University of Tennessee Press, 1969.

D'Espagnat, Bernard. *Conceptual Foundations of Quantum Mechanics*. 2nd ed. Reading, MA: W. A. Benjamin, 1976.

———. "The Quantum Theory and Reality." *Scientific American*, Nov. 1, 1979. https://static.scientificamerican.com/sciam/assets/media/pdf/197911_0158.pdf.

———. *Veiled Reality: An Analysis of Present-Day Quantum Mechanical Concepts*. Reading, MA: Addison-Wesley, 1995.

Devlin, Hannah. "Slaves to the Rhythm: Rats Can't Resist a Good Beat, Researchers Say." *The Guardian*, Nov. 11, 2022. https://www.theguardian.com/science/2022/nov/11/rat-instictively-move-time-music-ability-previously-thought-uniquely-human-study?CMP=Share_AndroidApp_Other.

DeVries, William, and Carl Sachs. "Wilfrid Sellars." In *Stanford Encyclopedia of Philosophy*. Stanford University, 1997–. Article published Aug. 9, 2011; last modified Oct. 1, 2024. https://plato.stanford.edu/entries/sellars/.

Diamond, Malcolm L. *Contemporary Philosophy and Religious Thought: An Introduction to the Philosophy of Religion*. New York: McGraw-Hill, 1974.

Diderot, Denis. *D'Alembert's Dream*. Translated by Leonard Tancock. Harmondsworth, UK: Penguin, 1976.

Ditz, Helen M., and Andreas Nieder. "Neurons Selective to the Number of Visual Items in the Corvid Songbird Endbrain." *PNAS* 112 (2015) 7827–32. https://www.pnas.org/doi/full/10.1073/pnas.1504245112.

Dorrien, Gary. *The Making of American Liberal Theology: Crisis, Irony, and Postmodernity, 1950–2005*. Louisville: Westminster John Knox, 2006.

Downing, Lisa. "George Berkeley." In *Stanford Encyclopedia of Philosophy*. Stanford University, 1997–. Article published Sept. 10, 2004; last modified July 6, 2021. https://plato.stanford.edu/entries/berkeley.

Duignan, Brian, et al. "Postmodernism and Relativism." *Britannica*, last modified Dec. 19, 2025. https://www.britannica.com/topic/postmodernism-philosophy/Postmodernism-and-relativism.

Eastman, Timothy E., and Hank Keeton, eds. *Physics and Whitehead: Quantum, Process, and Experience*. Albany: State University of New York Press, 2012.

Editors of the Encyclopædia Britannica et al. "Apologetics." *Britannica*, last modified Apr. 12, 2025. https://www.britannica.com/topic/apologetics.

———. "Career in the United States of Alfred North Whitehead." *Britannica*, last modified Dec. 26, 2025. https://www.britannica.com/biography/Alfred-North-Whitehead/Career-in-the-United-States.

Edwards, Paul. *The Encyclopedia of Philosophy*. 8 vols. New York: Macmillan, 1967.

Einstein, Albert. "Über einen die Erzeugung und Verwandlung des Lichtes betreffenden heuristischen Gesichtspunkt." *Annalen der Physik* 17 (1905) 132–48.

———. "Zur Elektrodynamik bewegter Körper." *Annalen der Physik* 17 (1905) 891.

Einstein, Albert, et al. "Can Quantum-Mechanical Description of Physical Reality Be Considered Complete?" *Physical Review* 47 (1935) 777–80. https://cds.cern.ch/record/405662/files/PhysRev.47.777.pdf.

Epperson, Michael. *Quantum Mechanics and the Philosophy of Alfred North Whitehead*. New York: Fordham University Press, 2004.

ESA/Hubble. "New Shot of Proxima Centauri, Our Nearest Neighbour." Oct. 28, 2013. https://esahubble.org/images/potw1343a/.

Evans, C. Stephen, and David Baggett. "Moral Arguments for the Existence of God." In *Stanford Encyclopedia of Philosophy*. Stanford University, 1997–. Article published June 12, 2014; last modified Oct. 4, 2022. https://plato.stanford.edu/entries/moral-arguments-god/.

Evans, Richard. "Science to the Grower: Are Plants Intelligent Enough to Earn a Scouting Merit Badge?" *UCNFA News* 20.3 (2016) 16–18. https://ucnfa.ucdavis.edu/sites/g/files/dgvnsk14416/files/inline-files/Download_UCNFA_News_as_PDF90389.pdf.

Faye, Jan, and Henry J. Folse. Introduction to *Causality and Complementarity*, by Niels Bohr. Vol. 4 of *Philosophical Writings of Niels Bohr*. Woodbridge, CT: Ox Bow, 1998.

Fermilab. "What Are Virtual Particles?" Uploaded Apr. 17, 2024. YouTube video, 10:28. https://www.youtube.com/watch?v=ayQhNLqbTFk.

Fernández, Lucia. "Nuclear Share in Electricity Generation in the U.S. 1975–2024." Statista, Nov. 28, 2025. https://www.statista.com/statistics/273208/nuclear-share-of-electricity-generation-in-the-us/.

Ferré, Frederick. *Basic Modern Philosophy of Religion*. New York: Scribner's Sons, 1967.

Ferrie, Chris. "Quantum Entanglement Isn't All That Spooky After All: The Way We Teach Quantum Theory Conveys a Spookiness That Isn't Actually There." *Scientific American*, Feb. 13, 2023. https://www.scientificamerican.com/article/quantum-entanglement-isnt-all-that-spooky-after-all/.

Feynman, Richard. *The Character of Physical Law*. With an introduction by James Gleick. New York: Modern Library, 1994.

———. *QED: The Strange Theory of Light and Matter*. Princeton: Princeton University Press, 1985.

Fidler, David. *Restoring the Soul of the World: Our Living Bond with Nature's Intelligence*. Rochester, VT: Inner Traditions, 2014.

Flew, Anthony. *A Dictionary of Philosophy*. Rev. 2nd ed. New York: Random House, 1999.

Flowers, Jane. "The Incredible Killer Tree That Broadcasts an SOS to Its Neighbours." Blasting News United Kingdom, last updated Oct. 19, 2015. https://uk.blastingnews.com/world/2015/10/the-incredible-killer-tree-that-broadcasts-an-sos-to-its-neighbours-00610971.html.

Focus on the Family. "Submission of Wives to Husbands." https://www.focusonthefamily.com/family-qa/submission-of-wives-to-husbands/.

Frankenberry, Nancy. *Religion and Radical Empiricism*. Albany: State University of New York Press, 1987.

Franklin, Benjamin. *Franklin's Autobiography*. New York: Macmillan Pocket Classics, 1928.

Freud, Sigmund. *Civilization and Its Discontents*. Translated by J. Strachey. New York: Norton, 1961.

———. *The Future of an Illusion*. Translated by J. Strachey. New York: Norton, 1989.

———. *Moses and Monotheism*. Translated by Katherine Jones. New York: Vintage, 1967.

Ghosh, Pallab. "Are Animals Conscious? How New Research Is Changing Minds." BBC, June 15, 2024. https://www.bbc.com/news/articles/cv223z15mpmo.

The Giffords. "Lord Gifford's Will." https://giffordlectures.org/will/.

Goodenough, Ursula. *The Sacred Depths of Nature*. New York: Oxford University Press, 1998.

Gopnik, Adam. "The Real Thing: Homer, Remington, and the American Empirical Tradition." In *Homer-Remington*. New Haven: Yale University Press, 2020.

Gorey, Matthew. "Atomism and the Receptacle in Plato's Timaeus." https://classicalstudies.org/atomism-and-receptacle-platos-timaeus.

Goudge, T. A. "Bergson, Henri." In Edwards, *Encyclopedia of Philosophy*, vol. 1.

Government of Ontario. "Walnut Toxicity." Last modified Aug. 22, 2022. https://www.ontario.ca/page/walnut-toxicity.

Grant, Richard. "Do Trees Talk to Each Other?" *Smithsonian Magazine*, Mar. 2018. https://www.smithsonianmag.com/science-nature/the-whispering-trees-180968084/

Gray, Harry B., and Gilbert P. Haight. *Principles of Chemistry*. New York: W. A. Benjamin, 1967.

Greenstein, George. *Quantum Strangeness: Wrestling with Bell's Theorem and the Ultimate Nature of Reality*. Cambridge, MA: MIT Press, 2019.

Grene, Marjorie G. *A Portrait of Aristotle*. Chicago: University of Chicago Press, 1963,

Gribbin, John. "Heisenberg, Hayfever, and Heligoland." Johngribbinscience, Apr. 20, 2020. https://johngribbinscience.wordpress.com/2020/04/20/heisenberg-hayever-and-heligoland/.

———. *History of Western Science: 1543–2001*. London: Folio Society, 2002.

Guerra, Silvia, et al. "Asymmetrical Distribution of Supports Affect Pea Plants Movement and Shape: Evidence of Quantity Discrimination?" *PloS One* 20 (2025) e0322859. https://doi.org/10.1371/journal.pone.0322859.

Gunter, Pete A. Y. "Henri Bergson." Petegunter.net. https://petegunter.net/philosophy/henri-bergson/.

Haas, Kathryn. "2.2.2: Quantum Numbers and Atomic Wave Functions." LibreTexts. https://chem.libretexts.org/Bookshelves/Inorganic_Chemistry/Inorganic_Chemistry_(LibreTexts)/02%3A_Atomic_Structure/2.02%3A_The_Schrodinger_equation_particle_in_a_box_and_atomic_wavefunctions/2.2.02%3A_Quantum_Numbers_and_Atomic_Wave_Functions.

Halewood, Michael. "Death, Entropy, Creativity and Perpetual Perishing: Some Thoughts from Whitehead and Stengers." *Social Sciences* 4 (2015) 655–67. https://doi.org/10.3390/socsci4030655.

Hall, Calvin S. *A Primer of Freudian Psychology*. New York: World Publishing, 1954.

Halpern, Paul. *Synchronicity: The Epic Quest to Understand the Quantum Nature of Cause and Effect*. New York: Basic, 2020.

Hamer, Ashley. "Empty Space Isn't Empty, and Quantum Researchers Now Have Direct Evidence." Discovery, Aug. 1, 2019. https://www.discovery.com/science/Empty-Space-Quantum-Researchers-Have-Direct-Evidence.

Hanson, Norwood Russell. "Quantum Mechanics, Philosophical Implications of." In Edwards, *Encyclopedia of Philosophy*, vol. 7.

Harris, William, ed. *Heraclitus: The Complete Fragments*. Middlebury, VT: Middlebury College, 2015. https://fliphtml5.com/jmfd/opdc/Heraclitus%3A_The_Complete_Fragments_-_Middlebury_College/.

Hartshorne, Charles. *Anselm's Discovery: A Re-Examination of the Ontological Proof for God's Existence*. La Salle, IL: Open Court, 1965.

Harvard Medical School. "Molecular Motor Struts Like Drunken Sailor." Uploaded Jan. 6, 2012. YouTube video, 0:30. https://www.youtube.com/watch?v=-7AQVbrmzFw.

Haskell, Thomas L. "The Curious Persistence of Rights Talk in the 'Age of Interpretation.'" *Journal of American History* 74 (1987) 984–1012. https://doi.org/10.2307/1902162.

Hättich, Frank. *Quantum Processes: A Whiteheadian Interpretation of Quantum Field Theory*. Münster: GmbH, 2004.

Hawking, Stephen. *The Grand Design*. New York: Bantam, 2010.

Hedrick, Lisa Landoe. *Whitehead and the Pittsburg School: Preempting the Problem of Intentionality*. Lanham, MY: Lexington, 2021.

Heisenberg, Werner. *Philosophical Problems of Quantum Physics*. Woodbridge, CT: Ox Bow, 1979.

———. *Physics and Philosophy*. New York: Harper, 1958.

———. *Physics and Philosophy*. Amherst, NY: Prometheus, 1999.

———. "Werner Heisenberg – Session VII." Interview by Thomas S. Kuhn, Feb. 22, 1963. Niels Bohr Library and Archive transcript, Dec. 18, 2024. https://doi.org/10.1063/nbla.wbnv.eibc.

Hendel, Charles W., Jr., ed. *Hume Selections*. New York: Scribner's Sons, 1955.

Henderson, Joseph L. "Ancient Myths and Modern Man." In *Man and His Symbols*, edited by Carl G. Jung. New York: Dell, 1964.

Hick, John. *Classical and Contemporary Readings in the Philosophy of Religion*. 2nd ed. Englewood Cliffs, NJ: Prentice-Hall, 1970.

Hirst, R. J. "Realism." In Edwards, *Encyclopedia of Philosophy*, vol. 7.

Historic Royal Palaces. "Henry VIII: Terrible Tudor?" https://www.hrp.org.uk/hampton-court-palace/history-and-stories/henry-viii/#gs.8sma4p.

Hofstadter, Douglas, and Emmanuel Sander. *Surfaces and Essences*. New York: Basic, 2013.

Holewinski, Britt. "Underground Networking: The Amazing Connections Beneath Your Feet." The National Forest Foundation. https://www.nationalforests.org/blog/underground-mycorrhizal-network.

Holton, Gerald. "Centennial Focus: Millikan's Measurement of Planck's Constant." Physics, Apr. 22, 1999. https://physics.aps.org/story/v3/st23.

Horgan, John. "A 25-Year-Old Bet About Consciousness Has Finally Been Settled." *Scientific American*, June 26, 2023. https://www.scientificamerican.com/article/a-25-year-old-bet-about-consciousness-has-finally-been-settled/.

Horsten, Leon. "Philosophy of Mathematics." In *Stanford Encyclopedia of Philosophy*. Stanford University, 1997–. Article published Sept. 25, 2007; last modified Nov. 6, 2023. https://plato.stanford.edu/entries/philosophy-mathematics/.

Hossenfelder, Sabine. "What Is Reductionism?" Back Reaction, Apr. 9, 2020. http://backreaction.blogspot.com/2020/04/what-is-reductionism.html.

———. "You Don't Have Free Will, But Don't Worry." Uploaded Oct. 10, 2020. YouTube video, 11:04. https://www.youtube.com/watch?v=zpU_e3jh_FY.

Howard, Don. "A Physicist Gets Philosophical." *Physics Today*, Mar. 1, 2023. https://physicstoday.scitation.org/doi/full/10.1063/PT.3.5199.

HyperPhysics. "Michelson Interferometer." http://hyperphysics.phy-astr.gsu.edu/hbase/phyopt/michel.html.

Irving, Michael. "James Webb Keeps Breaking Records for Most Distant Galaxies Ever Seen." New Atlas, Aug. 3, 2022. https://newatlas.com/space/james-webb-most-distant-galaxies-record/.

Isaacson, Walter. *Einstein: His Life and Universe*. New York, Simon & Schuster, 2007.

Ito, Shingo. "Japan Scientists Hope Slime Holds Intelligence Key." Phys.org, Dec. 28, 2011. https://phys.org/news/2011-12-japan-scientists-slime-intelligence-key.html.

Jabr, Ferris. "How Brainless Slime Molds Redefine Intelligence." *Nature* (2012). https://doi.org/10.1038/nature.2012.11811.

James, William. *The Principles of Psychology*. New York: Holt, 1893.

———. *Some Problems of Philosophy*. Edited by H. M. Kallen New York: Longmans, Green, 1921.

———. *The Varieties of Religious Experience*. New York: Collier, 1961.

———. "The Will to Believe." In *Essays on Faith and Morals*, edited by Ralph Barton Perry. New York: World, 1962.

———. *"The Will to Believe" and Other Essays in Popular Philosophy*. London: Longmans, Green, 1897.

———. "A World of Pure Experience." In *Essays in Radical Empiricism*. Lincoln: University of Nebraska Press, 1996.

Jankowiak, Tim. "Immanuel Kant." Internet Encyclopedia of Philosophy. https://iep.utm.edu/kantview.

Jung, Carl G. *Synchronicity: An Acausal Connecting Principle*. Translated by R. F. C. Hull. Princeton: Princeton University Press, 1969.

Jung, Carl G., and Wolfgang Pauli. *Atom and Archetype: The Pauli/Jung Letters, 1932–1958*. Edited by C. A. Meier. Translated by David Roscoe. Updated ed. Princeton: Princeton University Press, 2001.

Kant, Immanuel. *Critique of Pure Reason*. Translated by Norman Kemp Smith. New York: St. Martin's, 1965.

Kaplan, Irving. *Nuclear Physics*. 2nd ed. Reading, MA: Addison-Wesley, 1962.

Kauffman, Stuart. *At Home in the Universe: The Search for the Laws of Self-Organization and Complexity*. New York: Oxford University Press, 1995.

Kaufman, Gordon D. *God, Mystery, Diversity*. Minneapolis, Fortress, 1985.

———. *In the Beginning . . . Creativity*. Minneapolis: Fortress, 2004.

Keller, Catherine. *On the Mystery: Discerning Divinity in Process*. Minneapolis: Fortress, 2007.

King, Martin Luther, Jr. *A Testament of Hope: The Essential Writings and Speeches*. Edited by James M. Washington. New York: HarperOne, 1986.

———. "Where Do We Go From Here?" Speech, Aug. 16, 1967. Stanford, The Martin Luther King, Jr. Research and Education Institute, transcript. https://kinginstitute.stanford.edu/where-do-we-go-here.

Kittsley, Scott L. *Physical Chemistry*. 3rd ed. New York: Barnes & Noble, 1969.

Koberlein, Brian. "That's About the Size of It." Brian Koberlein (blog), Apr. 14, 2015. https://briankoberlein.com/blog/thats-about-the-size/.

Kravitz, Fran, and the ACS Committee on Ethics. "Dynamite and the Ethics of Its Many Uses." American Chemical Society. https://www.acs.org/education/outreach/celebrating-chemistry-editions/2021-ncw/dynamite-ethics.html.

Kuhlmann, Meinard. "Physicists Debate Whether the World Is Made of Particles or Fields—or Something Else Entirely." *Scientific American*, Aug. 1, 2013. https://www.scientificamerican.com/article/physicists-debate-whether-world-made-of-particles-fields-or-something-else/.

Lamont, Michele. "How to Become a Dominant French Philosopher: The Case of Jacques Derrida." *American Journal of Sociology* 93 (1987) 584–622. https://www.journals.uchicago.edu/doi/10.1086/228790.

Landauer, Rolf. "The Physical Nature of Information." *Physics Letters A* 217 (1996) 188–93. https://www.sciencedirect.com/science/article/abs/pii/0375960196004537.

Leibniz, Gottfried Wilhelm. "Third Reply." In Alexander, *The Leibniz-Clarke Correspondence*.

Lindbeck, George. *The Nature of Doctrine: Religion and Theology in a Postliberal Age*. Philadelphia: Westminster, 1984.

Loomer, Bernard M. "S-I-Z-E." *Criterion* 13 (1974) 5–8.

———. "The Size of God." In Dean and Axel, *Size of God*.

———. "Two Conceptions of Power." *Process Studies* 6 (1976) 5–32. https://www.religion-online.org/article/two-conceptions-of-power/.

Lowe, Victor. *Alfred North Whitehead: The Man and His Work*. 2 vols. Baltimore: Johns Hopkins University Press, 1985.

Malin, Shimon. "Whitehead's Philosophy and the Collapse of Quantum States." In *Physics and Whitehead: Quantum, Process, and Experience*, edited by Timothy Eastman and Hank Keeton. Albany: State University of New York Press, 2003.

Margenau, Henry. *The Nature of Physical Reality: A Philosophy of Modern Physics*. New York: McGraw-Hill, 1950.

Maron, Samuel, and Carl Prutton. *Principles of Physical Chemistry*. 4th ed. New York, Macmillan, 1965.

Marx, Karl. Introduction to *Critique of Hegel's Philosophy of Right*, edited by Joseph O'Malley. Translated by Annette Jolin and Joseph O'Malley. Cambridge, UK: Cambridge University Press, 1970. Marxists.org, transcribed by Andy Blunden. https://www.marxists.org/archive/marx/works/1843/critique-hpr/intro.htm.

Mathas, Carolyn. "The Basics of Quantum Computing." EDN, Aug. 13, 2019. https://www.edn.com/the-basics-of-quantum-computing-a-tutorial/.

Maturana, Humberto R., and Francisco J. Varela. *Autopoiesis and Cognition: The Realization of the Living*. Boston: Reidel, 1980.

McFague, Sallie. *Metaphorical Theology: Models of God in Religious Language*. Philadelphia: Fortress, 1982.

Meland, Bernard E. *Fallible Forms and Symbols: Discourses on Method in a Theology of Culture*. Philadelphia: Fortress, 1976.

———. "Grace: A Dimension Within Nature?" *Journal of Religion* 54 (1974) 119–37. https://doi.org/10.1086/486381.

———. "Introduction: The Empirical Tradition in Theology at Chicago." In *The Future of Empirical Theology*. Vol. 7 of *Essays in Divinity*. Edited by Jerald C. Brauer. Chicago: University of Chicago Press, 1969.

Miller, Johanna L. "Does Quantum Mechanics Need Imaginary Numbers?" *Physics Today*, Mar. 1, 2022. https://physicstoday.scitation.org/doi/10.1063/PT.3.4955.

Momaday, N. Scott. *The Way to Rainy Mountain*. Albuquerque, NM: University of New Mexico Press, 1976.

———. "We Are What We Imagine." Goodreads. https://www.goodreads.com/author/quotes/64204.N_Scott_Momaday.

Moore, Patrick. *Travellers in Space and Time*. Garden City, NY: Doubleday, 1984.

Moore, Ruth. *Niels Bohr: The Man, His Science and the World They Changed*. New York: Knopf, 1966.

Morgan, Matthew. "Turner: Painting *The Fighting Temeraire*." The National Gallery, uploaded Mar. 18, 2016. YouTube video, 25:55. https://www.youtube.com/watch?v=8O-fna8HrWw.

Morowitz, Harold. "Teilhard, Complexity, and Complexification." *Complexity* 2.4 (1999) 7–8. https://doi.org/10.1002/(SICI)1099-0526(199703/04)2:4<7::AID-CPLX2>3.0.CO;2-D.

Morris, Andréa. "Testing a Time-Jumping, Multiverse-Killing, Consciousness-Spawning Theory of Reality." *Forbes*, Oct. 23, 2023. https://www.forbes.com/sites/andreamorris/2023/10/23/testing-a-time-jumping-multiverse-killing-consciousness-spawning-theory-of-reality/.

Moskowitz, Clara. "Physicists Disagree over Meaning of Quantum Mechanics, Poll Shows." Live Science, Jan. 21, 2013. https://www.livescience.com/26444-quantum-mechanics-physicists-poll.html.

Nadeau, Robert, and Menas Kafatos. *The Non-Local Universe: The New Physics and Matters of the Mind*. New York: Oxford University Press, 1999.

Nature Education. "Rosalind Franklin: A Crucial Contribution." 2014. https://www.nature.com/scitable/topicpage/rosalind-franklin-a-crucial-contribution-6538012/.

NEI. "Nuclear Fuel." https://www.nei.org/fundamentals/nuclear-fuel.

Neiman, Susan. "Why the World Still Needs Immanuel Kant." *New York Times*, Apr. 17, 2024. https://www.nytimes.com/2024/04/17/arts/immanuel-kant-300-anniversary.html.

Neville, Robert Cummings. “Cosmological and Ontological Contingency.” *American Journal of Theology and Philosophy* 40 (2019) 54–61.

News Staff. “Newly Proposed Experiment Could Confirm That Information Is Fifth State of Matter.” Sci.News, Mar. 21, 2022. http://www.sci-news.com/physics/information-fifth-state-matter-10638.html.

Newton, Isaac. *Opticks: Or, A Treatise of the Reflections, Refractions, Inflections and Colours of Light.* 4th ed. London: Innys, 1730. https://archive.org/details/opticksoratreatoonewtgoog/page/n6/mode/2up.

New York University. “The New York Declaration on Animal Consciousness.” Apr. 19, 2024. https://sites.google.com/nyu.edu/nydeclaration/declaration#h.5e13vfkpjc7j.

NOAA Fisheries. “Common Bottlenose Dolphin.” Last modified Mar. 6, 2025. https://www.fisheries.noaa.gov/species/common-bottlenose-dolphin.

The Nobel Prize. “Nobel Prize in Physics 1918.” https://www.nobelprize.org/prizes/physics/1918/summary/.

———. “Nobel Prize in Physics 2022.” https://www.nobelprize.org/prizes/physics/2022/summary/.

Oberlander, Elana. “A New Theory in Physics Claims to Solve the Mystery of Consciousness.” Neuroscience News, Aug. 11, 2022. https://neurosciencenews.com/physics-consciousness-21222/.

Oldenberg, Otto, and Wendell Holladay. *Introduction to Atomic and Nuclear Physics.* 4th ed. New York: McGraw-Hill, 1961.

Pagels, Heinz R. *Perfect Symmetry: The Search for the Beginning of Time.* New York: Simon & Schuster, 1985.

Paley, William. *Natural Theology*. In *Philosophical and Religious Issues: Classical and Contemporary Statements*, by Ed L. Miller. Encino, CA: Dickenson, 1971.

Pangambam, S. “My Stroke of Insight by Jill Bolte Taylor.” TED Talk transcript. Singju Post, Sept. 1, 2014. https://singjupost.com/stroke-insight-jill-bolte-taylor-transcript/?singlepage=1.

Papatheodorou, C., and Basil J. Hiley. “Process, Temporality, and Space-Time.” *Process Studies* 26 (1997) 247–78.

Patterson, Nancy. *Quantum Physics and the Power of the Mind.* Self-published, Lulu, 2020.

PBS. “The Incredible Termite Mound.” Oct. 28, 2011. https://www.pbs.org/wnet/nature/the-animal-house-the-incredible-termite-mound/7222/.

Pegis, Anton C., ed. *Introduction to St. Thomas Aquinas: The Summa Theologica; The Summa Contra Gentiles.* New York: Modern Library, 1945.

Penrose, Roger. *The Emperor's New Mind: Concerning Computers, Minds, and the Laws of Physics.* Oxford: Oxford University Press, 1989.

Perry, Ralph Barton. *Philosophy and Psychology.* Vol. 2 of *The Thought and Character of William James.* Boston: Little, Brown, 1935.

Peters, Francis E. *Greek Philosophical Terms: A Historical Lexicon.* New York: New York University Press, 1967.

Petersen, Aage. “The Philosophy of Niels Bohr.” *Bulletin of Atomic Scientists* 19 (1963) 8–14.

Plato. *Timaeus.* In *The Collected Dialogues of Plato*, edited by Edith Hamilton and Huntington Cairns, translated by Benjamin Jowett. Princeton: Princeton University Press, 1961.

Polanyi, Michael. *Personal Knowledge: Towards a Post-Critical Philosophy*. New York: Harper & Row, 1958.

Pollan, Michael. "The Intelligent Plant: Scientists Debate a New Way of Understanding Flora." *New Yorker*, Dec. 15, 2013. https://www.newyorker.com/magazine/2013/12/23/the-intelligent-plant.

PoorLeno. "Hydrogen Density Plots." Wikimedia Commons, uploaded Aug. 17, 2008. Image. https://commons.wikimedia.org/wiki/File:Hydrogen_Density_Plots.png.

Potthoff, Harvey H. *God and the Celebration of Life*. Chicago: Rand McNally, 1969.

Price, Lucien. *Dialogues of Alfred North Whitehead*. Westport, CT: Greenwood, 1977.

Prigogine, Ilya, and Isabelle Stengers. *Order Out of Chaos: Man's New Dialogue with Nature*. New York: Bantam, 1984.

Prum, Richard O. *The Evolution of Beauty: How Darwin's Forgotten Theory of Mate Choice Shapes the Animal World and Us*. New York: Doubleday, 2017.

Puiu, Tibi. "Is Information the Fifth State of Matter? Physicist Says There's One Way to Find Out." ZME Science, Mar. 9, 2022. https://www.zmescience.com/science/news-science/information-energy-mass-equivalence/.

Quine, Willard Van Orman. *Word and Object*. Cambridge, MA: MIT, 1960.

Ramroop, Tara, and Kara West. "To the Ends of the Earth." *National Geographic*, last modified Jan. 10, 2024. https://education.nationalgeographic.org/resource/ends-earth/.

Raposa, Michael L. "The Divided Self of William James." *First Things*, Aug. 1, 2001. https://www.firstthings.com/article/2001/08/the-divided-self-of-william-james.

Reiff, Joseph T. *Born of Conviction: White Methodists and Mississippi's Closed Society*. New York: Oxford University Press, 2016.

Reppert, Steven M., and Jacobus C. de Roode. "Demystifying Monarch Butterfly Migration." *Current Biology* 28 (2018) R1009–22. https://doi.org/10.1016/j.cub.2018.02.067.

Richardson, Robert D. *William James: In the Maelstrom of American Modernism*. Boston: Houghton Mifflin, 2006.

Riebeek, Holli. "Planetary Motion: The History of an Idea That Launched the Scientific Revolution." NASA Earth Observatory, July 7, 2009. https://science.nasa.gov/earth/earth-observatory/planetary-motion/.

Robinson, Mary-Russell. "Mockingbirds Can Learn Hundreds of Songs, but There's a Limit." Cornell Labs, Apr. 12, 2016. https://www.allaboutbirds.org/news/mockingbirds-can-learn-hundreds-of-songs-but-theres-a-limit.

Rorty, Richard. *Philosophy and Social Hope*. New York: Penguin, 1999.

Ross, W. D. *Aristotle: A Complete Exposition of His Works and Thought*. New York: Meridian, 1959.

Rothenberg Gritz, Jennie. "These Gorgeous Photos Capture Life Inside a Drop of Seawater." *Smithsonian*, Jan./Feb. 2023. https://www.smithsonianmag.com/science-nature/these-gorgeous-photos-capture-life-inside-drop-seawater-180981297/.

Rovelli, Carlo. "Consciousness Is Irrelevant to Quantum Mechanics." Interview by Alexis Papazoglou. *IAI News*, July 19, 2022. https://iai.tv/articles/consciousness-is-irrelevant-to-quantum-mechanics-auid-2187.

Ruether, Rosemary Radford. *Sexism and God Talk: Toward a Feminist Theology*. Boston: Beacon, 1983.

Russell, Bertrand. *A History of Western Philosophy*. New York: Simon & Schuster, 1945.

Schleiermacher, Friedrich. *The Christian Faith.* Edited by H. R. Mackintosh and J. S. Stewart. Edinburgh: T. & T. Clark, 1928.

Schwarcz, Joe. "Aristotle: The First Real Scientist." McGill, Apr. 17, 2018. https://www.mcgill.ca/oss/article/general-science-history/aristotle-man-who-relied-observed-facts.

Sears, F. W., and Mark W. Zemansky. *University Physics.* 3rd ed. Reading, MA: Addison-Wesley, 1964.

Segall, Matthew David. *Physics of the World-Soul: Whitehead's Adventure in Cosmology.* Grasmere, ID: SacraSage, 2021.

Seigfried, Charlene Haddock. *William James's Radical Reconstruction of Philosophy.* Albany: State University of New York Press, 1990.

Shakespeare, William. *The Complete Works of William Shakespeare.* San Diego, CA: Canterbury Classics, 2014.

Sharp, Daryl. "Collective Unconscious." In *C. G. Jung Lexicon: A Primer of Terms and Concepts.* Toronto: Inner City, 1991.

Sherburne, Donald W., ed. *A Key to Whitehead's "Process and Reality."* Bloomington: Indiana University Press, 1966.

Sieczkowski, Cavan. "Loyal Dog Reportedly Prevents Owner's Suicide Attempt by Knocking Gun Away." *Huffington Post*, Mar. 7, 2013. https://www.huffpost.com/entry/loyal-dog-allegedly-prevents-owners-suicide_n_2829627.

Siegel, Ethan. "The Sun Wouldn't Shine Without Quantum Physics." Medium, Nov. 30, 2018. https://medium.com/starts-with-a-bang/the-sun-wouldnt-shine-without-quantum-physics-9a8b5b8bca39.

———. "Yes, the Universe Really Is 100% Reductionist in Nature." Medium, Aug. 16, 2022. https://medium.com/starts-with-a-bang/yes-the-universe-really-is-100-reductionist-in-nature-3d5aa4bd434f.

Smuts, Jan C. *Holism and Evoluton.* New York: Macmillan, 1926.

Sölle, Dorothee. *Thinking About God: An Introduction to Theology.* Translated by John Bowden. Philadelphia: Trinity International, 1990.

Souter, Alexander. *A Pocket Lexicon to the Greek New Testament.* London: Clarendon, 1916.

Stapp, Henry. "Minds and Values in the Quantum Universe." In *Information and the Nature of Reality: From Physics to Metaphysics*, edited by Paul Davies and Neils Henrik Gregersen. Cambridge, UK: Cambridge University Press, 2010.

Starr, Michelle. "Crows Can Actually Count Out Loud, Amazing New Study Shows." ScienceAlert, May 24, 2024. https://www.sciencealert.com/crows-can-actually-count-out-loud-amazing-new-study-shows.

Stokes, Philip. *Philosophy: 100 Essential Thinkers.* Kettering: Index, 2004.

Strogatz, Steven. "What Is Quantum Field Theory and Why Is It Incomplete?" *Quanta*, Aug. 10, 2022. https://www.quantamagazine.org/what-is-quantum-field-theory-and-why-is-it-incomplete-20220810/.

Strong, James. "936. Basileuó." In *Strong's Exhaustive Concordance* (1890). https://biblehub.com/greek/936.htm.

Suchocki, Marjorie. *The Fall to Violence: Original Sin in Relational Theology.* New York: Continuum, 1994.

Taylor, Jill Bolte. "My Stroke of Insight." Filmed Feb. 2008. TED Talk video, 18:24. https://www.ted.com/talks/jill_bolte_taylor_my_stroke_of_insight.

TED. "The Secret Lives of Plants." Video playlist. https://www.ted.com/playlists/356/the_secret_lives_of_plants.

Teilhard de Chardin, Pierre. *The Phenomenon of Man*. New York: Harper & Row, 1959.

Tennant, F. R. *Philosophical Theology*. Vol. 2. Cambridge, UK: Cambridge University Press, 1930.

Tennyson, Alfred. "In Memoriam, Epilogue [O True and Tried, So Well and Long]." Poets.org. https://poets.org/poem/memoriam-epilogue-o-true-and-tried-so-well-and-long.

Tillich, Paul. *Systematic Theology*. 3 vols. Chicago: University of Chicago Press, 1951–1963.

Tomforde, Mark. "The History of Calculus." https://marktomforde.com/academic/miscellaneous/calculus-history/calchistory.html.

Underhill, Evelyn. *Worship*. London: Collins, 1936.

University of California, Davis. "Caterpillars Foiled When Tomato Plants Summon Parasitic Wasps." ScienceDaily, June 22, 1999. www.sciencedaily.com/releases/1999/06/990622055654.htm.

USGS. "The Great 1906 San Francisco Earthquake." https://earthquake.usgs.gov/earthquakes/events/1906calif/18april/.

———. "How Do Salmon Know Where Their Home Is When They Return from the Ocean?" https://www.usgs.gov/faqs/how-do-salmon-know-where-their-home-when-they-return-ocean.

———. "How Long Was the 1906 Rupture?" https://earthquake.usgs.gov/earthquakes/events/1906calif/18april/howlong.php.

Vale, Ron. "1: Molecular Motor Proteins." Science Communication Lab, uploaded Feb. 2, 2016. YouTube video, 35:25. https://www.youtube.com/watch?v=9RUHJhskWoo.

Varsity Tutors. "Master Adding and Subtracting Vectors." https://www.varsitytutors.com/hotmath/hotmath_help/topics/adding-and-subtracting-vectors.

Veritasium. "Your Body's Molecular Machines." Uploaded Nov. 20, 2017. YouTube video, 6:20. https://www.youtube.com/watch?v=X_tYrnv_o6A.

Weisberger, Mindy. "Why Crows Hold Funerals." Live Science, Jan. 6, 2016. https://www.livescience.com/53283-why-crows-hold-funerals.html.

Welch Labs. "How the Bizarre Path of Mars Reshaped Astronomy [Kepler's Laws Part 1]." Uploaded May 8, 2024. YouTube video, 16:40. https://youtu.be/Phscjlou6TI?si=5UOF12NZMTHDpsCD.

Wheeler, Demian. *Religion Within the Limits of History Alone: Pragmatic Historicism and the Future of Theology*. Albany: State University of New York Press, 2020.

Wheeler, Demian, and David Conner, eds. *Conceiving an Alternative: Philosophical Resources for an Ecological Civilization*. Anoka, MN: Process Century, 2019.

Wheelwright, Philip. *Heraclitus*. New York: Atheneum, 1971.

White, Alan R. "Coherence Theory of Truth." In Edwards, *Encyclopedia of Philosophy*, vol. 2.

Whitehead, Alfred North. *Adventures of Ideas*. New York: Free Press, 1933.

———. *Essays in Science and Philosophy*. New York: Greenwood, 1968.

———. *Modes of Thought*. New York: Free Press, 1966.

———. *Process and Reality: An Essay in Cosmology*. Edited by Donald Sherburne and David Ray Griffin. Rev. ed. New York: Free Press, 1978.

———. *Religion in the Making*. New York: World, 1960.

———. *Science and the Modern World*. New York: Free Press, 1967.

———. *Symbolism: Its Meaning and Effect*. New York: Putnam's Sons, 1955.

Whitford, Margaret, ed. *The Irigaray Reader: Luce Irigaray*. Cambridge, MA: Blackwell, 1991.

Whittaker, Gavin. "Beevers Crystal Structure Model of Kaolinite." Wikimedia Commons, uploaded Jan. 10, 2019. Image. https://commons.wikimedia.org/wiki/File:Beevers_crystal_structure_model_of_Kaolinite.jpg.

Wieman, Henry Nelson. *The Source of Human Good.* Carbondale: Southern Illinois University Press, 1974.

Wigner, Eugene P. "The Unreasonable Effectiveness of Mathematics in the Natural Sciences: Richard Courant Lecture in Mathematical Sciences Delivered at New York University, May 11, 1959." *Communications on Pure and Applied Mathematics* 13 (1960) 1–14. https://doi.org/10.1002/cpa.3160130102.

Wikipedia. "Anthropic Principle." Wikimedia Foundation, last modified Dec. 17, 2025. https://en.wikipedia.org/wiki/Anthropic_principle.

———. "Apophatic Theology." Wikimedia Foundation, last modified Jan. 12, 2026. https://en.wikipedia.org/wiki/Apophatic_theology.

———. "Applications of Quantum Mechanics." Wikimedia Foundation, last modified Sept. 28, 2025. https://en.wikipedia.org/wiki/Applications_of_quantum_mechanics.

———. "Atomic Orbital." Wikimedia Foundation, last modified Oct. 20, 2025. https://en.wikipedia.org/wiki/Atomic_orbital.

———. "Bowerbird." Wikimedia Foundation, last modified Mar. 21, 2023. https://en.wikipedia.org/w/index.php?title=Bowerbird&oldid=1145944336.

———. "Cloud Chamber." Wikimedia Foundation, last modified Oct. 12, 2025. https://en.wikipedia.org/wiki/Cloud_chamber.

———. "Copleston–Russell Debate." Wikimedia Foundation, last modified Oct. 17, 2023. https://en.wikipedia.org/wiki/Copleston%E2%80%93Russell_debate.

———. "DDT." Wikimedia Foundation, last modified Jan. 18, 2026. https://en.wikipedia.org/wiki/DDT.

———. "Direct and Indirect Realism." Wikimedia Foundation, last modified Nov. 5, 2025. https://en.wikipedia.org/wiki/Direct_and_indirect_realism.

———. "Double-Aspect Theory." Wikimedia Foundation, last modified Mar. 9, 2025. https://en.wikipedia.org/wiki/Double-aspect_theory.

———. "Einstein–Podolsky–Rosen Paradox." Wikimedia Foundation, last modified Jan. 3, 2026. https://en.wikipedia.org/wiki/Einstein%E2%80%93Podolsky%E2%80%93Rosen_paradox.

———. "Emmy Noether." Wikimedia Foundation, last modified Jan. 12, 2026. https://en.wikipedia.org/wiki/Emmy_Noether.

———. "Field (Physics)." Wikimedia Foundation, last modified Nov. 21, 2025. https://en.wikipedia.org/wiki/Field_(physics).

———. "God of the Gaps." Wikimedia Foundation, last modified Oct. 8, 2025. https://en.wikipedia.org/wiki/God_of_the_gaps.

———. "Hard Problem of Consciousness." Wikimedia Foundation, last modified Dec. 23, 2025. https://en.wikipedia.org/wiki/Hard_problem_of_consciousness.

———. "Henri Bergson." Wikimedia Foundation, last modified Jan. 13, 2026. https://en.wikipedia.org/wiki/Henri_Bergson.

———. "Henry Stapp." Wikimedia Foundation, last modified Dec. 5, 2025. https://en.wikipedia.org/wiki/Henry_Stapp.

———. "Hidden-Variable Theory." Wikimedia Foundation, last modified Oct. 4, 2025. https://en.wikipedia.org/wiki/Hidden-variable_theory.

———. "Higgs Field." Wikimedia Foundation, last modified July 30, 2025. https://simple.wikipedia.org/wiki/Higgs_field.

———. "History of Entropy." Wikimedia Foundation, last modified Dec. 22, 2025. https://en.wikipedia.org/wiki/History_of_entropy.

———. "Humorism." Wikimedia Foundation, last modified Dec. 28, 2025. https://en.wikipedia.org/wiki/Humorism.

———. "Jacques Derrida." Wikimedia Foundation, last updated Jan. 6, 2026. https://en.wikipedia.org/wiki/Jacques_Derrida.

———. "James Watson." Wikimedia Foundation, last modified Jan. 10. 2026. https://en.wikipedia.org/wiki/James_Watson.

———. "Johannes Kepler." Wikimedia Foundation, last modified Jan. 14, 2026. https://en.wikipedia.org/wiki/Johannes_Kepler.

———. "Jungian Archetypes." Wikimedia Foundation, last modified Jan. 7, 2026. https://en.wikipedia.org/wiki/Jungian_archetypes.

———. "Kaolinite." Wikimedia Foundation, last modified Jan. 6, 2026. https://en.wikipedia.org/wiki/Kaolinite.

———. "Many-Worlds Interpretation." Wikimedia Foundation, last modified Dec. 27, 2025. https://en.wikipedia.org/wiki/Many-worlds_interpretation.

———. "Mass–Energy Equivalence." Wikimedia Foundation, last modified Jan. 15, 2026. https://en.wikipedia.org/wiki/Mass%E2%80%93energy_equivalence.

———. "Max Planck." Wikimedia Foundation, last modified Jan. 10, 2026. https://en.wikipedia.org/wiki/Max_Planck.

———. "Mind–Body Problem." Wikimedia Foundation, last modified Oct. 13, 2025. https://en.wikipedia.org/wiki/Mind%E2%80%93body_problem.

———. "Muon." Wikimedia Foundation, last modified Jan. 19, 2026. https://en.wikipedia.org/wiki/Muon.

———. "Oil Drop Experiment." Wikimedia Foundation, last modified Jan. 2, 2026. https://en.wikipedia.org/wiki/Oil_drop_experiment.

———. "Ontological Argument." Wikimedia Foundation, last modified Jan. 20, 2026. https://en.wikipedia.org/wiki/Ontological_argument.

———. "Parable of the Invisible Gardener." Wikimedia Foundation, last modified Jan. 16, 2026. https://en.wikipedia.org/wiki/Parable_of_the_Invisible_Gardener.

———. "*Philosophy and the Mirror of Nature*." Wikimedia Foundation, last modified Dec. 17, 2025. https://en.wikipedia.org/wiki/Philosophy_and_the_Mirror_of_Nature.

———. "Philosophy of Mathematics." Wikimedia Foundation, last modified Jan. 5, 2026. https://en.wikipedia.org/wiki/Philosophy_of_mathematics.

———. "Photoreceptor Cell." Wikimedia Foundation, last modified Sept. 25, 2025. https://en.wikipedia.org/wiki/Photoreceptor_cell.

———. "Pythagoras." Wikimedia Foundation, last modified Dec. 3, 2025. https://en.wikipedia.org/wiki/Pythagoras.

———. "Quantum Entanglement." Wikimedia Foundation, last modified Jan. 17, 2026. https://en.wikipedia.org/wiki/Quantum_entanglement.

———. "Quantum Field Theory." Wikimedia Foundation, last modified Jan. 18, 2026. https://en.wikipedia.org/wiki/Quantum_field_theory.

———. "Quantum Superposition." Wikimedia Foundation, last modified Dec. 3, 2025. https://en.wikipedia.org/wiki/Quantum_superposition.

———. "Quantum Tunnelling." Wikimedia Foundation, last modified Dec. 30, 2025. https://en.wikipedia.org/wiki/Quantum_tunnelling.

———. "Richard Rorty." Wikimedia Foundation, last modified Jan. 12, 2026. https://en.wikipedia.org/wiki/Richard_Rorty.

———. "Robert S. Mulliken." Wikimedia Foundation, last modified Jan. 10, 2026. https://en.wikipedia.org/wiki/Robert_S._Mulliken.

———. "Schrödinger Equation." Wikimedia Foundation, last modified Jan. 20, 2026. https://en.wikipedia.org/wiki/Schr%C3%B6dinger_equation.

———. "Schrödinger's Cat." Wikimedia Foundation, last modified Dec. 8, 2025. https://en.wikipedia.org/wiki/Schr%C3%B6dinger%27s_cat.

———. "Specious Present." Wikimedia Foundation, last modified Sept. 4, 2025. https://en.wikipedia.org/wiki/Specious_present.

———. "Synchronicity." Wikimedia Foundation, last modified Nov. 19, 2025. https://en.wikipedia.org/wiki/Synchronicity.

———. "Transuranium Element." Wikimedia Foundation, last modified Jan. 3, 2026. https://en.wikipedia.org/wiki/Transuranium_element.

———. "Uncertainty Principle." Wikimedia Foundation, last modified Jan. 11, 2026. https://en.wikipedia.org/wiki/Uncertainty_principle.

———. "The Unreasonable Effectiveness of Mathematics in the Natural Sciences." Wikimedia Foundation, last modified Dec. 16, 2025. https://en.wikipedia.org/wiki/The_Unreasonable_Effectiveness_of_Mathematics_in_the_Natural_Sciences.

———. "Waggle Dance." Wikimedia Foundation, last modified Nov. 20, 2025. https://en.wikipedia.org/wiki/Waggle_dance.

———. "Whitehead's Theory of Gravitation." Wikimedia Foundation, last modified Dec. 21, 2025. https://en.wikipedia.org/wiki/Whitehead%27s_theory_of_gravitation.

———. "Wolfgang Pauli." Wikimedia Foundation, last modified Jan. 13, 2026. https://en.wikipedia.org/wiki/Wolfgang_Pauli.

Wildman, Wesley. *Effing the Ineffable: Existential Mumblings at the Limits of Language.* Albany: State University of New York Press, 2019.

Williams, Bernard. "Descartes, René." In Edwards, *Encyclopedia of Philosophy*, vol. 2.

WIRED. "How This Blob Solves Mazes." Uploaded Oct. 25, 2019. YouTube video, 10:06. https://www.youtube.com/watch?v=7YWbY7kWesI.

Wolchover, Natalie. "Experiment Reaffirms Quantum Weirdness." *Quanta*, Feb. 7, 2017. https://www.quantamagazine.org/physicists-are-closing-the-bell-test-loophole-20170207/.

Wolf, Jessica. "The Truth About Galileo and His Conflict with the Catholic Church." UCLA Newsroom, Dec. 22, 2016. https://newsroom.ucla.edu/releases/the-truth-about-galileo-and-his-conflict-with-the-catholic-church.

Wood, Charlie. "Inside the Proton, the 'Most Complicated Thing You Could Possibly Imagine.'" *Quanta*, Oct. 19, 2022. https://www.quantamagazine.org/inside-the-proton-the-most-complicated-thing-imaginable-20221019/.

Yen, Jeannette, et al. "Sensory-Motor Systems of Copepods Involved in Their Escape from Suction Feeding." *Integrative and Comparative Biology* 55 (2015) 121–33. https://doi.org/10.1093/icb/icv051.

Zeyl, Donald, and Barbara Sattler. "Plato's *Timaeus*." In *Stanford Encyclopedia of Philosophy*. Stanford University, 1997–. Article published Oct. 25, 2005; last modified July 28, 2023. https://plato.stanford.edu/entries/plato-timaeus/.

Zilsel, E. "P. Jordans Versuch, den Vitalismus Quanten Mechanisch zu Retten." *Erkenntnis* 5 (1935) 56–64.

Zyga, Lisa. "Physicists Find Quantum Coherence and Quantum Entanglement Are Two Sides of the Same Coin." Phys.org, June 25, 2015. https://phys.org/news/2015-06-physicists-quantum-coherence-entanglement-sides.html.

Index

www.ingramcontent.com/pod-product-compliance
Lightning Source LLC
LaVergne TN
LVHW020519100826
845148LV00010B/1280

* 9 7 9 8 3 8 5 2 5 9 4 8 9 *